DISCOVERING ENZYMES

David Dressler

Huntington Potter

**SCIENTIFIC
AMERICAN
LIBRARY**

A division of HPHLP
New York

Library of Congress Cataloging-in-Publication Data

Dressler, David, 1941–
  Discovering enzymes / David Dressler.
    p. cm.—(Scientific American Library series, ISSN 1040-3213
  ; no. 34)
  Includes bibliographical references and index.
  ISBN 0-7167-5013-9
  1. Enzymes.   I. Potter, Huntington.   II. Title.
QP601.D69 1990
574.19′25—dc20   90-44448
                                                            CIP

Printed in the United States of America

Scientific American Library
A division of HPHLP
New York

Distributed by W. H. Freeman and Company
41 Madison Avenue, New York, New York 10010
20 Beaumont Street, Oxford OX1 2NQ, England

  2 3 4 5 6 7 8 9 0 KP 9 8 7 6 5 4 3 2 1

This book is number 34 of a series.

Contents

To our parents
Dr. Charles Dressler and Mrs. Gay Dressler,
Dr. Jane Huntington Potter and, in memory,
Dalton Potter

Acknowledgments

This book is based on experience gained teaching molecular biology to students at Harvard College and Harvard Medical School. It offers an integrated and self-contained approach to one of the most fascinating areas of biology and chemistry—the study of enzymes, the molecular machines that are responsible for carrying out the activities of living cells. The book begins simply and gradually becomes more complex as discoveries are introduced that represent the frontiers of modern chemistry. The whole of the material is arranged as a story—it leads from the discovery of enzymes to an exploration of their molecular mechanism of action to an understanding of their role in carrying out the physiological processes of the body.

It is a pleasure to acknowledge a number of colleagues who have made invaluable contributions to this book. Dr. Peter Atkins of Oxford University reviewed the entire manuscript with great care and insight. His advice was cheerfully given, and by suggesting several careful and almost painless excisions, he helped sharpen the focus of the book. We are also indebted to Dr. Richard Dickerson, who reviewed the chapters on enzyme physiology and evolution at galley stage. In addition, every sentence had the benefit of Gunder Hefta's keen eye and many years of experience as an editor. The book has gained enormously from the thoughtfulness of its reviewers. We, of course, remain responsible for any errors of fact or interpretation that may exist.

We are grateful to Dr. Christopher Ruggles of the Polygen Corporation, with whom the computer-generated enzyme molecules were created. These occur, for example, on pages 90 and 139 and were produced using Polygen's state-of-the-art Quanta software on a Silicon Graphics computer. Other scientific illustrations include the excellent photomicrographs provided by Dr. Mary Osborn of the Max Planck Institute in Göttingen, West Germany, and by Dr. Sanford Palay of the Harvard Medical School; the fine medical and molecular illustrations by Tomo Narashima; and the pictures collected by photo researcher Travis Amos.

Linda Chaput, Publisher of the Scientific American Library, made available to us the resources of a great publishing house. Indeed, the devotion and patience of the Scientific American Library staff were phenomenal, and they were willing to take the time to get things right. Susan Moran pored over the entire manuscript and guided it through the numerous and intricate stages of editorial development. Susan is as stubborn about removing bad ideas as we are about keeping good ones, and the book benefited greatly from our shared obsession. In several instances, Linda Davis, Director of Development, helped guide our discussions to a mutually acceptable compromise. Nancy Singer designed the physical book, and for the final layout, Michael Braunschweiger integrated text and illustrations page by page. As the book approached completion, Ellen Cash managed its production, successfully maintaining her wits and the schedule despite authorial improvements sufficient to upset both. The anchor of the team was Sonia DiVittorio, who moved the book through the final stages of editorial production, supervising the creation of the artwork and ironing out every remaining problem. Sonia worked Saturdays and some Sundays—even July 4th—to put the best possible finish on the book.

As we recall, every chapter in the book went through about ten typewritten drafts. Others put the number much higher. Sylvia Klun and Doreen Wong-Gideonse typed the early versions, and the remainder of the massive word-processing effort was carried through to completion by Stefan Cooke. Stefan's indispensable efforts also included constructing diagrams, both freehand and with a computer, and tracking down scientific papers—some ancient and some not yet published.

David Dressler
Huntington Potter
August 1990

Discovering Enzymes

*Science should be made as simple as possible but
not simpler.*

Albert Einstein

*We are searching for the essence that lies behind
the fortuitous.*

Paul Klee

*To see the World in a Grain of Sand . . .
And Eternity in an Hour.*

William Blake

*What we want is a story that starts with an earthquake
and works its way up to a climax.*

Samuel Goldwyn

Alchemists of the Living Cell

According to convention, there is fire, there is water, there is air, and there is earth. There is a sweet and a bitter, and a hot and a cold. According to convention there is inherent order in the universe. In truth, there are only atoms and a void.

Democritus, 400 B.C.

Democritus expressed these thoughts almost 2500 years ago. They bring into focus one of the most fundamental mysteries of Nature—how does a world full of complex, almost miraculous structures and phenomena arise out of nothing more than atoms and a void? The answer will prove to be that evolution came upon the molecular machines that today we call enzymes. These machines are of such immense power that they have summoned life out of lifeless atoms.

For billions of years, long before the advent of life on earth and enzyme-directed biological evolution, a lengthy period of chemical evolution shaped the world. Under the influence of gravity, the predominant element of the early universe, hydrogen, settled into huge collections of gaseous clouds. The pressure produced in these massive bodies by gravitational compaction led to temperatures in the millions of degrees—enough to sustain thermonuclear reactions. When these nuclear reactions ignited, the simple clouds of hydrogen gas became the first stars. As a

Chemical evolution gave rise to an impressive but stark inorganic world. It had oceans, an atmosphere, and endless rock formations—all built of simple inorganic compounds. The major forces that shaped this world were gravity, heat, and pressure.

consequence of the stellar nuclear reactions, hydrogen became fused into helium and progressively heavier elements such as carbon, nitrogen, oxygen, and iron—the elements out of which the universe is made. Thus, over millions of years, stars became very complex structures.

Character is fate, according to the Greeks, and this is true for stars as well as humans. While some stars simply burned off their fuel over billions of years, others, overlarge and burning too fast, became gravitationally unstable and exploded, scattering vast amounts of their complex mixture of elements into space. This material, when it condensed, became the starting matter for subsequent generations of stars and, from smaller condensations of the heavier elements, planets.

Earth was one such planet. But it proved to be a planet with a very special property. Its distance from the sun endowed it with a moderate temperature,

Biological evolution began with the advent of the first living cells. The force behind this development was the invention of a set of microscopic cellular machines called enzymes, which had the special ability to promote specific chemical reactions. By choosing the molecules they wanted to react, enzymes were able to deliberately build up the highly complicated structures upon which life depends. The result was a world of ever increasing complexity as organisms developed greater capabilities. In the photograph, plant forms cover the surfaces on a rockbound coast. The different colors represent pigmented molecules enzymatically produced in the plant cells to gather the energy in the various wavelengths of sunlight.

which was neither too hot nor too cold. Thus when heat-driven volcanic eruptions and the vaporization of small compounds led to the formation of a gaseous atmosphere, it was neither boiled off, nor frozen into an inert mass. Instead, the molecules of the primitive atmosphere experienced an intense bombardment by high-energy radiation that poured from the sun onto the young planet. Energized to a reactive state, these simple molecules combined to form somewhat larger compounds that were relatively stable. As the water vapor in the atmosphere condensed upon cooling, torrential rainstorms were unleashed, filling the cratered surface of the planet. Over hundreds of millions of years, the various compounds in the atmosphere were washed into the developing lakes and oceans. Their concentration steadily increased until the sea resembled a dilute broth containing a wide variety of dissolved substances. An impressive, if stark and lifeless, world arose. It had oceans, an atmosphere, and endless rock formations—

all built of simple inorganic compounds. None of these compounds contained more than 10 to 15 atoms.

It was at this stage that the great—and perhaps unique—step was taken. Collections of primitive molecules—lifeless in themselves, but composed of the same atoms as are found in living forms today—became assembled into larger aggregates that had the capacity to absorb additional similar material from the environment. At some point the ability of these aggregates to ''grow'' changed from a process of accretion to a more orderly and exact process of reproduction, and the first living cells were born. As if a great invisible force had swung into motion, the vast reserve of lifeless molecules in the oceans began to be converted into living protoplasm. But the molecular components of living protoplasm were

Ten weeks after conception, the human embryo is well along the journey of life. Thousands of enzymes, genetically inherited from both parents, are building the major structures of the body (the heart and parts of the brain are already functioning) and beginning to unfold the blueprint of a unique life.

of much greater size and vastly more organized than the molecules produced by chemical evolution, and the steps in their synthesis had to be links in a highly structured chain of chemical transformations. Thus, for life to exist, some method was needed to promote reactions that were not random but purposeful, and whose products were not merely aggregates of simple compounds but considerably larger molecules of much greater complexity.

The key to this second stage of creation was the invention of a remarkable set of microscopic machines called enzymes. Working in a way that contrasted with the powerful but random forces that drove reactions in the nonliving world, enzymes brought to nature a new approach to the problem of promoting chemical transformations. Rather than relying on gravitational, thermal, and electromagnetic energy, enzymes were able to bring molecules together in a more subtle way that took advantage of their natural tendencies to react with each other. Under their influence, the small molecules absorbed from the environment were led to participate in highly specific chemical transformations that, step by step, built up new and complex compounds within the cell. These contained thousands of atoms and, most important, displayed novel properties that were useful to the cell. Unlike anything that had existed in the prebiological world, here was a microscopic part of the universe that, under its own direction, was becoming more and more complex as time went on.

For the past three billion years, the original enzymes and their descendants have been transforming the inorganic world into the living world. Diversifying and evolving ever greater capabilities, they have generated the range and scope of life forms that fill the earth. The descendants of these enzymes still reside in the cells of every living organism and have passed the spark of life down through all the generations of living things.

This book is about enzymes and how they built an island of life within the vast ocean of the chemical universe. Working within a cosmos in which the immensely more powerful forces of gravity, temperature, and pressure drive chemical transformations and lead to the production of inanimate stars and planets, enzymes function in a completely different way. They have devised a method to promote reactions that are deliberate rather than random, and whose products are not merely aggregates of simple compounds but much more sophisticated molecules of virtually unlimited complexity.

As a consequence, we now live in two worlds—a macroscopic world dominated by random chemical and physical forces, and a microscopic world created and maintained by invisible enzymes. It is not an overstatement to say that the fundamental difference between the living and the nonliving worlds is the presence of enzymes. *In essence, it is enzymes that have created life out of nothing more than atoms and a void.*

We will begin our discussion with a series of vignettes that are designed to illustrate some of the processes enzymes carry out. Not to put too fine a point on it, enzymes are responsible for all of the chemical transformations in the living world.

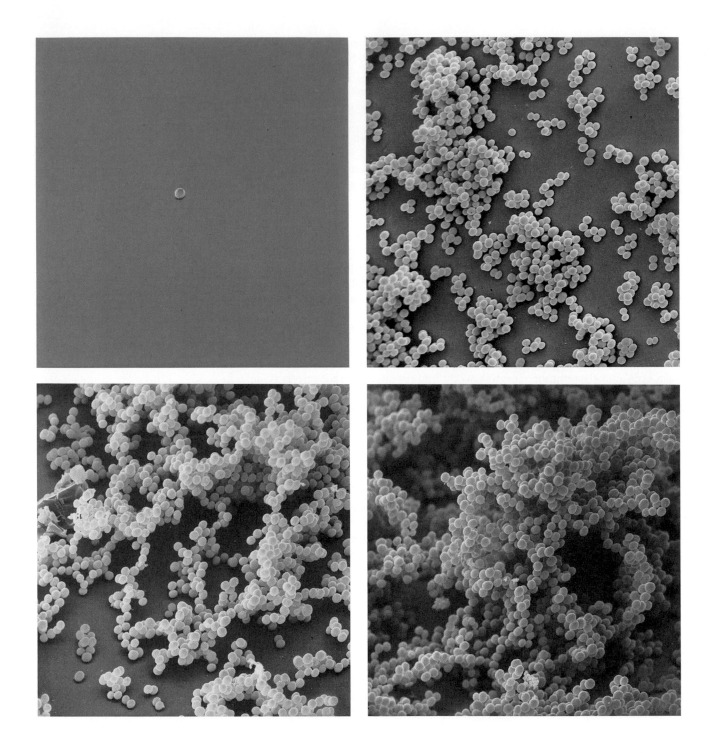

ENZYMES AND THE GROWTH OF CELLS

On the facing page we see a single bacterial cell in laboratory culture. The solution in which the cell is growing consists of nothing more than water, a common sugar, glucose, and a few simple salts, such as sodium sulfate and ammonium chloride. From these basic substances, the bacterium is able to construct all of the complex molecules required for life.

That cells would be able to survive, and even to prosper, in such a bleak chemical world may be surprising. And yet they do. If we were to look at the cell an hour later, we would find two cells; 10 hours later, a thousand cells; and 20 hours later, a population of over a million cells. Since a bacterium can divide roughly every 60 minutes, such growth, if it could continue unchecked, would theoretically produce more than 200,000,000,000,000,000,000,000,000,000,000, 000,000,000,000,000,000 (that is, 2×10^{44}) cells in the short span of six days. Each of these cells would weigh only a trillionth of a gram, but their total mass would exceed 2×10^{32} grams—more than the mass of the earth. Of course, such unlimited growth would never actually occur, but it does bring into focus a point that is of interest to us here. So great is the "life force" of the cell that a single bacterium, given an unlimited food supply, has the inherent capacity to undo the whole of creation in just six days.

In essence, the growth of cells represents the transformation of the simple molecules of the environment into living protoplasm—a transformation no less remarkable than the medieval alchemist's attempt to change lead into gold. There is, however, an important difference. In the case of the alchemist, the tool for the transformation was often a fraudulent wand with a hollow tip, from which gold could be secretly dispensed. But in the case of the cell, the agents of change are not at all illusory. They are very real, extraordinarily powerful microscopic machines called enzymes. *Collectively, enzymes are responsible for carrying out all of the chemical reactions of the living cell, and, as such, they represent the life force of the cell.*

Enzymes work by altering and rearranging the simple molecules of the environment one step at a time to form the many complex molecules upon which the life of the cell depends. For example, while many of the sugar molecules taken from the growth medium are completely dismantled (to release energy), others are only partially broken apart, and the remaining pieces are recombined to generate a family of related small molecules. Often these new molecules are further diversified by the addition of nitrogen or sulfur atoms, taken from the ammonium and sulfate salts in the surrounding medium. This process generates a set of about a hundred small organic molecules that serve as universal building

The growth of a single bacterial cell can, over a period of hours, give rise to millions of cells. These most primitive of living cells construct the complex molecules of protoplasm directly from simple substances they absorb from the environment.

blocks—the amino acids, the nucleotides, the sugars, and the lipids. These building blocks are then used in a second cycle of synthesis to construct all the larger, more complicated *macromolecules* of living protoplasm—an assembly process that is again guided by enzymes.

Because of the complexity with which the building blocks can be put together, macromolecules have the potential for displaying special, physiologically useful properties. For instance, one group of basic building blocks, the lipids, are useful in forming membranes that create a boundary around the cell and also divide it into compartments. Another group of building blocks, the nucleotides, are assembled into the DNA of genes. It is through the synthesis of such macromolecules that enzymes develop the components of protoplasm that will carry out the functions of the living cell. The situation is not unlike the making of a clock. The clock begins as a collection of small defined pieces—gears, wheels, weights, dials, and pointers—all of which can easily be mass-produced and none of which display any individual complexity. These pieces can be used by a skilled artisan, who assembles them in such a way that the final product is more than the sum of the parts. What emerges is a machine that measures time. The cell is just such a synthetic invention. But whereas human hands put together the clock, enzymes build the cell—they are the workers in the living factory.

In broad overview, the role of enzymes in the life of the cell is to mediate the transformation of the simple chemical world of the environment into living protoplasm. In effect, they guide molecules across the threshold of life.

The cells of higher organisms are not unlike the single bacterial cell. Here, too, enzymes are involved in converting small molecules into living protoplasm. But in higher organisms, the individual cells are not in direct contact with the environment and must therefore be supplied with a special nutrient solution created for them by the organism as a whole. This "growth medium" is generated through the process of digestion. As celebrated in the Brueghel painting on the facing page, higher organisms have developed a complex feeding pattern designed to allow them to follow a predatory form of existence. The strategy of predation entails breaking down the ingested food particles and using their subunits to produce new, somewhat different sets of macromolecules.

To carry out digestion, the organism devotes a considerable percentage of its cells to making a complex internal tube, in which various outpocketings become specialized structures, such as the salivary glands, the liver, and the pancreas. Food particles—relatively complex aggregates of once-living tissue, and sometimes even whole plants or animals—are collected within the tube and progressively broken down into their constituent building blocks. At each level, the cells that line the tube produce different enzymes designed to break down specific components in the ingested material. To take just two examples: near the beginning of the tube, the salivary glands supply enzymes that break down the carbohydrates of grains and fruits into free sugars; farther along, certain cells associated with the stomach and intestine secrete enzymes that break down the proteins of meat into their building blocks—the amino acids.

After the food particles have been broken down enzymatically, the released

The elaborate feeding patterns of higher organisms as chronicled by the Flemish painter Jan Brueghel.

building blocks are absorbed by cells lining the digestive tube and passed into the circulatory system. From the bloodstream they are taken up into the individual cells of the body and reassembled into the macromolecules that characterize the predator organism. Enzymes are involved at every stage of this process. After one set of enzymes have broken down the original food particles, others transport the released building blocks through cellular membranes, and still others mediate the restructuring of the building blocks in the recipient cells.

The growth of microorganisms and individual cells on the one hand, and the digestion of food by higher organisms on the other, illustrate two complementary facets of enzyme function. Enzymes involved in the buildup of cellular molecules are said to be carrying out the process of *anabolism*. Enzymes involved in breaking down molecules, as in digestion, are carrying out the process of *catabolism*. The sum total of all these chemical changes is termed *metabolism* (from the Greek *metaballein*, ''to throw about or change'').

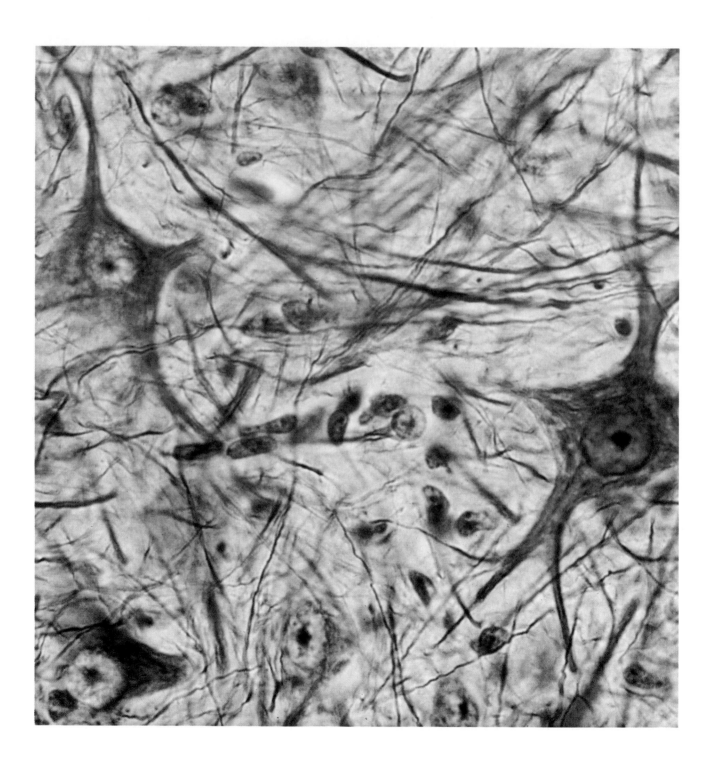

ENZYMES AND CELL SPECIALIZATION

Modern laboratory procedures allow one to grow not only bacterial cells in culture but also the individual cells of higher organisms. Basically, one removes an organ, dissociates it into individual cells, and places the cells in an artificial chemical solution in a petri dish.

We have already seen that bacteria in laboratory culture have an essentially unlimited capacity for cell division and growth. In contrast, animal cells in culture generally stop dividing within a short time (5 to 50 generations). The differing abilities of bacteria and animal cells to sustain growth reflect their different roles. Whereas the sole activity of bacteria is to grow—to convert as much of the environment as possible into protoplasm—such uncontrolled growth would be ultimately lethal to a higher organism. The component cells of higher organisms are members of a cellular community, with each type of cell having its own assigned responsibility, and to preserve this balance, cell growth is strictly limited. Indeed, it is the specialization and diversity of the cells of higher organisms, not their growth potential, that is their important characteristic.

The key to cell specialization is enzyme specialization. In addition to the basic set of enzymes needed for maintaining the life of the cell, each cell type in a higher organism also manufactures its own unique set of enzymes, which are designed to achieve the specific goals of the cell. As these specialized enzymes carry out their assigned tasks, some first construct the diverse structures of the cells we find in a typical multicellular organism, and others then direct these cells as they make their special contribution to the life of the organism.

The important point is that, in the end, the function of individual cells is determined by the enzymes they contain. This is readily illustrated by considering the type of cells shown in the facing photograph, which are nerve cells, or neurons. As part of the system that integrates all of the functions of higher organisms (ranging from sensory perception to muscle movement, and from heartbeat to logic), individual neurons extend a complex network of information-carrying fibers to all parts of the body. As is readily apparent in the photograph, most of the mass of a nerve cell is concentrated in a central area, the *cell body*, where the basic synthetic reactions that maintain the cell protoplasm are carried out. Projecting outward from this area is a long fiber, or *axon*. Here, the most characteristic function of the nerve cell is accomplished—its transmission of a signal.

The basis of nerve cell function is to be found in certain specialized enzymes located in the cell's membrane. These enzymes generate a difference in electrical charge between the outside and the inside of the cell by pumping ions

A collection of neurons that have been rendered visible by staining with a dye. Highly branched fibers, or dendrites, collect stimulatory information from sensory organs or other neurons. These stimuli are then summed at the cell body and, if a certain threshold is reached, a wave of electrochemical change is sent down the axon and then on to the next cell.

across the cell membrane. This creates a higher concentration of positive ions (particularly sodium ions) in the fluid surrounding the cell and leaves behind a relative excess of negatively charged ions inside the cell. Thus, in the resting neuron there exists a difference in electrical charge across the cell membrane—a condition that prepares the cell to function.

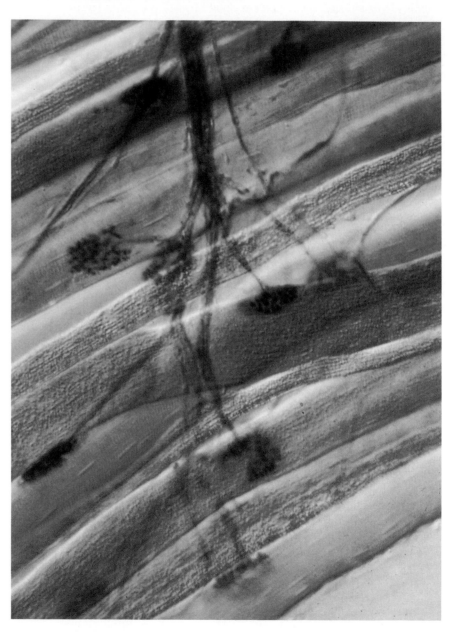

A nerve fiber with several axons descends into an area of muscle. The branching ends of each axon secrete a chemical signal to cause the target muscle to contract.

When a nerve cell is stimulated (either by another neuron in the brain or by a sensory organ), the permeability of the membrane changes. Ions begin to move more freely into and out of the cell, and as a result there is a decay (or *depolarization*) in the charge differential across the cell membrane. When this decay reaches a critical level, a further change in the membrane occurs: channels for sodium ions suddenly open. Sodium ions rush into the axon, and the charge differential totally collapses. A self-propagating wave of electrochemical change is generated that sweeps rapidly down to the end of the axon.

The continued propagation of the signal after it reaches the end of a nerve fiber poses another problem—and again, one that is solved by enzymes. The next cell in the pathway—for example, a muscle cell—is not generally in direct contact with the activated neuron, and therefore the signal can no longer be propagated in precisely the same way. Instead, the electrical impulse is converted into a chemical signal. When the wave of depolarization reaches the end of the nerve fiber, it causes the rupture of vesicles filled with a small, signaling molecule. This enzymatically made chemical serves as a *neurotransmitter* to carry the signal to the next cell. For example, in the photograph on the facing page, the branching ends of a nerve fiber are seen permeating ribbons of skeletal muscle. In this case, the impulse that sweeps down the nerve axon causes the release of millions of neurotransmitter molecules, which spill into the space between the neuron and the adjacent muscle cell. The neurotransmitters then bind to specific receptor sites on the surface of the target cell. Through this transfer of the original signal across the intercellular space (the *synapse*), the muscle is galvanized into action and contracts.

Within a short time after the target cell has been activated by the neurotransmitter, it is necessary for the system to cease signal transmission and return to its original state. This aspect of the signal transmission process is also accomplished by enzymes. First, the charge differential across the original nerve cell membrane is restored, as the ion channels close and the enzymatic ion pumps redistribute sodium ions to the outside of the cell. At the same time, the neurotransmitter that has flooded the neuromuscular junction is cleared away by a specific enzyme. This enzyme, which is needed by no other cell, is secreted into the intercellular space, where it inactivates the neurotransmitter molecules, in this case by cleaving them into two pieces. The system returns to its original state, prepared to respond again to future stimuli.

It is interesting that a variety of toxic substances—both natural and synthetic—exert their influence by interfering with the enzymes involved in signal transmission in the nervous system. For example, the neurotoxins of the cobra and the puffer fish block two major ion channels, and many nerve gases inhibit the enzyme that inactivates one of the body's major neurotransmitters.

The important point is that the entire process of nerve impulse transmission is controlled by enzymes. Not only do enzymes build all of the components of the nerve cell, but they also play a continuing role in the function of the neuron. The same dependence on specialized enzymes holds true for all cells of the body.

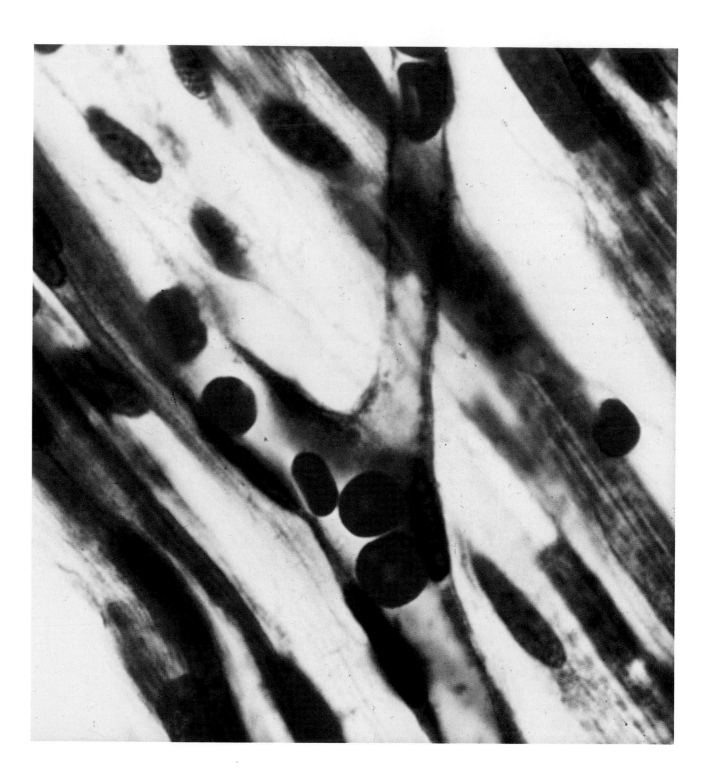

ENZYMES AND PHYSIOLOGY

Not only are the growth of cells and the development of their specialized functions under the control of enzymes, but so too are the major physiological processes that characterize the organism as a whole. As an example of the role of enzymes in physiology, we will consider the process of blood clotting.

With the evolution of large multicellular organisms, individual cells could no longer rely on direct contact with the environment to obtain nutrients and oxygen and to carry away toxic waste products. The solution to this problem was the development of the circulatory system. Through this internal network of ramifying tubes, all of the hundred or so types of molecules upon which the life of the cell depends can be carried to or transported away from the isolated cell (see the micrograph on the facing page). However, this development brought with it a serious problem: a single break in the system could easily result in an organism bleeding to death in a matter of minutes.

The defense against this mechanical disaster is a highly sophisticated set of enzymatic reactions leading to the formation of a blood clot. Stated briefly, a group of enzymes working together change the character of blood from a liquid to a solid in the localized area of the circulatory system where damage has occurred. Although the full sequence of events involved in clot formation is highly complex, the key event is well understood. It involves the enzyme-controlled conversion of an inactive blood component, *fibrinogen,* to an active form, *fibrin.*

Before clotting occurs, fibrinogen is simply a large, inert molecule floating freely in the blood. However, its properties are dramatically changed if a portion of the molecule is clipped off, as shown in the diagram on the next page. This trimming leads to the release of material with a "blocking" function, and the modified molecules—now called fibrin—are able to interact with each other in a side-by-side and end-to-end fashion to form long fibers. In the region of the circulatory system where fibrinogen has been converted into fibrin, the long fibers form a mesh, and, as this three-dimensional lattice traps large numbers of blood cells, a strong, gelatinous blood clot forms. The effectiveness of the fibrin network in promoting clot formation is indicated by the fact that the final clot can contain as little as one-twentieth of 1 percent of fibrin, with the rest of the clot coming from the cells that have become trapped in the network of fibers. The electron micrograph on page 17 shows a red blood cell becoming entangled in a network of fibrin filaments.

The trimming process that converts fibrinogen into fibrin is the major "switch" thrown in the blood-clotting reaction, but it takes 15 distinct enzymes working in sequence to bring about the apparently simple change in fibrinogen. Certain of these enzymes become active in the presence of damaged tissue, others transmit the signal forward, and ultimately numerous copies of the final,

Capillaries delivering oxygen (via red blood cells) and nutrients to an area of muscle tissue.

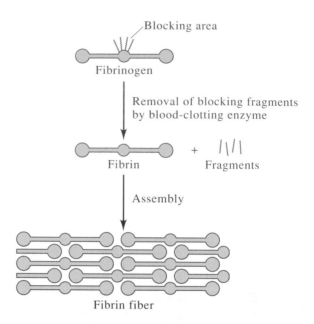

The conversion of inactive to active fibrin.

fibrinogen-trimming enzyme are activated. Because each enzyme in the series activates hundreds of copies of the succeeding enzyme, an ever-expanding "cascade" develops, which culminates in the nearly simultaneous conversion of large numbers of fibrinogen molecules into active fibrin. The whole process proceeds very rapidly, and the formation of the blood clot begins within a fraction of a second.

As in any complex mechanical process, there is the potential for malfunction. This is exactly what happens in *hemophilia,* a serious human disease that results in the failure of blood to clot normally. Individuals with this disease have difficulty converting fibrinogen into fibrin, and the time required for blood clotting is thus significantly increased. Serious injuries result in uncontrolled bleeding and are, of course, life threatening. In addition, hemophiliacs experience periodic episodes of internal hemorrhaging because the ordinary small ruptures that routinely occur in the circulatory system due to wear and tear cannot easily be repaired.

The powerful techniques of modern biochemistry and molecular biology have succeeded in identifying the molecular basis for hemophilia. The disease is most commonly caused by the lack of a single protein that participates in an enzymatic reaction. Affected individuals lack one of the 15 components in the blood-clotting system, and as a result their blood-clotting process falters.

In addition to introducing the relationship between enzyme function and human disease, hemophilia provides another important insight. This comes from the observation that the disease does not occur randomly—as is true for many other illnesses, such as those due to viral infection. Rather, hemophilia occurs repeatedly in certain families; the inability to clot blood is passed down in an orderly way from one generation to the next. Evidently, a defective "particle of inheritance"—that is, a defective *gene*—underlies the inability to produce the missing blood-clotting component. The important point is that *enzymes are produced by genes, and a mutation in a gene can lead to the alteration or absence of a specific enzyme, resulting either in death or in serious disease.*

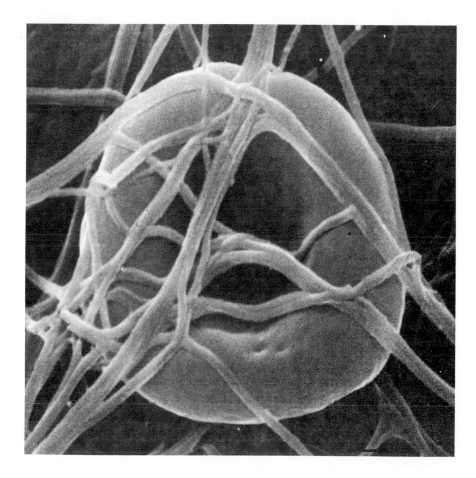

Fibrin filaments wrap around and entrap a single red blood cell.

ENZYMES IN DEVELOPMENT

Thus far we have seen that enzymes play a key role in cell growth, in cell differentiation, and in basic physiological processes. We now turn to a new level of enzyme responsibility—in the development of the organism. As a particularly simple example of the role of enzymes in development, we will consider an enzyme made by the silk moth.

To become a moth, every caterpillar must first spin a cocoon. This is an elaborate structure, made primarily from a single macromolecule—the protein silk—which is spun out in a rhythmic, more or less circular pattern as the caterpillar bends its body back and forth, while remaining attached to a twig. Altogether, about a mile of silk is synthesized to make the cocoon. Within this protective structure, the caterpillar completely reorganizes the tissues of its body and undergoes the dramatic developmental changes that transform it into an adult moth. At the end of this metamorphosis, the adult is still trapped within the cocoon. Its escape depends upon the timely production of a specific enzyme that digests the cocoon.

The picture on the facing page shows a silk moth emerging from its cocoon. At the end of its developmental cycle, the salivary glands of the maturing moth produce *cocoonase,* an enzyme whose sole purpose is to digest a hole in the cocoon. The enzyme attacks the silk fibers until enough protein threads have been cut to allow the adult moth to escape from the cocoon. This enzyme illustrates, in a very simple way, the fact that *each stage in the development of an organism is guided by specific enzymes.*

We have now seen a number of interesting, important, and very different physiological processes—ranging from basic cell growth to the transmission of nerve impulses and from blood clotting to the digestion of a cocoon. These processes illustrate the wide range of essential chemical transformations that are carried out by enzymes.

In the next chapter we will see how, in the nineteenth century, enzymes were discovered to be the cellular agents of change. Then, in the following six chapters, we will continue our discussion of enzymes through several levels of increasing complexity. Our major goal will be to determine the chemical nature of these substances and explore how they work at the molecular level, focusing on a particularly interesting group of enzymes as a "case study." In the last chapters, we will consider further the role of enzymes in carrying out physiological processes. Indeed, we will find that the processes discussed in this prologue were not chosen at random. Although they are very different in appearance, in fact they are all carried out in whole or in part by members of one family of enzymes, which has evolved from a single ancestral enzyme.

A silk moth emerging from its cocoon.

1

The Discovery of Enzymes

Each science arises in its own special way. Classical genetics and the concept of the gene, for example, arose fully born in the brilliant analysis of one scientist, Gregor Mendel. Biochemistry, on the other hand, struggled its way into existence amidst fragments of data and endless debate. The story we trace in this chapter is a brief recapitulation of the hard-fought intellectual wars of the nineteenth century out of which the modern science of biochemistry arose. The core concept of the new science was that living systems prevail against the randomness and simplicity of their external environment through the use of *enzymes*—molecules they make to promote specific chemical transformations.

As history, our story is highly interesting, but even more important is the perspective it provides about the nature of scientific progress. Progress is made through an interactive process in which data and ideas are exchanged and debated in the search for a unifying context in which both can be understood. Generally, this dialectic involves many scientists working over a period of years. As we will see, the road is not always smooth or clearly lit. But once a few basic principles have been established and scientists know how to design specific experiments to extend their knowledge, progress can be very rapid indeed, and astounding new insights can be achieved.

The origin of biochemistry lies in the pioneering research of nineteenth-century scientists.

This Egyptian relief sculpture from the tomb of a Fifth Dynasty ruler (about 2400 B.C.) shows several steps in the brewing of beer, which is an alcoholic fermentation process analogous to the making of wine. The essential difference is that the sugars of grains rather than of grapes are used as the starting material. The figures in the top panel are engaged (right to left) in pounding, winnowing, and grinding the grain. Those in the middle panel are soaking the coarse-ground material in water, allowing some of the grains to sprout, or malt (left). In the bottom panel, the resulting mash with its released sugars is being poured into a fermenting vat, which rests on a stand resembling a coiled rope. During the next few days fermentation occurs, and the finished beer is then transferred from the vat to pottery jars, which are capped, sealed with clay, and placed in storage.

An entire world of physiological processes is mediated by enzymes, as we saw in the Prologue, but it is convenient that the discovery and early history of enzymes was intimately bound up with one particular process. This is alcoholic fermentation—the key step in the making of wine. It was during the study of this process that enzymes were discovered and the modern science of biochemistry emerged.

Fermentation: The Process That Led to the Discovery of Enzymes

Perhaps 10,000 years ago—certainly more than 6000 years ago—our ancestors observed that the juice of crushed grapes, if allowed to stand for extended periods, underwent a remarkable transformation, giving rise to a liquid with very special properties. Over a period of weeks, what was originally a sweet and sticky mash was transformed into a somewhat turbulent, frothing mixture by an invisible force that, when it subsided, left behind a calm liquid bearing the gift of euphoria. So impressive were the effects of this process, which came to be known as *fermentation,* that early man never ceased to study it. The relief sculpture on the facing page is an ancient Egyptian record of the basic steps involved in fermentation—steps that have remained essentially unchanged to the present day.

The making of wine begins when grapes are brought in from the fields and lightly crushed to release their juice. Fermentation sets in within a few hours, and during the next several days there is a gradual conversion of the grape juice into wine. Some of the changes that occur in the fermenting grape juice, or *must,* are readily visible. One observes, for example, a slow and constant production of bubbles, which continues for a period of days. Indeed, this process is so active that the emerging bubbles become enmeshed in the drifting fragments of skins, seeds, and stems to form a foaming layer across the surface of the fermenting juice. Heat is also generated during the fermentation process, and the must may become warm to the touch. *But the most significant observation is that in every fermentation mixture a fine, cloudy suspension begins to build up, which eventually settles to the bottom of the fermentation vat as a thin layer of viscous sediment.* This material, known since antiquity as *yeast,* has always been associated with the fermentation process: it is present at the end of every successful fermentation and, moreover, its addition to fresh grape juice will accelerate a new round of fermentation.

Not nearly as apparent as the effervescence, the heat, and the buildup of yeast is the central event that underlies fermentation—the gradual conversion of the grape juice into wine. Only relatively recently could this event be identified and explained scientifically. We now understand the fermentation process to involve the conversion of one of the components in grape juice, sugar, into alcohol, with the attendant release of a gas, carbon dioxide. This transformation is brought about by the yeast, which is actually composed of living microorganisms that are present on the surface of grapes in the vineyard and come into contact with the juice when the grapes are crushed.

However, when our story begins, all of this has yet to be discovered, and the process of wine making is understood only as a change in the character of grape juice "upon aging." That this change is brought about by yeast—and, even more important, by machinelike substances within yeast that would eventually come to be called enzymes—was completely unknown.

The Vital Force

In the ancient world there were several speculations about the driving force behind the fermentation process. One of the earliest hypothesizers, Aristotle, worked from the premise that the changes that occurred in inanimate objects were analogous to those experienced by living beings. Aristotle believed that every living thing had a natural end or goal toward which it tended. The acorn's natural tendency, for example, was to become an oak tree. The warrior's natural goal was to become a hero. When this teleological view was applied to fermentation, grape juice was seen as an infantile form of wine, and the act of fermentation was analogous to maturation. The final change from wine to vinegar (which occurs as wine ages) was the equivalent of death.

In the Middle Ages, it was discovered that the vapors of alcohol could be captured from heated wine by using the sorts of distilling apparatus shown scattered in this painting of an alchemist's workshop.

In the Aristotelian system, change resulted from an object's inherent tendency to move toward a specific goal. This movement was presumed to be under the control of an inner driving force—a *vis viva,* or *vital force*. Though undefined in material terms, the vital force was thought of as a very real part of the organism. It drew its existence from the context or organization of matter, although it was not simply matter in the ordinary sense. Instead, it was somehow built into the fabric of an organism's component parts, endowing them with life. Indeed, the basic distinction between the animate and inorganic worlds was presumed to be the presence of the vital force.

At first glance, we are tempted to dismiss the concept of the vital force as nonscientific and primitive. In fact it

is. But its disguises are often so sophisticated that vital forces persist unnoticed. For example, our current, rather vague concepts of ''will power'' and ''creative thinking'' often closely approximate this general type of idea— activities that are above and beyond the world of molecules. Thus, vitalism is by no means a thing of the past; it is faced by every generation of scientists. In the study of biology, from Aristotle's time through the succeeding 2000 years, the *vis viva* remained the driving force in bringing about changes in organic matter. In one form or another this pattern of thinking persisted, offering an appealing, almost religious, alternative to more physical ideas, and blocking the development of biochemistry until nearly the end of the nineteenth century. *The emergence of modern biochemistry was, in large measure, the gradual replacement of the vital force by the enzyme.*

The Alchemists Identify Alcohol as the Product of Fermentation

The first real progress in understanding the changes that occur during the fermentation process—and in the world of organic matter in general—appeared with the rise of alchemy in the Middle Ages. New facts became available and, as is so often the case in science, they stimulated the development of new ideas—not initially the right ideas, but better ideas.

Much of ancient Greek chemistry had been based on the principle that the world was made up of four fundamental substances—Earth, Air, Fire, and Water. In addition, the idea had gradually arisen from philosophical considerations that, along with these physical elements, there must be a *fifth* element (the *quinta essentia*, or quintessence) that reflected the vital force and determined the specific nature and activity of each animate thing. Since such essences were often thought of as having a volatile character, the early chemists heated various materials and saw in the expulsion of fumes and vapors the liberation of the spirit characteristic of each substance. And, as the techniques of alchemy became more highly developed, the conviction grew that it might be possible to isolate the quintessences of various substances.

By the twelfth century, distillation methods had become sufficiently refined to allow the condensation and recovery of the vapors released from heated substances by using cooled receiving flasks. When this procedure was applied to wine, it gave an important insight into the nature of fermentation. The key observation came when the alchemists compared the distillation of wine to the distillation of grape juice. It was apparent that a specific new substance was formed during fermentation—a substance that could not be detected in the starting grape juice. The new substance was a clear liquid that could burn, and it was given the name *aqua ardens,* or "burning water." Today we know this substance as alcohol (or, more properly, ethanol or ethyl alcohol). Thus, from the work of the

early chemists it became clear that fermentation was not simply a "maturation," or change in the state or organization, of the starting grape juice; rather, it was the formation of an entirely new substance.

Fermentation as a Specific Chemical Reaction

The understanding of fermentation took another leap forward in the eighteenth century through the work of Antoine Lavoisier (1734–1794), one of the founders of

Antoine Lavoisier and Mme. Lavoisier. Lavoisier's brilliant career in science was cut short by the French Revolution. He was brought to trial because of his job as a member of the Ferme Générale, which collected taxes on a commission basis for the monarchy. Although he argued that his work was nonpolitical and that he was predominantly a scientist, the president of the tribunal nonetheless rejected his plea for mercy: "The Republic has no need of chemists and savants. The course of justice shall not be interrupted." Lavoisier was guillotined the same day.

modern chemistry. Lavoisier approached the problem from the point of view of a chemist who had spent a lifetime studying the chemical changes that occur in the nonliving world. From the analysis of such reactions, he had arrived at two fundamental conclusions: (1) substances undergoing a reaction combine (or break apart) in specific quantitative amounts—an observation that eventually led to *the law of combining ratios,* and (2) the amount of matter present at the beginning of a reaction is equal to the amount of matter present at the end—*the law of the conservation of mass.*

In one of the reactions for which he is most famous, Lavoisier studied a mixture of hydrogen and oxygen gas that, when it was ignited, explosively combined to form an entirely new substance—water. His analysis of the amounts of reactants and products showed both that the

interacting substances had combined in a specific ratio and that there had been a strict conservation of matter. Today we would write such results in the form of a *chemical equation:*

$$2H_2 + O_2 \rightarrow 2H_2O$$

Combining ratio: $\quad 2 + 1 \rightarrow 2$

Conservation of matter: $\quad 4H + 2O \rightarrow 4H + 2O$

From this and numerous other reactions, Lavoisier concluded that, when a substance was detected at the end of a reaction, it was not a starting substance whose properties had been altered or "matured" but rather an entirely new

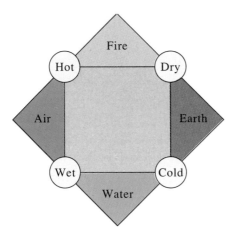

About 2500 years ago, Greek scientists developed a system of chemistry based on the idea that all substances are composed of fundamental building blocks, or "elements." The original elements, described by Empedocles, were Earth, Air, Fire, and Water, and they embodied pairs of the common properties of hot or cold, wetness or dryness. For example, Earth was cold and dry; water was cold and wet. To Empedocles, wood was a complex substance containing all four elements. Thus, when wood burns to yield ash (Earth), smoke (Air) rises from it, followed by flame (Fire), and a cool surface held near the flame will have moisture (Water) condense upon it.

By 1800, the Greek system and related types of alchemy had given way to the new science of chemistry, which was better grounded empirically and rationally. Several "true" elements had been discovered—for example, nitrogen, hydrogen, oxygen, and a number of metals. After Lavoisier had discovered the law of the conservation of mass, his near contemporary John Dalton devised the "atomic theory," which showed how compounds could be understood as assemblies of integral numbers of elements. Dalton's original symbolism for three reactions involving carbon, hydrogen, and oxygen is shown above.

The experimental equipment of Lavoisier enabled him to recover and measure the products generated during chemical reactions and to compare them with the starting materials. These experiments gave rise to the law of the conservation of mass. Lavoisier's quantitative approach to science may have come from his other occupation as a member of a firm that handled large tax accounts. The apparatus that keeps track of substances produced during a chemical reaction is not unlike the bookkeeping ledgers of Lavoisier's professional life.

substance created by a chemical transformation of the original material. The general interpretation that came to be placed on these results was that the products of chemical reactions were derived by a rearrangement of the components—or, as we would now say, atoms—in the starting material.

These concepts of modern chemistry were just beginning to become established when Lavoisier turned his attention to the process of fermentation. A recent chemical analysis of grape juice had shown that sugar was one of its principal components, accounting for some 25 percent of the material present. Combining this new fact with the older observation of the alchemists that *aqua ardens,* or ethanol, was the primary product of fermentation, Lavoisier proposed that *fermentation was a chemical reaction in which the sugar of the starting grape juice was converted into the ethanol of the finished wine.*

To test his idea about fermentation, Lavoisier devised an artificial, and very clever, experimental system: he found that it was possible to use a pure sugar (glucose) instead of fruit juice as the starting material for the fermentation process. When the sugar was mixed with water, it underwent fermentation in the laboratory—provided that, as in natural fermentations, one added a small amount of the undefined material, yeast, that had sedimented to the bottom of a previous fermentation mixture.

Using this simplified system, which avoided the need to analyze the more complex mixture of organic substances found in whole grapes, Lavoisier was able to monitor the fate of the original carbon atoms in the sugar. By burning a sample of either the starting sugar or the product ethanol in oxygen, he was able to determine the amount of carbon present before and after the fermentation reaction. The addition of yeast did not appear to complicate the analysis because it was needed in such small amounts and was present at the end of the fermentation in only slightly greater quantity than at the beginning. Thus the yeast could simply be ignored and the overlying fluid examined. Lavoisier determined that, at the beginning, his artificial fermentation reaction contained 26.8 pounds of carbon— initially present in pure sugar. At the end, the sugar had entirely disappeared. There were still, however, 27.2 pounds of carbon (the same amount, within the limits of experimental error), but it was now present partly in ethanol and partly in carbon dioxide. Thus, in accordance with his formulation of the problem in terms of the law of the

conservation of matter, the atoms of the starting sugar had indeed appeared in the products, ethanol and carbon dioxide. Matter had been conserved—as in the reactions of inorganic chemistry—but specifically *transformed* during the fermentation process. This was the first time a chemical equation expressing the law of the conservation of matter had been shown to apply to an organic material (that is, to a substance, sugar, formed by a living organism). The ability to apply the same analytic approach to the reactions of both the inorganic and organic worlds was important because it served to unify the chemistry of the two worlds.

In the early part of the nineteenth century, Joseph Louis Gay-Lussac adjusted Lavoisier's initially imprecise measurements and firmly established that, during alcoholic fermentation, sugar is converted into equal parts of the two products, ethanol and carbon dioxide:

$$\text{Sugar} \rightarrow \text{ethanol} + \text{carbon dioxide}$$

Grape sugar (glucose)   Ethanol   Carbon dioxide

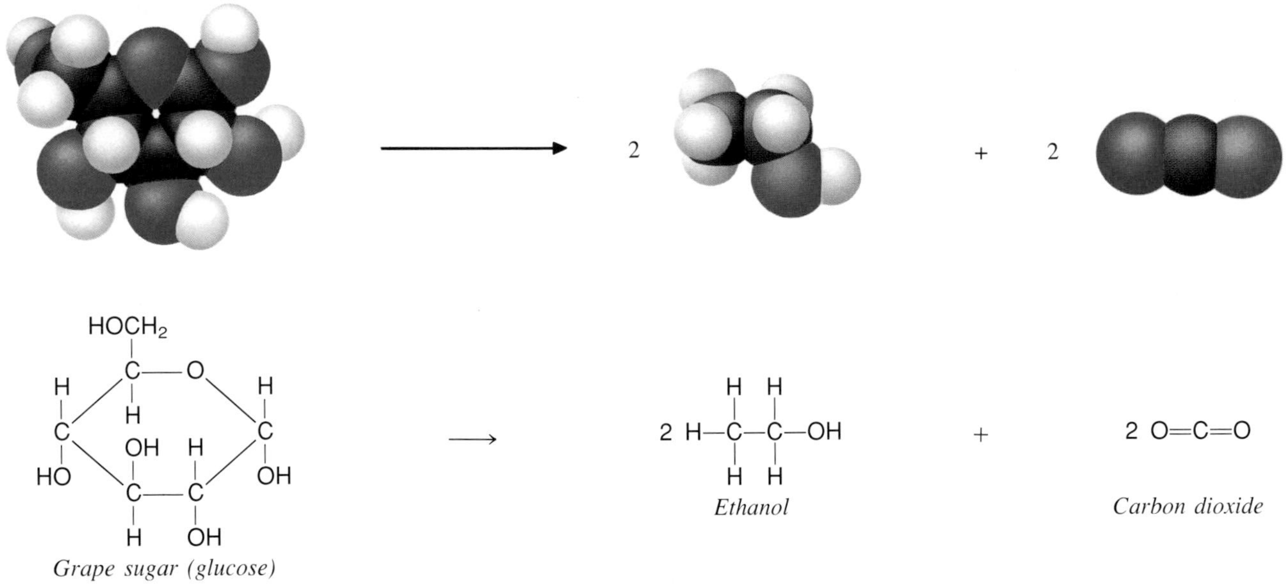

We now know the specific arrangements of atoms in the molecules that participate in the fermentation process and can thus represent the transformation of glucose into ethanol and carbon dioxide in the realistic representation shown above. Because the molecular structures can be hard to interpret (in part because we cannot see all the atoms), it is often convenient to write them in the planar schematic way shown at bottom.

Because of the limitations of the chemistry of the day, Lavoisier and Gay-Lussac were not able to assign molecular structures to such compounds as sugar or ethanol. Gradually, however, as the structures of molecules were elucidated, organic chemistry became a well-defined discipline, and it became possible to treat the chemical aspects of living systems with the same precision that was applied to inorganic reactions. In terms of modern structural chemistry, Lavoisier and Gay-Lussac's equation can be written:

$$C_6H_{12}O_6 \rightarrow 2C_2H_5OH + 2CO_2$$

Or, in terms of chemical bookkeeping:

Starting material			*Product material*				
	Sugar	$\xrightarrow{\text{Fermentation}}$	Ethanol	Carbon dioxide			
Carbon	6		4	+	2	=	6
Hydrogen	12		12	+	0	=	12
Oxygen	6		2	+	4	=	6

Thus, by 1800, inorganic chemistry and the new chemistry of living material were united in obeying the rule of the conservation of matter when changes occurred. Lavoisier and Gay-Lussac had shown that the ancient process of fermentation was actually a chemical reaction—specifically, the transformation of sugar into ethanol and carbon dioxide. But what *caused* the transformation to occur? The answer to this question was as obscure as it had been in Aristotle's time. Lavoisier and Gay-Lussac had determined the essential characteristics of the fermentation reaction so that one now understood the chemical nature of the starting material and the end products, and could thus monitor the reaction quantitatively. This, however, did not help to explain the nature of the driving force that brought about the fermentation reaction. In particular Lavoisier and Gay-Lussac had ignored any significant role for the yeast. They knew it was needed in some way, but it appeared to be a minor component of the reaction mixture and not much changed during the fermentation process. It was the yeast, however, that was the key to solving the next part of the puzzle.

The Role of Yeast in Fermentation

The fact that fermentation was not understood did not prevent the leading scientists of the day from having their own strongly held hypotheses. Although they no longer spoke openly in terms of vital forces, they had, in fact, developed equally metaphysical ideas.

There had always been a considerable debate as to the nature of the yeast, or "ferment"—the vague name given to the reaction-promoting material in the fermentation mixture. The controversy centered primarily on the question of what specific role the yeast played in the fermentation process.

Liebig's Chemical Theory of Fermentation

In 1839 the eminent German chemist Justus von Liebig (1803–1873) developed a mechanistic explanation for the role of yeast in the fermentation process. He viewed the yeast that was present in the fermentation mixture as decomposing matter that emitted certain vibrations:

The atoms of a putrefying body, the ferment, are in a ceaseless movement; they change their positions and form new combinations. These moving atoms are in contact with the atoms of the sugar, which is held together only by weak forces. The movement of the atoms of the ferment cannot be without effect on the atoms of the sugar mixed with it; either their motion will be abolished or the atoms of the latter will move too. Thus the sugar atoms suffer a displacement; they rearrange themselves in such a way as to form alcohol and carbon dioxide.*

---

*In this and other excerpts of papers cited in this book, I have edited out various passages and have taken slight liberties with the translations where this seemed appropriate. For instance, sentences have sometimes been rearranged and terms and words updated to their modern equivalent. However, I have tried to remain true to both the style and substance of the original text. The sources of the quotations appear in the back of the book under "Further Readings"; see especially the books by Friedmann and Fruton.

Justus von Liebig was one of the great nineteenth-century leaders in the field of chemistry. Through his teaching and the research carried out in his laboratory, he made the University of Giessen one of the foremost centers of chemistry in the world. Frustrated by his own experience as a student—when, having no access to scientific apparatus, he was forced to learn entirely from books and lectures—Liebig instituted the first chemistry laboratories for university students.

Schwann and Cagniard-Latour's Biological Theory of Fermentation

At almost the same moment, a fundamentally different view of the role of yeast in fermentation was put forward by a different group of researchers—the biologists. A young German physiologist, Theodor Schwann (1810–1882), and a noted French inventor, Charles Cagniard-Latour (1777–1859), independently reported the results of their microscopic examination of fermentation mixtures. Both men advanced the view that the yeast that built up in the fermentation vats was actually composed of living material—''a mass of globules that reproduce by budding, and not merely a simple chemical or organic substance.'' Moreover, they thought it was very probable that

Theodor Schwann is remembered today primarily for his proposal of the cell theory, which states that the basic structural and functional unit of both plant and animal organisms is the cell. In addition, during his investigation of muscle contraction, Schwann discovered the sheath of insulating material that covers nerve fibers. Both the sheath and the cells that produce it now bear his name. Schwann's position that yeast are living organisms brought him into conflict with the leading German chemists of the day, and their ridicule led to his early retirement.

In other words, under the vibrational impact of the decomposing yeast, the sugar is literally shaken to pieces, whereupon it forms the reaction products ethanol and carbon dioxide. In Liebig's theory, not only did the sugar molecules decompose, but the yeast, through its vibrations, also brought about its own decomposition.

The general idea of an energy transfer from decomposing yeast to sugar molecules seemed intuitively reasonable and was widely accepted by contemporary chemists.

the decomposition of the sugar and its conversion into carbon dioxide and alcohol are a consequence of the growth of the yeast.

Cagniard-Latour and Schwann had thus made the important suggestion that the ''force'' that drove the fermentation reaction came from a living organism as a result of its growth. They believed that, by seeding a sugar-containing solution with yeast, one was adding a microcosm of living cells, which in the course of their growth produced the alcohol. The first fermentation reactions—those observed without the deliberate addition of yeast—would have been due to yeast already present on the skins of the grapes or to yeast carried to the fermentation mixture through the air.

This proposal was, in fact, very close to the truth. But Schwann and Cagniard-Latour had only circumstantial evidence to support their conclusion: the correlation of fermentation with the presence of apparently growing yeast. What they did not have was a specific experiment to show a direct cause-and-effect relationship between yeast and the fermentation process. Perhaps fermenting grape juice was simply a favorable environment for the growth of yeast, and that accounted for its presence.

Because the proposal that yeast was alive was antithetical to their view of the fermentation process, chemists quickly rejected the idea. The introduction of living yeast cells into the discussion about fermentation was viewed as an unwarranted intrusion of biology—to them, a regrettably inexact science—into the higher domain of chemistry. And, to make matters worse, it was an intrusion based on a scientific instrument that, at that time, had a dubious reputation. For a hundred years, refocusing images without distortion had proved to be extremely difficult, and during much of the period before 1830 microscopists had reported observing many objects that, in retrospect, can be attributed more to their imagination than to the resolving power of their lenses.

It was therefore not surprising that the proposal that yeast was a living organism was vigorously opposed by the most influential chemists of that generation, who, while in disagreement as to the exact chemical mechanism of fermentation, were united in opposing the biological explanation of Schwann and Cagniard-Latour. As Liebig wrote:

Certain scientists have been misled into explaining the ferment (yeast) as living organisms—as plants or as animals that incorporate the components of sugar for their development, and then expel these as excrements in the form of alcohol and carbon dioxide. In this manner they seek to explain sugar decomposition and the increase in the amount of the added yeast during fermentation. This theory, however, is self-contradictory . . . [since yeast is not a living organism].

The chemists continued to maintain that yeast was lifeless decaying matter and that fermentation should be considered strictly as a chemical reaction. To think otherwise and to seek explanations for the phenomenon in terms of the murky life-associated activities of such a poorly defined material as yeast would only be a return to the vitalism of an earlier era. Schwann and Cagniard-Latour's characterization of yeast as living cells that decompose sugar through their life processes fell into relative obscurity.

Almost a quarter of a century was to pass before a successful attack against Liebig's theory refocused thinking about fermentation on the essentially correct view of Schwann and Cagniard-Latour. This was largely the work of the French chemist Louis Pasteur. Pasteur would demonstrate that yeast were indeed living cells, and that fermentation was an integral part of their growth and metabolism. It would, after all, be necessary to take into account the ''life forces'' of growing yeast cells. But, as we will see, Pasteur placed too much emphasis on the special life forces that might exist within living cells. In the end, it would be necessary to leave even this greatest of French scientists behind, and to understand the internal processes of yeast within the context of the chemical world and without reference to vital forces. *The working out of this dialectic was to be one of the greatest accomplishments of chemistry in the nineteenth century, for, as we will see, it led step by step to the discovery of enzymes and the founding of modern biochemistry.*

Pasteur on Fermentation

Louis Pasteur (1822–1895) was one of the towering scientific figures of the nineteenth century. In addition to his contributions to the fields of microbiology and immunology—both of which he helped to found—he is famous for the statement that "Chance favors the prepared mind," and his own career illustrates this point.

Pasteur began his career as a chemist, and his success in studying small organic molecules, such as sugars, led to his early fame. The German chemist Eilhard Mitscherlich had pointed out that there was something odd about the crystals of the sugarlike compounds tartaric acid and racemic acid. Although they had the same atomic composition

$(C_4H_6O_6)$ and the same physical properties (for example, their melting points), a solution of tartaric acid would rotate the plane of polarization of a beam of light passing through, whereas a solution of racemic acid would not. Pasteur was convinced that this difference in the ability to interact with light must be due to some chemical or physical difference between the two substances. To test this idea, he prepared crystals of the two compounds and examined their tiny, distinctive facets under the microscope. In tartaric acid crystals, the facets were all oriented in the same direction—we might say they were "right-handed." In the case of racemic acid, in contrast, Pasteur found *two* kinds of crystals: right-handed and left-handed. The distinction was so clear that he could separate the two types of crystals by hand.

When Pasteur dissolved the *right-handed* crystals of racemic acid in water, he found that they were identical to the tartaric acid crystals in that they rotated a beam of polarized light in the same direction and to the same extent. The solution of *left-handed* racemic acid crystals, however, rotated the incoming light to an equal extent but in the opposite direction. Pasteur immediately grasped the significance of these results: racemic acid was optically inactive because it was a mixture of two chemically identical—but mirror-image—substances that rotated polarized light in opposite directions and thus in combination canceled each other out.

Pasteur made a further observation about racemic acid, which would prove to be especially valuable when he later took up the study of fermentation. A mold that had contaminated a solution of racemic acid in his laboratory changed the solution from optically inactive to optically active. This effect became more pronounced with time. Pasteur deduced that the mold was a microorganism that was growing by using a specific compound that it obtained from the surrounding environment, preferentially selecting the right-handed constituent of the racemic acid mixture and rejecting the left-handed one. Perhaps it was because of this experience that Pasteur, unlike Liebig, was able to accept the idea that yeast was a living organism. In any case, with this background, Pasteur's mind was prepared when the opportunity to study fermentation arose.

By Pasteur's time, managing the process of fermentation—developed into an art and raised to an industrial

Pasteur in his laboratory. Undeterred when his university diploma labeled him "mediocre" in chemistry, Pasteur believed that will power and hard labor were the key ingredients in success. The dying Pasteur's final advice to his students was "One must work." Pasteur is best remembered for formulating the germ theory of disease, which put the practice of medicine on a firm scientific basis.

level—had become a mainstay of the French economy. The wine-producing centers of Eppernay and Reims were as prosperous as any cities in France. But the control of alcoholic fermentation had always been an entirely empirical affair. Certain temperatures and time periods had been found to be optimal, but no one knew why. If something went wrong and the wine went sour, there was nothing to be done but to start again. Thus it was not unusual when, in 1856, a manufacturer of alcohol in the northern French town of Lille encountered repeated failures in a set of fermentations that were intended to produce commercial alcohol from beet sugar. Time and again the contents of some of the fermentation vats turned acid, and at the end of the process there was no alcohol, only a substance that smelled like sour milk. The man's son happened to be a student at the local college, and the dean of this institution was Louis Pasteur, then a young professor of chemistry. Pasteur was approached and asked if he would investigate the source of these unsuccessful fermentations. He agreed, and thus began the work that established his international reputation.

Pasteur first carried out a chemical analysis of the contents of the vats that had failed to produce alcohol. He found that they contained, instead of ethanol, a considerable quantity of lactic acid. His next step was to examine the sediment from the vats in which alcoholic fermentation had proceeded normally and to compare it with the sediment in those vats in which fermentation had failed. Pasteur had only a student's microscope, allowing a magnification of about 500-fold, but this was sufficient to obtain the crucial result: the sediment from the vats that had produced alcohol contained large globules of yeast, some of which showed buds, suggesting the possibility of active growth. On the other hand, there were no yeast globules in the sediment from the vats that produced lactic acid, but there were "other globules much smaller than those of yeast."

Pasteur drew the straightforward but significant conclusion: that the production of ethanol resulted from one microbe and the production of lactic acid resulted from the other. In other words, the two types of fermentation reactions were carried out by two distinct types of living, growing cells—yeast and the other smaller microbe. The important point to Pasteur was that in each case the "fer-

ment" was a living cell whose growth led to the accumulation of one or another chemical substance.

To reinforce this interpretation, Pasteur sought to demonstrate the two types of fermentation experimentally. Taking a minute sample of the sediment from each fermentation mixture, he inoculated them into fresh flasks that contained sugar solutions supplemented with a variety of inorganic salts and minerals necessary for life. The addition of sediment from the vats that had produced alcohol resulted in a typical alcoholic fermentation, in which the sugar disappeared and yeast globules accumulated at the

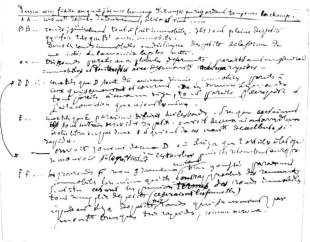

A page from Pasteur's notes concerning yeast and alcoholic fermentation. Pasteur considered the presence of bacteria in fermentation mixtures to represent a "disease of wine." This interpretation may have led to his most important accomplishment, the germ theory of disease, in which he proposed that microorganisms are the causative agents of a number of illnesses in higher organisms. Indeed, Pasteur provided part of the initial proof of this theory by identifying the pathogenic microorganisms responsible for several diseases.

bottom of the flask. In a second flask, the other sediment gave rise to the other type of fermentation: lactic acid was formed, and Pasteur observed an increase in the number of the smaller globules. As a consequence of this work, the problem of the fermentation industry was solved. Pasteur was able to advise manufacturers that, to achieve successful fermentations, vats must be carefully protected from accidental contamination by unwanted microorganisms.

But there were scientific dividends as well. Pasteur's finding represented a distinct advance over the work of Schwann and Cagniard-Latour. He now had more that just a correlation between alcoholic fermentation and the presence of yeast. He had developed an experimental system in which either of the two fermentations could be made to occur at will by the addition of either of two microorganisms.

In subsequent studies, Pasteur went on to show that no one chemical equation (such as the one derived from the work of Lavoisier and Gay-Lussac) was entirely satisfactory for describing a fermentation reaction. On the contrary, Pasteur reported that he always found additional, if minor, products in the fermentation mixture—products such as glycerol, butyric acid, and succinic acid. Which organic substances were produced and in what proportions depended on the specific microorganism that constituted the "ferment" and on the exact experimental conditions.

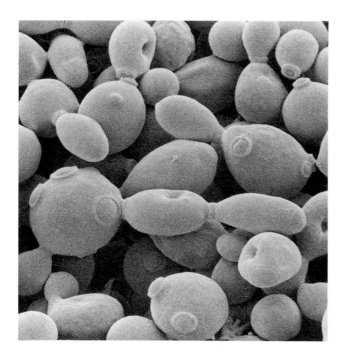

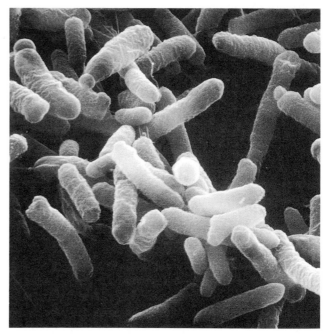

A scanning electron microscope view of a budding yeast cell of the type recovered from an alcoholic fermentation reaction. In the scanning electron microscope objects are observed in three dimensions at a magnification of about 10,000-fold. The portal-like structures are scars where daughter yeast cells have budded off.

In his studies, Pasteur found that some fermentation reactions are overtaken by minute microbes, now known as bacteria. He correctly interpreted the failure of these fermentations to be due to the rapid growth of the bacteria, which led to the production of a variety of unwanted compounds.

From all of these findings Pasteur obtained the first insight into the important fact that living cells possess the ability to transform, or *metabolize,* numerous organic compounds to create a variety of other products. In his view, *the growth of the cell was dependent on such chemical reactions—and, conversely, such reactions were an expression of the life of the cell.*

Pasteur had thus developed a general theory about fermentation—or, more accurately, fermentations—based on the nutritional activation of specific microorganisms. In 1858, he published the first of a series of classic papers. He began by evaluating the role of yeast in alcoholic fermentation:

> In 1835 Cagniard-Latour and Schwann, by employing a more perfect microscope, discovered that yeast consists of cells that grow and multiply. From that time on, the physical and chemical phenomena associated with fermentation have been postulated to be acts connected with the life processes of a little cellular plant—and our subsequent researches have confirmed this view.

> In introducing yeast into a sugar solution one is sowing a multitude of minute living cells, representing innumerable centers of life, each capable of growing with extraordinary rapidity. The globules of yeast are true living cells, and may be considered to have as the physiological function correlative with their life the transformation of sugar, somewhat like the cells of the mammary glands transform the elements of the blood into the various constituents of milk in connection with their life functions.

Pasteur then came to his most important conclusion:

> I see in the act of fermentation a phenomenon correlative with life—a physiological act—giving rise to multiple products, all of which are necessary for the cell. I believe that there is never any fermentation without there being simultaneously the organization, development, and multiplication [that is, the growth] of yeast or other globules, or at least the continued life of the

globules that are already present. The totality of the results seem to me to be in complete opposition to the opinions of M. Liebig.

Pasteur's evidence greatly strengthened the earlier suggestions of Cagniard-Latour and Schwann and made him the leading opponent of the prevailing chemical view of fermentation championed by Liebig. The stage was now set for one of the great debates in biochemistry. Liebig and Pasteur fought out their disagreement publicly in the open literature, and neither would budge from his position. The argument is especially interesting because, as in many good scientific debates, both participants were in fact partly correct and both were partly wrong. What had happened was that both Liebig and Pasteur had sensed part of the overall picture and each had drawn somewhat distorted conclusions. The final irony of the debate was that Liebig was perceived as losing—although he had correctly focused on the most fundamental aspect of the problem (the need for a specific mechanism to explain fermentation), while Pasteur was perceived as winning—although, to achieve his victory, he relied on a quasi-vitalist position that overemphasized the importance of living cells to physiological processes and in the end prevented him from participating in the final, correct solution to the fermentation problem.

Pasteur versus Liebig

By the 1860s, Pasteur was able to force Liebig to concede that yeast was, in fact, a living organism. What had been learned by the microbiologists was now believed by the chemists. Liebig, however, would *not* concede that the life processes of yeast were directly responsible for the chemical reaction that occurred during fermentation. He continued the argument by proposing that the living yeast cells *contained* or *produced* some substance that, when some of the yeast cells died, began to decompose and act in essentially the way he had originally proposed—by transmitting vibrational energy to the surrounding sugar molecules, causing them to change into molecules of alcohol and carbon dioxide:

Liebig in his study.

There seems to be no doubt as to the part that yeast plays in the phenomenon of fermentation. It is through it that some unstable substance is produced that can manifest an action on the sugar. Should the yeast cease to grow, the bond that unites the constituent parts of its cellular contents is loosened, and it is through the motion produced therein that yeast brings about a disarrangement or separation of the elements of the sugar to form other organic molecules.

It was too much like the old theory, and Pasteur's response was scathing. Liebig, he said,

has been compelled to renounce his original opinions concerning the nature of fermentation [that yeast consisted entirely of lifeless, decaying matter], and has now devised an equally unacceptable theory. But what purpose is served by the gratuitous hypothesis of communicated motion? It would be easy to believe that the printer of the Journal has made some mistake, so great is the obscurity of his writing.

Whether we take the new form of Liebig's theory or the old one, neither takes account of the growth and increase of yeast during the fermentation process. So, let us briefly see what Liebig thinks of our experiment in which fermentation is produced by the impregnation of a sugar-salt solution with a minute quantity of living yeast. After deep consideration he pronounces our experiment to be inexact, and the result ill-founded. Liebig, however, is not one to reject a fact without grave reasons for his doing so, or with the sole object of evading a troublesome discussion. "I have repeated Pasteur's experiment," he says, "a great number of times, with the greatest possible care, and have obtained the same results as M. Pasteur, *excepting as regards the formation and increase of the yeast*." It was, however, the very formation and increase of the yeast [its obligatory *growth* during fermentation, as opposed to its use only as a source of decomposing matter] that constituted the point of our experiment. The situation, therefore, is this: Liebig has denied that yeast undergoes growth in a fermenting sugar solution, whilst we have asserted that this growth does actually take place, and is comparatively easy to prove. In 1871 we replied to M. Liebig before the Paris Academy of Sciences and offered to prepare in a solution of sugar and salt, in the presence of a commission to be chosen for the purpose, as great a weight of yeast as Liebig could reasonably demand. Liebig did not accept our proposal, nor did he even reply to our Note—and up to the time of his death, on April 18th, 1873, he wrote nothing more on the subject.

In fact, at the time Pasteur wrote this paper, Liebig had already been dead for three years. But so great was Pasteur's sense of invective that he kept Liebig alive throughout the dialogue of the paper, and the reader would never have suspected, until near the end, that Liebig was no longer an active participant in the debate.

Had Liebig still been alive, he would undoubtedly have responded that Pasteur had misrepresented his position. To Liebig, the real question had become not so much whether yeast was alive but *whether or not the fermenting ability of yeast was intimately bound up with its living state*. From Liebig's point of view, there was a much more important passage in his paper than the one Pasteur had so vehemently attacked. He had written:

It is possible that the only correlation between the phenomenon of fermentation and the "physiological act" is the production, in the living cell, of a substance that, by some special property, may bring about the decomposition of the sugar into other organic molecules. In this view, the "physiological act" [provided by the yeast] would only be necessary for the *production* of this substance, but would have nothing else to do with fermentation.

Liebig could justifiably have asked why Pasteur had not addressed himself to this important question. And, in fact, it was Pasteur's unwillingness to consider the problem from this point of view, and his reliance instead on the special properties that might be inherent in the organization of living protoplasm, that left him unprepared for the final stage of the debate, in which data began to accumulate that cells did indeed produce specific substances that promote chemical transformations.

A Reason for Stubbornness: The Soluble Ferments

There was good reason for Liebig and his fellow chemists not to concede to Pasteur that fermentation depended on the living state of yeast. During the very period in which the debate over fermentation was taking place, several other chemical transformations involving organic matter were being discovered. *And, unlike alcoholic fermenta-*

tion, these transformations were clearly not associated with living cells.

In 1833, Anselme Payen and Jean François Persoz reported that the water solution surrounding germinating barley grains (malt) contained a material that could decompose the starchy interior of seeds into free sugar. The two French chemists then went on to achieve a partial purification of the "active principle"—by allowing the split-open barley grains to settle out, skimming off the overlying fluid, and adding alcohol. This procedure yielded a flaky white precipitate that, when redissolved in water, efficiently converted starch into sugar. Payen and Persoz named their material *diastase* (from the Greek, meaning "to make a break"). Diastase was a remarkably powerful substance. It could transform more than 1000 times its own weight of starch into sugar in a few minutes. *And, most important, diastase differed from the ferment involved in wine making since, as a purified substance, its ability to promote a chemical transformation clearly did not depend on the presence of living cells.*

At the same time that Payen and Persoz were discovering diastase in plant tissue, Theodor Schwann was studying the digestive process in animals. In 1836, Schwann found that the fluid obtained from the gastric mucosa (the lining of the stomach) was capable of bringing about "artificial digestion" in a test tube. Filtration of the initial preparation through cloth and paper yielded a cell-free solution of the "digestive principle." Schwann demonstrated that small amounts of his digestive principle were capable of degrading muscle tissue, the fibrous material of blood clots, and coagulated egg whites. He named this substance *pepsin* and concluded that it digested a variety of animal substances.

Diastase and pepsin were clearly ferments in the broad sense of the word, inasmuch as they promoted chemical transformations. However, they differed from the more widely studied ferments involved in the formation of alcohol or lactic acid, which Pasteur had shown to be living cells. In contrast, diastase and pepsin clearly did not require the higher "organizing" power of intact cells, and, because they were noncellular and found free in solution, they were given the name *unorganized* or *soluble ferments*.

Traube's Hypothesis and Berthelot's Proof

Neither Payen and Persoz nor Schwann had generalized their observations. Diastase and pepsin were considered merely as agents involved in two isolated and unrelated chemical changes. They were not thought of as examples of any more broadly based theory of cellular chemistry. This, however, was exactly the view taken by the Polish chemist Moritz Traube.

Traube (1826–1894) was a gifted scientist who brought together the salient facts known about ferments in 1860 and framed them in terms of one unified theory. Traube believed that neither Liebig nor Pasteur, the major theorists of the day, had fully appreciated the significance of the soluble, or unorganized, ferments. Each had chosen instead to emphasize a different (and, to Traube, misleading) aspect of the fermentation problem. Liebig had been overly concerned with the specific way that ferments might bring about chemical transformations, but he had no actual evidence to support his particular mechanism. Hence, Traube took the view that "Liebig was wrong in saying that ferments are substances in a state of decomposition, which can transmit to ordinarily inert substances their chemical vibrations." Moreover, Traube realized that, although Pasteur had shown that many fermentations were carried out by living cells, there was no direct evidence supporting the view that fermentation was indissolubly linked to the living state of the cells, and therefore "Pasteur's hypothesis that fermentations are to be regarded as the expressions of the vital forces of lower organisms is unsatisfactory." Traube thus felt free to propose that *substances analogous to the known soluble ferments existed inside cells* and were responsible for carrying out such processes as alcoholic fermentation. The proposed "intracellular ferments" would be specific, reaction-promoting particles—just like diastase and pepsin, which functioned outside cells. In his view, there was no need to ascribe alcoholic fermentation to any special, unexplained forces in living cells. There was really no fundamental difference between such processes as the fermentation of glucose to produce ethanol and carbon diox-

Moritz Traube made the farsighted proposal that soluble ferments exist within cells, where they carry out specific chemical transformations. His work in science was all the more impressive because it was accomplished in a laboratory in his attic; the death of his brother had obliged Traube to leave behind his training in physiology and medicine in order to take over the family wine business.

ide and the transformation of starch into free sugar molecules—other than the *location* of the particlelike ferments.

Traube went on to emphasize that this was a broadly applicable idea—chemical transformations by extracellular and intracellular ferments could explain not only "the most important vital–chemical processes in lower organisms, *but in higher organisms as well.*"

Traube's proposal was the first formulation of the idea that *chemical transformations in all life forms—not just microorganisms—might be traced back to the actions of specific, reaction-promoting substances (analogous to diastase and pepsin) that would function like a set of cellular machine tools.*

Berthelot Isolates a Soluble Ferment from Living Cells

Within two years, Traube's hypothesis had received strong support from the experiments of one of Pasteur's countrymen, Marcelin Pierre Eugène Berthelot (1827–1907). In 1860, Berthelot macerated yeast—the very organism that was at the heart of the controversy surrounding alcoholic fermentation—and obtained a "soluble" ferment capable of breaking apart a particular compound sugar, *sucrose* (cane or table sugar), into its two constituent simpler sugars, glucose and fructose. He named his substance *invertase.* Like diastase and pepsin, invertase could be purified away from all living cellular material by filtration and precipitation with alcohol. *But invertase was special in that it was a ferment extracted from the living yeast cell itself.*

In discussing his findings, Berthelot noted the work of Payen, Persoz, and Schwann—and then drove home the important point:

In the case of invertase (and other soluble ferments, such as diastase and pepsin) it is seen clearly that the living cell is not the ferment, but gives rise to it. Moreover, once the soluble ferments are produced, they act independently of any further vital act; their activity shows no necessary correlation with any physiological phenomenon. I insist on these words in order not to leave any ambiguity about my way of regarding the action of soluble ferments.

Berthelot's experiments provided the first example of a ferment that was *normally* contained within cells but could be released and still function, thus acting long after the cell was no longer intact and alive. It was clear that the living cell itself could not be the ferment. Rather, the cell was the producer of the ferment, which was then merely one of its internal components. When considering the mechanism of the chemical transformations carried out by these ferments, the most important point to Berthelot was that, because ferments such as diastase, pepsin, and invertase were capable of acting outside cells, there was no need to consider any physiological phenomenon or "vital life force" that might be associated with living cells.

Berthelot now made the proposal that brought him into conflict with Pasteur:

If a deeper study leads to the extension of the view that I propose here, and to its application to the insoluble [cellular] ferments, as well as to the soluble ferments, then all fermentations would be brought back to the same general concept.

The interests of Marcelin Berthelot, who recovered the first soluble ferment (enzyme) from cells, extended beyond chemistry into higher education and politics. He served the French government as minister of public instruction and went on to briefly become minister of foreign affairs.

In other words, it was possible that all chemical transformations in living systems would eventually be found to be promoted by ferments such as diastase, pepsin, and invertase—ferments that, instead of being secreted, were retained within cells. It was only one step further to suggest that the cell was simply a collection of ferments, each of which would ultimately be extracted and purified, like any other chemical substance.

Pasteur and the Vital Force

Pasteur could not fail to respond to Berthelot and Traube. He had a clear choice. On the one hand, he could accept the proposal that alcoholic fermentation was explainable under a general theory in which chemical transformations were caused by specific ferments that reside within cells—and thus retreat from his position that the living state of yeast was essential for the fermentation process. Alternatively, he could attempt to stand by his dictum "No fermentation without life."

In the end he decided to accept Berthelot's data—but to reject their general applicability.

It is true that there are circumstances other than the fermentation of grape sugar to yield ethanol under which yeast can bring about modifications in organic substances. It has been found that yeast imparts to water some material (invertase) that cleaves cane sugar into glucose and fructose. And M. Berthelot has shown that this substance may be isolated by precipitating it with alcohol, in the same way as diastase can be precipitated from its solutions. These are remarkable facts, *but they are at present only vaguely connected with the alcoholic fermentation of sugar by means of yeast.*

To deal with such substances as diastase, pepsin, and invertase, Pasteur preferred to view the world of organic transformations as divided into two parts: "fermentations properly designated as such"—the type of reaction he studied, and which, as far as the available evidence was concerned, required living and growing cells—and "other fermentation-type reactions," which were carried out by the soluble or unorganized ferments. The two realms were not equal. Pasteur regarded alcoholic fermentation as representative of the more important life functions of the cell, such as growth, assimilation, and respiration. These "higher" life processes were presumed by the majority of biologists at that time to require the full complexity of living protoplasm. In contrast, the "other" types of fermentations (such as those carried out by diastase, pepsin, and invertase) were simple cleavage reactions that occurred outside of cells and, as such, were seen as "lesser" activities that did not reflect the essence of the living state.

To bolster his distinction between the two types of fermentations, Pasteur offered one line of experimentation.

The researches in which we have proved the existence of living ferments in many fermentations have established beyond doubt that there are profound differences between these and the phenomena connected with non-cellular "soluble ferments." For instance, the ferments of Fermentations Proper multiply and reproduce themselves during the process, whilst the others do not. Moreover, it has been shown that poisons like chloroform, ether, cyanide, salts of mercury, and the oils of turpentine all stop alcoholic fermentation, whilst in no way interfering with the action of cell-derived substances such as diastase.

We consider such results as a proof that alcoholic fermentation is dependent on the life of the yeast cell, and that a distinction should be made between the two classes of fermentations.

Pasteur then reiterated his most basic conclusion:

The chemical act of fermentation is essentially a phenomenon correlative with a vital act, beginning and ending with the growth of a living cell.

Pasteur believed that the chemical transformation of sugar into ethanol and carbon dioxide could not possibly occur in the absence of intact, *living* cells. Destroy the cell (as with "hydrogen cyanide or salts of mercury") and the special nonisolable force woven into the fabric of its components would also be destroyed. Chemical transformations that might occur free in solution, such as the conversion of cane sugar into glucose and fructose by the soluble ferment called invertase, were fundamentally different and should not be allowed to influence thinking about more important processes such as alcoholic fermentation.

Indeed, the German biochemist Wilhelm Friedrich Kühne (1837–1900) supported Pasteur's view by giving the soluble ferments their own name: *enzymes*. This term was initially invented to segregate the soluble ferments into their own small area of biological chemistry. It would eventually, of course, come to encompass *all* chemical transformations in the biological world, including those "higher" cell functions held in reserve by Pasteur, but the time was not yet right.

Thus, in 1878, calcified in his point of view, Pasteur wrote:

It is possible for someone to agree with me when, first, it is accepted that Fermentations Proper require as an absolute prerequisite the presence of *living* microorganisms. Will M. Berthelot or will he not contradict this position—not with *a priori* points of view, but with serious facts? If yes, let our fellow member of the Academy have the kindness to say so; if no, there is nothing for us to discuss.

Thus, at the height of the fermentation debate, the most important point of all eluded Pasteur—that all types of fermentation do indeed reflect an underlying theme. As we will see, cells contain substances that promote various reactions, and these substances can either function inside the cell *or* be released and function in a suitable external environment. And most important, these substances, while truly remarkable, have nothing to do with the vital, living forces of the cell. On the contrary, they give rise to what appears to be the vital force of the cell. But it would

be a quarter of a century before this became evident, and in the meantime Pasteur's viewpoint largely prevailed.

Pasteur in Retrospect

In making the artificial distinction between types of transformations in the organic world and refusing to comment seriously on anything but "fermentations properly designated as such," Pasteur had taken a serious wrong step in the fermentation debate. Basically, he had failed to deal with the issue explicitly raised by Traube and Berthelot and implicit in the studies of invertase and the other soluble ferments. The intriguing question is why.

First, of course, there are Pasteur's stated reasons. But these were not entirely compelling even at the time, and, in retrospect, they can be seen to be invalid. For example, we now know that the interaction of cells with many poisons is complex and can result in the creation of a hostile environment in which intracellular ferments will not work.

Only after his death did it become known that, beyond the reasons specifically given by Pasteur for his attitude about the requirement for living cells in fermentation, there was perhaps an even more important reason. Pasteur had broken open yeast cells in an attempt to find an intracellular ferment that would transform sugar into alcohol and carbon dioxide. He had, however, been unsuccessful. This *negative result,* in the hands of such a gifted experimenter, could only have reinforced his belief that special forces or patterns of protoplasmic organization (in essence, vital forces) might be operating in the living yeast cell and are destroyed as soon as the cell is disrupted. To draw such a conclusion on the basis of negative data is, as Pasteur well knew, extremely dangerous, because one can easily obtain a negative result if the conditions chosen for an experiment are not exactly right. He therefore did not publish this experiment. Nonetheless, Pasteur may have been influenced by his inability to release a soluble alcoholic ferment from yeast.

Even more subtle pressures may have acted on Pasteur to determine his position in the fermentation debate.

Like many other highly motivated scientists, he was probably somewhat reluctant to accept a theory that was not his own, and that was at such variance with his own way of thinking. A leading scientist is, by definition, in front, and loses his position if he accedes to a rival. This happens, of course, but not easily—and only when sufficient evidence has accumulated to force a change. Making matters worse, in this case the change required the acceptance of a rival theory that was for the most part associated with the chemical tradition of Germany, a country that had only recently initiated and won an aggressive war against France.

It would, of course, be a distortion of history to view the fermentation debate between Pasteur, on the one hand, and Liebig, Traube, and Berthelot, on the other, as a battle between right and wrong in which those scientists who held the ultimately correct view were the heroes and their opponents, the villains. Science is an interactive exercise in which the very process of discussion leads to further experiments and to progress. Pasteur was only upholding a long-standing scientific tradition by not immediately accepting a general theory that had been proposed solely on the basis of a special situation (the discovery of the soluble ferments and their arguable applicability to intracellular processes). The proof that Pasteur demanded was to come, but not in his lifetime.

Berzelius: The Concept of Catalysis

Liebig, Pasteur, Berthelot, and Traube had all been struggling to understand the nature of the forces that brought about alcoholic fermentation and, more generally, the chemical transformations of living systems. The fundamental question, however, remained at issue: whether cellular reactions followed the laws of ordinary chemistry or might involve special life forces.

Liebig's proposal had been conservative and somewhat reasonable—a Newtonian transfer of vibrational energy. Although his specific ideas about the cause of fermentation, based on decomposition, had been largely discredited by Pasteur, there was a sense in which Liebig

had been on the right track. The notion that the sugar molecule was vibrated to pieces under the influence of an energy transfer was the beginning of thinking about what we would today call a *reaction mechanism*. At least Liebig was worrying about the crucial point—that is, how an ordinarily stable sugar molecule could be transformed into a different set of molecules.

Pasteur's biological intuition about reaction mechanisms had led him to expect that special forces or patterns of organization would exist in living cells. He rejected the simple mechanical proposal offered by Liebig, preferring instead that "the chemical act of fermentation should be considered as a phenomenon correlative with a vital act, beginning and ending with the growth of the living cell." To Pasteur, there might well need to be new laws of nature to explain the chemical reactions of living cells—and not necessarily just "the ordinary laws of chemistry."

Thus, in 1875, the most basic question about the forces underlying chemical transformations in living systems was still unanswered: How were biochemical reactions brought about?

The Swedish chemist Jöns Jakob Berzelius (1779–1848) had actually published the essence of the correct answer in an elegant study in 1835. This work had been largely overlooked during the battles between Liebig and Pasteur and then between Pasteur, Berthelot, and Traube. But now, forty years later and long after Berzelius's death, the time was right for this work to become fully appreciated.

Berzelius was one of the greatest chemical intellects of his time. He was the discoverer of no less than three elements (cerium, selenium, and thorium). He analyzed more than 2000 compounds with his own hands and determined the "combining ratios" of 43 elements. As the author of five editions of a multivolume textbook, he maintained an extensive grasp of the state of knowledge in all aspects of his science. His laboratory consisted of two unheated rooms, in which he worked 12 to 14 hours a day. From this laboratory, he published hundreds of important papers, not the least of which was his treatise on the mechanism of chemical reactions. Some important scientific papers, such as Gregor Mendel's on the existence of genes, are models of experimental clarity and deduction. Berzelius's is a model of the other type of great scientific

Among the many achievements of Jöns Berzelius was the introduction of the present system of writing chemical formulas—using the first letter of the Latin (or Greek) name of the element as its symbol and adding a subscript giving the number of atoms of that element in the compound.

ity was discovered. It soon became apparent that electrical and chemical relationships are one and the same thing, and that the affinities that are finally selected during chemical reactions merely result from the movement of atoms toward more perfect reciprocal electrical relationships. From that time on, the formation of new compounds could be explained in terms of an encounter of substances whose electrical relationships could be neutralized better by a transposition of their constituents.

We also learned from the experiments of that period that rearrangements leading to new chemical combinations could often be promoted by such external factors as heat, and sometimes light.

Berzelius then considered how chemical transformations might occur in living systems.

If we turn with the experience gained from inorganic Nature toward the study of the chemical processes that occur in living Nature, we find that the organs of the body produce substances with highly differing compositions. In animals the raw material is blood, and one sees an uninterrupted uptake of this fluid by the organs, and the discharge of milk, bile, urine, and so on from their openings. But the more we consider such processes the clearer it becomes that here is something for which inorganic Nature has not yet provided a key— [for such chemical reactions could not possibly be driven by, for example, heat, which would not be tolerated in a living system].

paper—that is, based on synthetic reasoning and intuitive insight outside the context of a specific proving experiment. In its careful intellectual searching from beginning to end, one sees the essence of a classic theoretical paper.

Berzelius entitled his paper "On a So Far Rarely Observed Force Which Is Probably Active in the Formation of Organic Substances." He began by discussing the way in which chemical reactions occur in the inorganic world.

New compounds are formed in inorganic Nature by the mutual interaction of substances. Up to 1800 one did not understand the basis for the affinity of substances for one another. At that point the influence of electric-

Because of his comprehensive knowledge of the literature, Berzelius was able to see the relevance of several previously unrelated findings and draw them together. The first was the work, fifty years earlier, of Gottlieb Sigismund Constantin Kirchhoff, the Russian royal apothecary in St. Petersburg. "Kirchhoff," wrote Berzelius,

has discovered a new driving force for some chemical reactions. He found that when starch [a chain of sugars] is placed in dilute acid, it is broken down into free sugar molecules.

The amazing finding, both from the work of Kirchhoff and also in experiments repeated by Berzelius, was that

> when we determined what the acid had given to or taken from the starch, so that it could form sugar, it was found that during this chemical reaction nothing gaseous escaped, nothing was found combined with the acid, all the acid originally present could be removed again with alkali [base], and in the final liquid only sugar was found.

The intriguing observation was that acid promotes a clearly defined reaction, yet is not one of the reactants and is not used up.

Berzelius recognized in Kirchhoff's finding a similarity to another set of experiments carried out by a French chemist, Louis Jacques Thenard.

> Thenard recently discovered a liquid—hydrogen peroxide—in which only a weak force holds the constituents in combination with each other. Under the influence of acid, the combination of the constituents remains undisturbed. Under the influence of alkali, however, the tendency of the constituents to separate is stimulated, and a type of slow fermentation occurs during which oxygen gas escapes, and in the end only water remains.

Today we know this reaction as the decomposition of hydrogen peroxide:

$$2H_2O_2 \xrightarrow{\text{Alkali}} 2H_2O + O_2$$

Berzelius went on to note that

> not only substances that can be dissolved in hydrogen peroxide (such as alkali) promote this decomposition;

but even some solids bring it about—for example, gold, silver, platinum, and manganese dioxide. The substances that cause the decomposition of H_2O_2 do not achieve this goal by being incorporated into the new compounds (H_2O and O_2); in each case they remain *unchanged* and hence act by means of an inherent force whose nature is still unknown, although its existence can be readily observed in this manner.

That a solid, particle-like substance could promote a chemical reaction without itself being changed

$$2H_2O_2 \xrightarrow{\text{Gold, silver, platinum, or manganese dioxide}} 2H_2O + O_2$$

was the most intriguing fact of all. Intuitively, Berzelius began to move toward the idea that the chemistry of living cells might be based on such particles—and presumably these particles, in contrast to heat or ordinary acids and bases, would be acceptable to physiological systems.

> Turning to the living world, let us think about the conversion of sugar to carbon dioxide and alcohol, which occurs during fermentation under the influence of a substance known by the name of ferment [yeast]. When compared with phenomena known in inorganic Nature, fermentation most closely resembles the decomposition of hydrogen peroxide under the influence of gold, platinum, or silver. It is very natural to imagine that an analogous activity exists on the part of the ferment involved in alcohol production.

The basis for the analogy lies in the fact that, in both cases, there is a conservation of matter

$$2H_2O_2 \rightarrow 2H_2O + O_2$$
$$C_6H_{12}O_6 \rightarrow 2CH_3CH_2OH + 2CO_2$$

and the causative agents—gold or platinum during the decomposition of hydrogen peroxide, and the ferment or yeast, in the case of fermentation—therefore cannot be considered to participate in the reaction by donating or accepting atoms. In this sense they are unchanged. Hence:

We have here a new force, belonging both to inorganic and to organic Nature, for bringing about chemical reactivity through the action of certain materials, simple as well as complex, solid as well as liquid. They accomplish this by bringing about a reordering of the components of a substance into other relationships *without a necessary change of their own components* and therefore act by means of an influence very different from ordinary chemical reactivity.

This force is almost certainly more widely distributed than had been thought up to now, and its nature is still hidden. When I call it a new ''force,'' however, I do not at all mean to imply that it is a capacity which exists independent of matter; on the contrary, I assume that it represents a special way by which some forms of matter can function.

Berzelius is clearly not prone to vitalism; he no more believes that a reaction-promoting agent such as yeast employs a vital force when it brings about alcoholic fermentation than he considers gold to use a vital force when it promotes the decomposition of H_2O_2. The new reaction-promoting reagents might be mysterious, but they are part of the known world of atoms and molecules.

Having clearly described the new force, Berzelius goes on to give it a name—a name that has become almost synonymous with biochemistry.

So long as the nature of the new force remains hidden, it will help our researches and discussions about it if we have a special name for it. I hence will name it the catalytic force of the substances, and I will name decomposition by this force ''catalysis.'' *The catalytic force is reflected in the capacity that some substances have, by their mere presence and not by their own reactivity, to awaken affinities that are slumbering in molecules* at a given temperature. As a consequence of this force, the elements in a complex substance become organized into other relationships.

Berzelius, in an impressively prescient passage, then posed two of the most important questions about catalysts in a way that forecast the coming century of research:

On the whole, catalytic substances behave in somewhat the same way as heat—in that they promote changes in otherwise stable and unreactive molecules. One can hence ask whether, like heat, a substance with catalytic properties is able to bring about a change in many compounds, or whether it catalyzes changes only in some substances without acting on others. Another question is whether substances with catalytic force yield various products from a given compound, or whether, in contrast to heat, they are specific in their catalysis, with each promoting just one type of chemical change.

In sum, if we turn with the idea of catalysis from the inorganic world to the chemical processes in living Nature, we see things in a completely new light. We find solid reasons to assume that, *in living plants and animals, thousands of catalytic processes occur within the tissues and fluids, generating a multitude of substances of differing chemical compositions* whose formation from the common raw material, the plant sap or the blood, we could never understand before. In the future we will perhaps discover these processes to be due to catalytic substances in the tissues and organs of which the living body is composed.

Both Pasteur and Liebig were aware of Berzelius's work. It had been republished in 1837 and had received considerable attention and approval among inorganic chemists. Further, the analogy to the reaction-promoting soluble ferments, such as diastase, pepsin, and invertase, is, at least in retrospect, readily apparent. However, Pasteur never mentioned Berzelius's work, even though it had explicitly raised the possibility that catalysts might underlie the fermentation process. As for Liebig and his followers, they failed to see anything useful in Berzelius's analogies and concepts:

To call a phenomenon catalytic is not to explain it; it is nothing but the replacement of a common word by a Greek word.

In the nineteenth century, chemists were cantankerous—and great chemists were greatly cantankerous.

Fermentation Outside the Living Cell

As the nineteenth century drew to a close, all of the facts and ideas necessary for the emergence of modern biochemistry had become available. These were:

- the knowledge that living cells actively promote specific chemical reactions (the work of Schwann, Cagniard-Latour, and Pasteur);

- the fact that—at least in some instances—substances *released* from cells have the ability to promote reactions (the work of Payen and Persoz, Schwann, Berthelot, and Traube);

- the realization of the need for precise chemical reaction mechanisms to explain the transformations that occur in living systems (the work of Liebig); and

- the subtle concept of catalysis: the ability of some substances, including particle-like substances, to promote chemical reactions—without themselves being used up during the process (the work of Berzelius).

The moment of synthesis, when all of this work would be drawn together, was now at hand. Remarkably, the facts and ideas that would be merged had come from scientists who, in their own lifetimes, found in them mostly a basis for disagreement. Of the major protagonists, only Berthelot lived to see the final synthesis. The rest had all died—Berzelius long before, in 1848; Liebig in the midst of the battle in 1873; and Pasteur just in time to escape a major scientific defeat, in 1895.

What stood in the way of the final resolution was Pasteur's artificial subdivision of the reactions in the living world into two spheres: the higher realm of "fermentations properly designated as such," which required living cells—and "other fermentations." This was an essentially arbitrary compartmentalization that did little more than provide a refuge for keeping vitalism alive for an-

other twenty years. Further progress occurred only when Pasteur's two-part world collapsed in the face of a key new experimental finding.

Fermentation in a Test Tube: The Discovery of Zymase

The debate between Pasteur, Liebig, Berthelot, and Traube had been silent for nearly twenty years. The force of Pasteur's personality and the lack of evidence showing that any important intracellular processes (such as fermentation) were actually carried out by soluble ferments—enzymes—left in doubt any general role for such substances in the life of the cell. Instead, an intricate theory of living protoplasm had arisen in the second half of the nineteenth century, and it held sway among a large number of physiologists. According to their theory, protoplasm had an elaborate structure, being widely regarded as a single complex molecule, or "biogen," with a multitude of reactive appendages attached to a stable core. Each appendage carried out a specific vital function made possible by the special complexity of the biogen molecule. The protoplasmists found only limited use for the known soluble ferments (such as pepsin and invertase) and consigned them to an extracellular realm where they carried out simple cleavage reactions involved in digestion.

In 1897, however, the smoldering ashes of the old fermentation debate were raked into flames by an accidental discovery that proved within a few years to be the cornerstone of modern biochemistry. It is a credit to the versatility of the scientific method, if not to its dignity, that this epoch-making achievement was not the outcome of an orderly intellectual process but the unplanned result of an experiment designed for a totally different purpose.

The story begins in the laboratory of Hans Buchner (1850–1902), a Munich bacteriologist who was attempting to prepare immunologically reactive substances from yeast and other microorganisms. It was in this context that he and a colleague, Martin Hahn, were developing new ways to break open yeast while maintaining their contents in an active form. The problem was that the extracts decayed within hours of their preparation, and Buchner and

Hans and Eduard Buchner laid the cornerstone of modern biochemical research with their achievement of fermentation in a test tube. The ability to recover enzymes from cells and study them is central to the understanding of modern biology.

Hahn therefore decided to try using a number of different preservatives. One of these was sucrose—cane or table sugar—added to the cell extract until it reached a concentration of 40 percent. The addition of a high concentration of sugar to prevent degradation of organic mixtures was a traditional method for making preserves—it is still used today in making jams and jellies. The sucrose exerts a lethal effect on any living cells that might fall into the mixture, such as yeast or bacteria, by drawing water from them.

Just such a yeast cell extract supplemented with sucrose had been prepared when Hans's brother Eduard Buchner (1860–1917) arrived in Munich to spend his vacation. Eduard saw, and was instantly intrigued by, a steady stream of bubbles generated in the preparation. As a chemist, he was familiar with the problem of fermenta-

tion and realized the significance of the bubbles. The complex mixture of substances derived from the broken yeast cells was evidently attacking the preservative, sucrose, cleaving it into free glucose and fructose, and then further processing the glucose to produce ethanol and carbon dioxide.

Here is an excerpt from Eduard Buchner's classic paper, which was to have such a profound influence on the course of subsequent events. It is so clearly written that the reader has the impression of virtually standing next to the researcher at the laboratory bench:

Previously it has not been possible to separate the process of fermentation from living yeast cells. However, this goal has now been achieved.

If one mixes brewer's yeast with an equal weight of sand and then grinds the mixture, the mass becomes moist and pliable. Now, if water is added, and the paste—wrapped in cheesecloth—is gradually subjected to several hundred atmospheres of pressure in a hydraulic press, one breaks open the cells and obtains "press juice." To remove any residual unbroken cells, the press juice is passed through a paper filter. The "cell extract" obtained in this way is a clear, slightly yellow liquid with a pleasant, yeasty smell. It is a collection of substances derived from the cell interior.

The new pressure procedure, unlike many others, had left the cell contents relatively unharmed, and this enabled the Buchners to make their seminal observation.

The most interesting property of the press juice is its capacity to bring about the fermentation of added carbohydrates. When it is mixed with sugar, one finds that beginning about 15 to 60 minutes later, there is a continuous evolution of carbon dioxide, which may go on for days. After three days of fermentation one may conveniently determine the amount of alcohol formed. In one experiment with 50 milliliters of yeast juice, 1.5 grams of ethanol was formed. [That is, in the end, the solution had become 3 percent ethanol.]

The fact that the fermentation reaction had occurred *in vitro* (literally "in glass," that is, in a test tube) allowed the important conclusion:

With respect to the theory of fermentation, the following statement can now be made: it is *not* necessary to have an apparatus as complicated as the living yeast cell for fermentation to take place. Rather a soluble ferment, or enzyme, is to be considered as the carrier of the fermentative activity of the press juice. We will name this subcellular agent of fermentation "zymase."

Thus two brothers, one a microbiologist and the other a chemist—neither one setting out to solve the problem of

fermentation—had succeeded where Pasteur (and many others) had failed. They had demonstrated that an extract of yeast, freed from intact cells by filtration, nonetheless retained the ability to convert sugar into alcohol. The process of alcoholic fermentation could now be understood outside the context of the living cell. At the same time, the soluble ferments, or enzymes, that had previously been relegated to a few simple, extracellular cleavage reactions, suddenly became much more important—taking on an *intra*cellular and much broader role.

The Response to Buchner's Work

One does not create an earthquake without moving boulders, and boulders cannot easily be moved. Although the publication of Eduard Buchner's paper created an immediate response in the scientific community, five years were to pass before the work and its interpretation gained general acceptance. The problem was that the apparently straightforward experiment was too important to accept at face value.

Carl Voit, a well-known Munich physiologist, was typical of the scientists who participated in the vigorous debate that ensued. He was willing in principle, he said, to accept the idea of an alcohol-producing enzyme, but he had reservations about the interpretation of the new study.

Ever since it became possible to extract soluble ferments from certain parts of the animal body [for example, pepsin from the digestive juices], it has been tempting to think that the various chemical transformations in muscle, liver, yeast, etc., were also due to extractable ferments. But soluble ferments for these cellular processes have not been found, and cautious workers have had to conclude that enzymes were not involved, but rather unknown agents in the organized cell. I have often stated that, as soon as substances were extracted from cells having the activity of the cell, I would accept these substances as the cause of the vital reactions.

Voit, however, was not at all sure that Buchner had accomplished this feat. He raised two major objections.

First, the extract was far less active (only a few percent, in fact) in producing ethanol and carbon dioxide than an equivalent amount of live yeast. This, Voit suggested, might mean that the observed production of ethanol *in vitro* was occurring by some minor mechanism, different from the way in which living cells carried out fermentation. Buchner had a perfectly reasonable answer: the low activity of the press juice could be explained by assuming that much of the zymase was destroyed upon breaking open the cells.

But Voit's second objection was more troublesome. He questioned how sure Buchner could be that the active agent in the extract was not just pieces of living protoplasm. This possibility could not be lightly dismissed. The absence of microscopically visible bits of protoplasm was not, in itself, sufficient proof that the juice pressed out of living yeast was completely lifeless. To Buchner, however, the argument that the cell extract still contained bits of intact living protoplasm was pure speculation. He maintained that he had established a *prima facie* case for the simpler explanation that alcoholic fermentation could be completely dissociated from the vital principle of the living cell:

> When specific chemical reactions can be made to occur *in vitro,* such advances should not be denied by saying that the reactions in the cell are due to the whole protoplasm, unless there are compelling reasons to do so. The presence of such special mysterious agents in press juice—these supposed pieces of living protoplasm— must be demonstrated before the simpler enzyme theory, which agrees with all the facts, is discarded.

This application of Occam's razor probably satisfied those scientists who worked on such substances as diastase, pepsin, and invertase, and therefore may have been predisposed to believe in the new soluble alcoholic ferment. But to others the question of interpretation remained open. *Nonetheless, if the biogen molecule could break off pieces that retained their functionality outside the cell, then the protoplasm theory had begun to merge with the enzyme theory.*

But no sooner had the theoretical debate reached this stable state than the worst possible event occurred. Reports began to appear that Buchner's experiments could not be repeated. Max Delbrück, Sr., and Heinrich Will, both of them leading figures in brewing research in Germany and authorities on the physiology of yeast, and a British professor of botany, Joseph Reynolds Green, reported that they were unable to prepare an active press juice. To this, the German chemist Alfred Stavenhagen added insult to injury. Even though he had obtained results that were not inconsistent with Buchner's findings, he nevertheless rejected zymase on the grounds that it "was in complete contradiction to Pasteur's theory." Buchner could only respond that Pasteur's authority was no argument against new experimental findings.

These negative results did not go unnoticed, and another German chemist, Carl Wehmer, summarized the case for the disbelievers in 1898. In a decidedly hostile attack on Buchner's work, Wehmer cited the various negative reports that had appeared and concluded that the claim of fermentation by a cell-free enzyme was nonsense:

> The hypothesis of zymase is completely unsubstantiated, and it deserves to be publicized no further. This reawakened version of Traube's old fermentation theory will, in all appearances, enjoy none too long a life.

No more serious a matter can arise in a scientist's career than for his peers to be unable to reproduce his experiments. The more important the scientific claim that is made, the more certain it is that other researchers will want to confirm the result and build upon it. Their failure even to reproduce the initial experiment casts doubt on the original investigator's competence and integrity—his most important scientific possessions. For what must have seemed to Hans and Eduard Buchner an interminable period, but was in reality less than eighteen months, the fate of zymase hung in the balance.

The reports of the death of zymase proved to be premature. Even as Wehmer was making his pronouncement, the tide was already turning. Delbrück's assistant Lange reported that he had obtained an active—in fact, very active—alcohol-producing extract. Within months, the two other doubters, Will and Green, also began to report obtaining active extracts. In retrospect, the problem had been that the Buchners' experiment was not as easy as it seemed; details of technique, and the particular strain of yeast used, made the difference between success and failure. One researcher was to remark that "at first we too got no positive result, while later we never failed." Every experimental scientist has had a similar experience.

The View from the Pasteur Institute

In the end, zymase also picked up important support from an unexpected source, the Pasteur Institute, which had been established in 1888 to honor France's premier scientist. The Institute's director, Émile Duclaux, gave the work his enthusiastic approval:

> This report is a considerable event in the history of science, and opens up a rich field of research. For a long time enzymes were believed to carry out only simple cleavage reactions. Buchner's alcoholic enzyme is more complex—it is the first to cause substantial changes and rearrangements of chemical groups within molecules, that is, the restructuring of molecules.

> I am aware that attempts have been made to explain the action of enzymes by a vital force derived from the organisms that make them. But it is becoming increasingly clear that this is pure mysticism.

In a moment of bittersweet reflection, Émile Roux, one of Pasteur's most famous successors, oscillates between loyalty and reality:

> The discovery of the alcoholic enzyme is unquestionably one whose importance will never be dimmed. We can ask whether, as some say, it begins a new era by ruining Pasteur's theory of fermentation. To be sure, fermentation by zymase is a purely chemical act, but

the *formation* of the enzyme may still be considered, in one sense, to be a vital act, inasmuch as the living cell makes the enzyme.

With the acceptance of Buchner's work, the protoplasm theory collapsed, and with it the outer wall of the fortress of vitalism. It was now clear that living cells with their presumed vitalistic properties were not a requirement for the chemical processes of fermentation, and also, by implication, that subcellular particles that could survive the death of the cell were responsible for carrying out all of the major chemical transformations in living cells.

Most important of all from the point of view of further experimentation was the fact that the cell's fermentation-producing enzymes, because they could now function *in vitro*, were no longer hidden within the black box of the cell and were therefore accessible to direct laboratory analysis.

Of course, not everyone was completely satisfied. Adolf von Baeyer, Eduard Buchner's former professor, commented: "This will bring him fame, even though he has no chemical talent." This sour judgment of a harsh and perhaps jealous teacher proved to be only partly correct. In 1907, Eduard Buchner was awarded the Nobel prize in chemistry "for his biochemical researches and his discovery of cell-free fermentation." Even if Baeyer did not realize it, Buchner had Pasteur's "prepared mind." It may seem strange that Hans Buchner, in whose laboratory much of the actual work was done, was not included in the Nobel prize. This, however, is almost certainly due to a particular requirement of the Nobel committee—that the prize be awarded only to living scientists. Hans Buchner had died in 1902. Eduard's career continued in a successful vein for many years thereafter. He became a professor at the University of Würzburg. In 1917, at the age of 57, he volunteered for active duty in the German army, his second tour of duty during World War I. He was wounded on the eastern front and died two days later. The picture shown at the top of the following page is taken from his obituary, and shows Captain Buchner, the Nobel laureate and founding father of biochemistry, in military uniform surrounded by the officers of his regiment.

Eduard Buchner is surrounded by members of his regiment.

The Enzyme Theory of Life

Long before Eduard Buchner died in 1917, the study of the chemistry of living organisms had been completely transformed, so great was the impact of the discovery of zymase. The demonstration that an isolable subcellular substance could promote a process such as fermentation was the turning point away from the vitalistic belief that the reactions of living cells were carried out by the whole cell protoplasm. Those who realized the importance and generality of the zymase work saw it as the triumph of new ideas over old. Using zymase as a paradigm, an expanding number of chemists—now calling themselves *bio*chemists—began to base their research on the idea of soluble enzymes as the explanation for cellular physiology.

And if fermentation could be explained in terms of enzymes, why not all the activities of living cells and of the organism as a whole? As ambitious as this idea was, it was true.

The ''enzyme theory of life'' was formally enunciated at the dawn of the twentieth century by Franz Hofmeister, who predicted that all cellular reactions would be found to be controlled by specific enzymes.

There are now a number of cases in which "intracellular ferments" have been brought to light from within the cell and their significance for vital processes made clear. Indeed, these discoveries, increasing almost daily, have revealed so universal an occurrence of ferments in organisms, and such a variety of modes of action, that we may be almost certain that sooner or later each vital cellular reaction will be discovered to be carried out by a particular specific ferment.

Thus, after half a century, Traube's original, tentative proposal had emerged as the central dogma of a new branch of science. In contrast to the old protoplasmic theory, which was barren in its ability to generate new experiments, the great advantage of the enzyme theory was that it provided an intellectual framework within which a new program of biochemical experimentation could develop. The touchstone of the new science was the idea that the whole metabolism of the living cell was guided by an array of very real, definable substances that could now be isolated and studied. Indeed, within a few years, hundreds of enzymes were discovered in the cells of all life forms, from microorganisms to plants and animals.

Fermentation, which had done so much to generate the enzyme theory, continued to play a central role in the further development of the new science. It turned out that even the apparently simple process of fermentation was more complex than originally thought. Harden, Embden, Meyerhof, and Warburg and their students showed that the transformation of glucose into ethanol was accomplished not in a single step by a single enzyme, as Buchner had believed, but in numerous small steps that had to occur in a precise order with each requiring its own proper enzyme. The main obstacle to progress had been the fact that although alcoholic fermentation appears to be simple, in fact it is very intricate. Thus, the search for *the* fermentation agent turned out to be a search for a concatenation of agents. Although years were to pass before all of the compounds en route from glucose to ethanol, and the various enzymes that formed them, were completely described, the principle of a connected chain of reactions—a *metabolic pathway*—quickly became established. Compounds were seen to be changed by the stepwise addition or removal of one atom or group of atoms at a time, and each such modification required its own enzyme.

During these studies, it became clear that the real function of fermentation was not to make ethanol, but to contribute to an even larger chain of reactions whose primary purpose was to use glucose to generate a variety of carbon compounds and energy-rich molecules for the cell. The buildup of ethanol (or other compounds, such as lactic acid) occurs only under special circumstances—for instance, when the whole cycle cannot be completed, owing to the absence of sufficient oxygen. In retrospect, what the early investigators had found were isolated reactions that turned out to be part of a larger network of intracellular transformations. In fact, by the middle of the twentieth century, so many enzymes and metabolic pathways had been discovered that it was indisputably clear that *enzymes are capable of generating all the compounds found in living tissue.*

While there are many hundreds of enzymes and each is interesting and important in its own right, we will focus our attention upon certain, exemplary enzymes, which will serve as a "case study" to answer fundamental questions about the mechanism of enzyme action. In particular, we will pursue two goals. The first is to understand, in precise molecular terms, how enzymes bring about chemical transformations. The second is to understand how these transformations, each a small step in itself, nonetheless culminate in such complex physiological processes as digestion, the clotting of blood, and the transmission of signals in the nervous system.

2

Lock and Key

The spirited ''discussion'' between Liebig, Pasteur, Berthelot, Traube, and Buchner led to the general hypothesis that cells manufacture special substances they use to promote chemical reactions. Although these substances had been given a name—*enzymes*—little was known about them, and the enzyme, as it emerged from the nineteenth century, was essentially a ''black box.'' One knew only that complex and undefined substances derived from cells were able to carry out various chemical transformations, such as fermentation. In the beginning, only a few chemical transformations could be attributed to enzymes, but the general theory quickly took hold: the entire life of the cell could be understood on the basis of the collective activities of a large number of enzymes, each responsible for promoting a particular chemical reaction. The work we will discuss in this chapter begins to break open the enzymatic black box, providing our first insights into how enzymes work.

At the outset, we begin with a somewhat paradoxical fact. The atoms of inorganic nature and those of the living world are the same—carbon, hydrogen, oxygen, nitrogen, sulfur, and so forth. Yet we recognize two different worlds with respect to chemistry. One is a somewhat mechanical world in which relatively simple compounds predominate and in which chemical transformations are brought about by increased tem-

Lock-and-key interactions are illustrated in this woodcut by M. C. Escher. This type of interaction underlies the activity of enzymes.

The pulp of the sugarcane plant contains a high concentration of dissolved sucrose, which can be purified commercially and used as common table sugar.

perature and pressure. In the other world—the world of living organisms—molecules tend to be much more varied and complex, and the reactions they undergo are guided by enzymes. It is, in fact, the activity of enzymes that generates the complexity of the compounds in the living world. Step by step, they build up molecules with biologically useful properties, using simple compounds they obtain from the nonliving world. In addition to building and rearranging complex molecules, enzymes also bring to the living world its high degree of organization. By accelerating certain reactions and not others, they selectively catalyze the very chemical transformations that form the basis of life.

In essence, the enzyme is an agent for change that has been superimposed during the course of evolution upon an already existing inorganic world of molecules and chemical reactions. The important and intriguing question is how these substances work.

A "Model" Reaction

To explore the way in which enzymes work, our strategy will be to choose a particular chemical transformation and consider it first as an ordinary chemical reaction occurring slowly without the help of an enzyme and then as a reaction proceeding enormously more rapidly under the influence of one of these powerful catalysts. By exploring the differences inherent in this contrast, we will begin to uncover the special power that enzymes bring to the reactions they catalyze.

As our example, let us consider more deeply a reaction involving a compound we are already familiar with— sucrose, or common table sugar. The sucrose molecule is simply a compound structure containing one molecule each of glucose and fructose, the two most common simple sugars. Sucrose is formed in many types of plants, where it is used as a temporary storage and transport form for sugar. After its production in the leaves during photosynthesis, the sucrose is distributed throughout the plant via the sap. When it reaches individual cells, it is broken down and its component simple sugars are used as starting materials for metabolic processes.

Glucose   *Fructose*

SUCROSE

The reaction we will consider entails the cleavage of sucrose into its two component parts, a reaction that can occur with or without the assistance of an enzyme:

$$\text{Sucrose} \rightarrow \text{glucose} + \text{fructose}$$

This reaction consists of a *hydrolysis*—literally, from the Greek, "breakage by water"—so named because the cleavage is brought about by incorporating the components of water (H and OH) into the starting compound. Although water consists almost entirely of H_2O, a small fraction of the molecules are dissociated into their chemically reactive components (H^+ and OH^-), and it is these ions that directly participate in the hydrolysis reaction. The incorporation of water is seen explicitly when the

complete molecular structures are drawn out, as in the figure at the bottom of this page. In chemical shorthand, one writes

$$C_{12}H_{22}O_{11} + H_2O \rightarrow C_6H_{12}O_6 + C_6H_{12}O_6$$
$$\text{Sucrose} \qquad \text{Water} \qquad \text{Glucose} \qquad \text{Fructose}$$

Compared with the ethanol fermentation reaction we considered in Chapter 1, this reaction is much easier to understand. In fermentation, glucose is broken down into several ethanol and carbon dioxide molecules—clearly a multistep process entailing the breakage of several carbon–carbon bonds and the rearrangement of various groups of atoms. In the decomposition of sucrose, only a single bond (marked with the wide arrow in the figure) has to be broken in order for the sugar to separate into its two con-

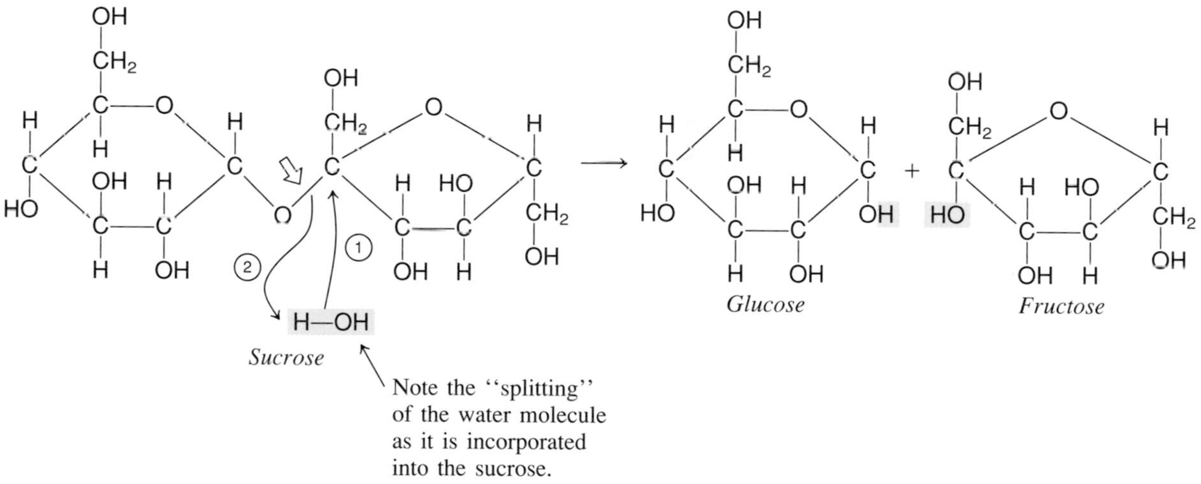

Note the "splitting" of the water molecule as it is incorporated into the sucrose.

The cleavage of sucrose into glucose and fructose. The straight lines between the atoms represent the covalent bonds that hold the molecule together. These bonds consist of pairs of electrons shared between the constituent atoms. During a chemical reaction, one or more covalent bonds undergo rearrangement. The curved arrows (in this and all such diagrams in the book) show the movement of electrons to achieve this end. Thus, in the cleavage of sucrose, one arrow shows the electron pair connecting the two sugars becoming associated with an H^+ derived from a water molecule. Simultaneously, an electron pair in the water molecule shifts from holding hydrogen to forming a new bond with one of the released sugars. The net result of these bonding rearrangements is the hydrolysis of sucrose.

stituent pieces. This simplicity makes the reaction ideal for discussing the basic mechanisms underlying chemical change.

The Nonenzymatic Breakdown of Sucrose

The low inherent reactivity of sucrose provides the starting point for our discussion. If sucrose is dissolved in water, it will undergo a nonenzymatic decomposition reaction in which glucose and fructose are formed in equal amounts as the sucrose disappears. To a traditional chemist, the most striking feature of this reaction is its incredible slowness. At room temperature and neutral pH, years would pass before even a few percent of the sucrose molecules decomposed. This is because the sucrose molecule is a stable arrangement for the atoms it contains; it is not easily disrupted to liberate its two constituent simple sugars. In

fact, most of the molecules found in nature have this characteristic stability—otherwise, they would have long ago fallen apart into simpler molecules. Thus, although the processes of life depend upon the constant changing of one molecular species into another, all the species involved are, like sucrose, somewhat resistant to change.

Promoting Chemical Change in the Laboratory: Heat

If sucrose is normally so stable, the question immediately arises as to how the molecule ever undergoes a chemical reaction. An important clue comes from one of the ways that laboratory chemists have found to increase the rate of this reaction (and many others)—the application of heat. This treatment essentially constitutes a form of environmental stress, which coaxes the electrons that hold the constituent atoms of the molecule together to undergo one or more rearrangements. Thus, the spontaneous break-

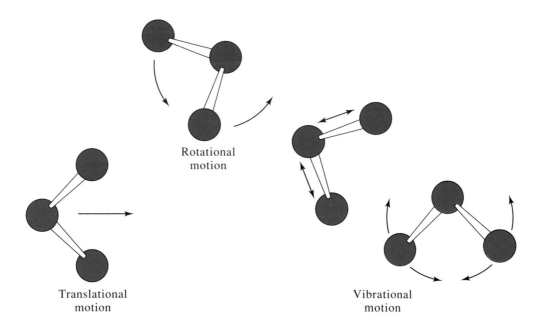

The translational, vibrational, and rotational motions of molecules contribute to their energy content.

down of sucrose is greatly accelerated if the sucrose solution is heated—and the higher the temperature, the faster the breakdown proceeds. What is the basis for the accelerating effect of heat? The answer lies in an understanding of molecular energy.

To a first approximation, the energy of a molecule is carried in the orbitals of the electrons that bond its constituent atoms together. Additional amounts of energy are contained in the velocity of the molecule as it moves through space (translational energy) and in the various types of vibrational and rotational motions within the molecule. If we could examine each of the sucrose molecules in a solution, we would find that their energies are not all identical; some have slightly more energy than the average, others less. For example, one sucrose molecule might be moving faster than another—that is, it will have a greater translational energy. This energy content, however, is not fixed: it can be changed if the molecule collides with another sucrose molecule, or with a molecule of the solvent. The situation is much like the interaction of two billiard balls: if a moving ball collides with a stationary one, it will transfer all or part of its translational energy to the latter, which will then move off in some new

direction depending on the angle of contact. Moreover, in the same collision, the interacting billiard balls can gain (or lose) various amounts of internal rotational or vibrational energy. For our purposes, molecules are not all that different from billiard balls, and we can consider them as miniature spheres moving through solution and colliding in large numbers—with each individual molecule having its own changing translational, rotational, and vibrational energy content. The graph below depicts the distribution of energies in a random population of sucrose molecules.

The basic rule of chemical change is that only those molecules that have sufficient energy are able to undergo a reaction. Moreover, this energy must become localized in a particular place in the molecule so that an unstable arrangement of atoms arises at the potential point of change (in the case of sucrose, the junction between its two component simple sugars). Those molecules that attain the more energized, strained state will either dissipate their extra energy through collision and again become stable sucrose molecules, or they will incorporate the components of water and break apart into glucose and fructose. The reason sucrose normally decomposes so slowly is that very few molecules exist in this unlikely, high energy state

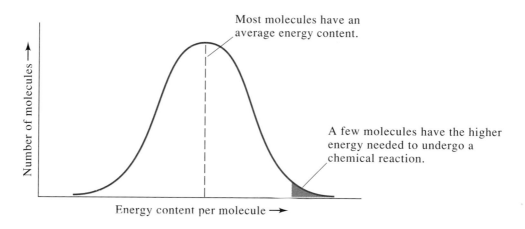

The distribution of energy among a population of sucrose molecules. Owing to their different amounts of translational, rotational, and vibrational motion, individual molecules have different energies. Only a small percentage of molecules have sufficient energy to undergo reaction.

at any particular time. But even if the *average* sucrose molecule does not have sufficient energy to react, nonetheless a few do and, over time, the chemical breakdown of sucrose proceeds spontaneously.

We can now understand why heat is generally so effective in promoting chemical reactions in the nonliving world. Temperature is a measure of the average energy content of a population of molecules. When one heats a reaction mixture and raises its temperature, more energy is pumped into the system, causing the average molecule to have a higher energy, as shown in the first graph below. The molecules of both compound and solvent then move more rapidly and participate in more frequent collisions. As this giant three-dimensional billiard game proceeds, energy is transferred between the colliding molecules, and a higher percentage have the energy necessary to react. The result is that more molecules undergo chemical transformation per unit time.

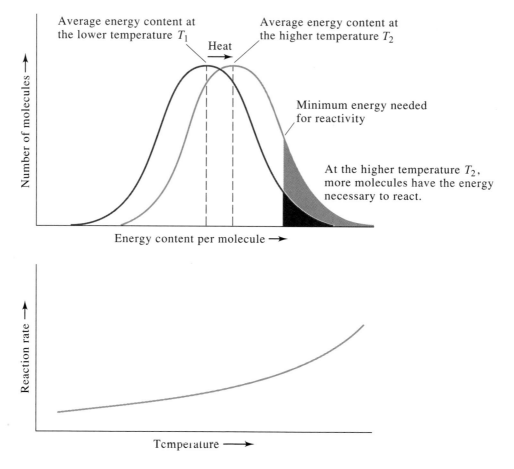

Heating a substance raises the average energy content of its molecules. More molecules acquire the energy necessary to undergo reaction (top), and the reaction rate is correspondingly increased (bottom).

Promoting Chemical Change in the Laboratory: Acid or Base

In addition to raising the temperature, laboratory chemists commonly employ one other method to accelerate chemical reactions: the addition of acid or base. The functional definition of an acid is a compound that can donate a hydrogen ion (H^+) to another molecule, and, correspondingly, a base is any compound that can accept an H^+ ion. Thus, adding acid gives the solution an excess of H^+, whereas adding base takes H^+ from water, releasing excess OH^- into solution. Acid, in particular, is effective in promoting the breakdown of sucrose. The decomposition of a few percent of the sucrose molecules (which would take years at neutral pH) will occur in a matter of hours in the presence of acid. If more acid is used, the reaction proceeds even more rapidly.

What is the role of the acid in promoting the hydrolysis reaction? Recall that the chemically reactive components of water, H^+ and OH^-, must be incorporated into the sucrose molecule if it is to be cleaved into glucose and fructose. The addition of acid releases extra H^+ into the surrounding solution and makes it more probable that one of these ions will become associated with a sucrose molecule and initiate cleavage (as shown in red in the reaction shown below). After the H^+ has become attached to the appropriate oxygen, a nearby water molecule, which has

The Use of Acid to Promote Sucrose Hydrolysis

Minutes of incubation	Percentage of sucrose hydrolyzed	
	No treatment	Strong acid
0	0	0
30	0	17
60	0	32
90	0	43
120	0	53
150	0	62
300	0	86

dissociated into H^+ and OH^-, contributes its OH^- component to the sucrose, completing the cleavage reaction. Thus, in an acid environment, more H^+ is available to participate in hydrolysis, and more sucrose molecules will react per unit time.

Elevated temperature and the presence of acid or base are two of the main factors that influence the rates of chemical reactions in the nonliving world. *The most notable point is that, in the absence of these factors, reactions tend to be infrequent and slow, reflecting the low inherent reactivity of molecules.*

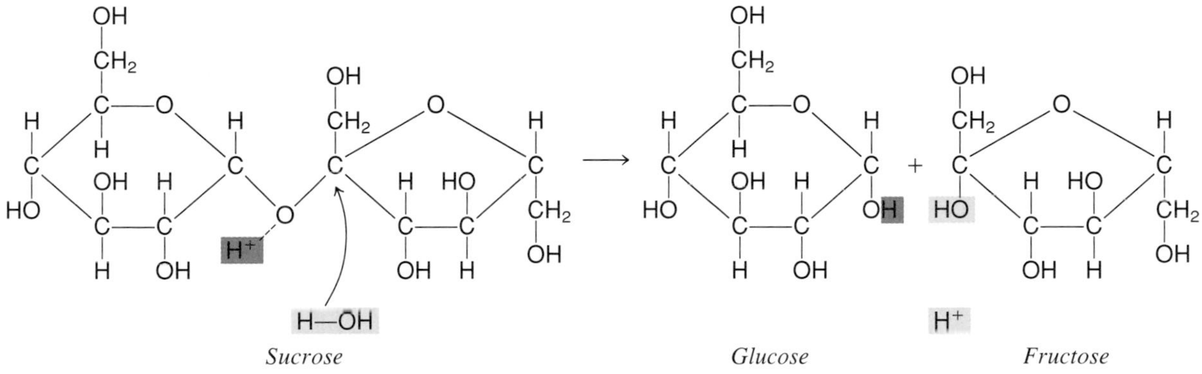

The accelerated hydrolysis of sucrose in the presence of acid.

Enzymatic Breakdown of Sucrose

Having considered the breakdown of sucrose as an ordinary chemical reaction, we will now consider the same reaction as it occurs under the influence of an enzyme.

The enzyme that nature has invented to promote the cleavage of sucrose is called *invertase* (or *sucrase*). This is the very enzyme that was studied by Berthelot and that played such an important role in the discovery of enzymes (Chapter 1). It will be recalled that in 1860 Berthelot broke open yeast cells and obtained a ''ferment'' capable

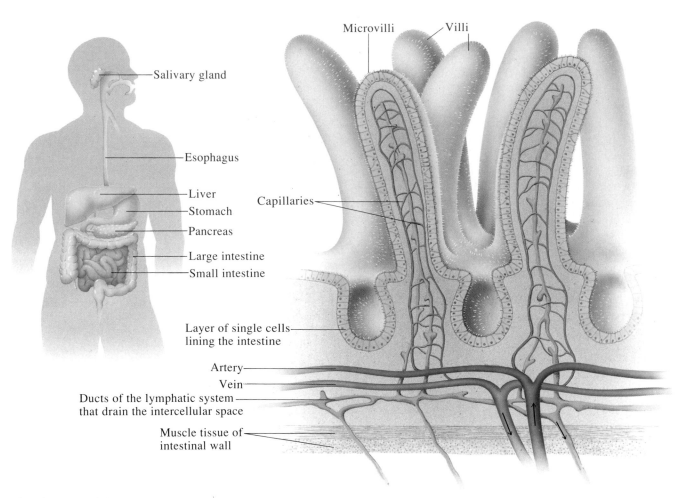

An enlargement of the wall of the small intestine showing the various structures involved in the absorption and utilization of sucrose. The fingerlike projections (villi) of the intestinal wall along with their hairlike cellular extensions (microvilli) increase the surface area available for nutrient absorption. Ingested sucrose (and other complex sugars) are first degraded by intestinal enzymes and then cleaved into individual sugars by enzymes bound to the outer membrane surface of the microvilli. Thus an invertase-like enzyme cleaves the sucrose into glucose and fructose, which then pass into the underlying capillary bed. After transport through the bloodstream, these smaller sugars are absorbed by other cells of the body.

of splitting cane sugar into glucose and fructose. This was the first instance of a reaction-promoting ferment (or "enzyme") that was normally contained within cells but that could be isolated from them without loss of function. It was thus clear that the living cell itself was not the ferment; rather, the cell was the producer of the ferment, which then became one of its internal components.

Berthelot originally isolated invertase from yeast, where it serves as a digestive enzyme. Like many other fungi, yeast lives primarily off decaying plant material, whose fluids can be a rich source of sucrose. The invertase is located between the cell membrane of the yeast and the surrounding cell wall, where it cleaves the sucrose into smaller glucose and fructose units that can easily be carried through the membrane and into the cell. The sugar molecules are then used to fuel the enzyme-directed chain of reactions leading to the formation of cell compounds (including, under certain growth conditions, ethanol). Invertase is widely distributed in nature, and it plays an important role in digestion in many organisms more complex than yeast. For example, in animals, sucrose is split into glucose and fructose by an invertase-like enzyme that is anchored to the outer membrane surface of the cells lining the small intestine. The released simple sugars are then absorbed into the bloodstream and ultimately transported to cells throughout the body. There, as in yeast cells, they serve as the starting point for cellular metabolism.

Invertase is a powerful catalyst for the cleavage of sucrose: if a tiny amount is added to a spontaneously decomposing solution of sucrose, there is a dramatic increase in the rate of hydrolysis. Measuring the amount of glucose and fructose that arises, we find that what once took years to detect now occurs in a matter of minutes. As the table shows, even a trace of enzyme has a great effect.

How does invertase compare, as a catalyst, with the use of heat, acid, or base—the standard chemical methods for promoting a reaction? In fact, the enzyme is orders of magnitude more effective. Whereas heat or the use of acid or base gives a ten- to a hundredfold increase in reaction rate at best, an enzyme will typically accelerate a reaction a billionfold or more.

The products of the invertase-catalyzed breakdown of sucrose—glucose and fructose—are, of course, exactly the same as for the previously considered uncatalyzed re-

The Enzymatic Hydrolysis of Sucrose

Minutes of incubation	Percentage of sucrose hydrolyzed		
	No treatment	Strong acid	Trace amount of enzyme
0	0	0	0
20	0	*	26
30	0	17	38
40	0	*	48
50	0	*	57
60	0	32	65
75	0	*	74
90	0	43	*
120	0	53	*
150	0	62	95
300	0	86	100

*Reaction mixture not assayed at this time.

action. *This illustrates a general principle: enzymes do not invent new reactions— rather, they accelerate the velocity of ordinary chemical reactions.* But does this necessarily mean that the way enzymes work is related to the methods laboratory chemists have developed for promoting reactions—the application of heat or the use of acid or base? In fact, what makes enzymes so interesting is that, upon first analysis, they do *not* appear to use these two standard mechanisms. On the contrary, the direct application of heat or acid/base is decidedly counterproductive.

Increased temperature will soon halt the invertase-catalyzed breakdown of sucrose, as shown in the graph at the top of the following page. At first, the reaction velocity increases as the temperature is raised—just as for the nonenzymatic breakdown of sucrose (see page 58). However, as the temperature continues to rise, a point is reached where the increase in reaction velocity begins to fall off. Then, a few degrees higher, the reaction stops altogether, and the production of glucose and fructose falls to the same low level observed in the absence of enzyme.

The most straightforward interpretation of this experiment is that the higher temperature has effectively destroyed the enzyme. Evidently, just as heat will promote the breakdown of sucrose, it can also cause the breakdown of the invertase enzyme itself. And, of these two sub-

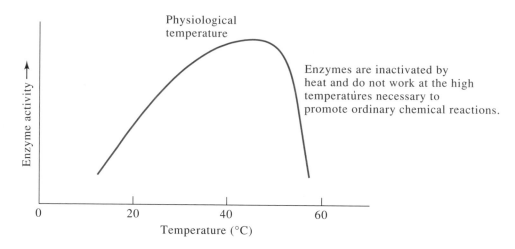

The sensitivity of enzymatically catalyzed reactions to heat. The heat sensitivity of enzymes is the primary reason why there are limits to the temperature at which life can exist.

stances, the enzyme is apparently more sensitive to thermal damage. We can conclude that enzymes are relatively fragile structures and therefore cannot benefit from the simple expedient of applying heat to promote a reaction. On the contrary, the living cell must seek to maintain its ensemble of enzymes at a moderate temperature.

Nor is the direct use of acid or base helpful in promoting enzymatic reactions. While the decomposition of sucrose is accelerated by the addition of acid—with higher concentrations of acid causing increasingly faster decomposition—such is not the case for the invertase-catalyzed reaction. Instead, there is an optimal concentration of acid at which most enzymes work—generally pH 7. As the pH moves away from this optimum, either by the addition of acid or base, the reaction velocity falls off to zero.

Evidently, just as sucrose is broken down by acid, the enzyme is also fragile and unstable in the presence of acid (or base). As a result, far from being able to use free acid or base to promote chemical transformations, cells must seek to maintain a particular pH at which their enzymes will work with optimal efficiency.

Is it the case, then, that enzymes accelerate reactions by a mechanism that is markedly different from those

available to the laboratory chemist? Despite first appearances, perhaps we should not be too quick to rule out a potential connection between traditional methods of accelerating reactions and enzymatic catalysis. An analogy can be made to the generation of power from nuclear reac-

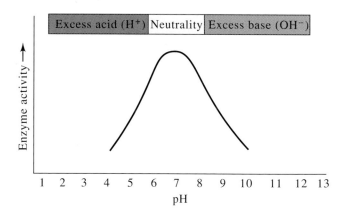

Because enzymes are inactivated by acids and bases, they do not work at the high concentrations of these reagents necessary to promote ordinary chemical reactions.

tions. The extraordinary temperatures achieved in the run-away nuclear reaction of an atomic weapon creates an explosive release of energy, with enormous destructive consequences. In contrast, in the controlled environment of a nuclear power generator, the same reaction can be constrained in order to harness the power inherent in the nuclear transformation without widespread and generalized damage. As a result, controlled nuclear reactions can be used to generate energy without pollution and to manufacture radioactive compounds for use in scientific experiments and medicine. The intriguing possibility remains that, in an analogous way, enzymes have found a mechanism to harness the principle that underlies the reaction-promoting power of heat or acid and base without subjecting the cell to uncontrolled, generalized damage.

Whether enzymes use an old method of catalysis in a new way or use an entirely new method, to make further progress in understanding how enzymes work, we must now set aside our thoughts based on general principles and study the enzyme more directly. We will begin by focusing on the interaction of the enzyme with its target molecule, searching for clues about the mechanism that enzymes actually use in promoting reactions.

Fischer's Lock-and-Key Concept of Enzyme Action

The first insights into the mechanisms used by enzymes to promote reactions were obtained at the turn of the century by the eminent German chemist Emil Fischer (1852–1919).

Fischer's interest was organic chemistry, the study of the compounds associated with living tissue. One of his many accomplishments during a long and productive career stands out as so fundamental that it is hard to realize it was not always known. Up until Fischer's time, molecules had been considered, somewhat abstractly, as undefined assemblages of atoms not rigidly fixed in space. It was Fischer who established that molecules have different and precisely defined three-dimensional structures. He came to this conclusion through studies of the common sugars with which we are all familiar—glucose, galactose, fruc-

Emil Fischer extended the field of organic chemistry into the realm of biologically important molecules. Fischer was awarded the Nobel prize in chemistry in 1902—the second recipient of the prize, which, by design, was timed to begin at the dawn of the new century.

tose, sucrose, lactose and so forth. By chemical treatments that converted one sugar into another, Fischer was able to deduce the precise geometrical arrangement of atoms within these sugars. For example, a five-carbon sugar could be reacted with a one-carbon compound to give a mixture containing a pair of two six-carbon sugars—one of which had the chemical properties of glucose and the other of the sugar mannose. It thus became apparent that glucose and mannose must differ only in the orientation ("up" or "down") of the OH group attached to carbon atom number two (where the one-carbon compound had reacted and created the difference). Through an extensive network of such interconversions, Fischer built up the structural relationships that characterize the different members of the family of sugar molecules. The important conclusion was that, while sugars may have the same chemical composition—six carbons, twelve hydrogens, and six oxygens, and hence the same chemical formula, $C_6H_{12}O_6$—in each sugar the three-dimensional configuration of the carbon atoms and their attached H and OH groups with respect to one another is different. These structural differences underlie the distinctive chemical and biological properties associated with each molecule—for instance, the solubility of a particular sugar in water, or its characteristic sweetness.

Fischer's analysis of the structure of various sugars laid the foundation for his seminal experiments in the field of enzymology. He knew of Berthelot's discovery, some forty years earlier, that invertase cleaves sucrose into glucose and fructose, and from his own work he knew the exact arrangement of the atoms in all three sugars. He then seized upon the idea of using this knowledge of molecular structure to probe the mechanism of action of an invertase-like enzyme to a depth no previous enzyme had ever been studied. He began by using the methods of laboratory chemical synthesis to produce modified forms of sucrose. These *structural analogs* were designed to be used, in addition to natural sucrose, as target molecules for the enzyme. How closely could the enzyme discriminate between one structure and another?

One of Fischer's modifications is shown in the figure on this page. It is sucrose in which the fructose half of the molecule has been replaced by a simple methyl group (—CH_3). When this synthetic compound, *methyl-glucose,* was offered to invertase, the enzyme was able to work on it very much as if it were ordinary sucrose. That is, methyl-glucose was cleaved into glucose and methanol (CH_3—OH), and it was therefore, like sucrose, a *substrate* (or target) for the enzyme. Apparently, the enzyme recognizes some localized area in its natural substrate, sucrose, and this area is duplicated in the artificial methyl-glucose molecule.

This result showed that individual enzymes are somewhat liberal with respect to the substrates they are willing to work on. There is nothing very unusual about a particular substrate—any closely related molecule, even one made in an organic chemistry laboratory, might have a similar internal area and also be a potential substrate.

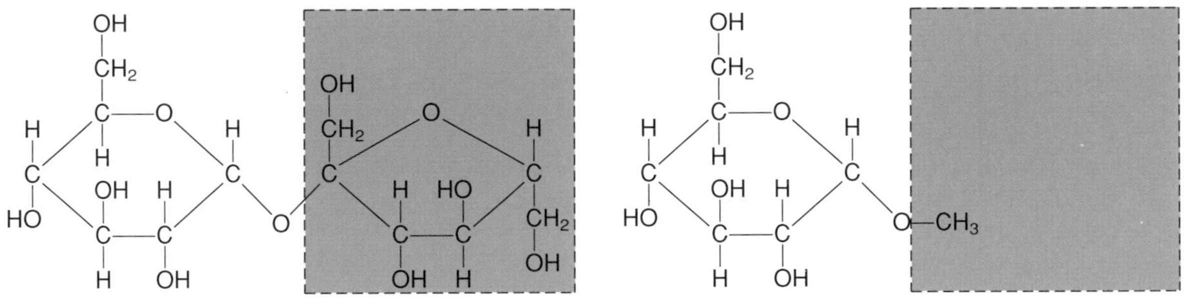

Normal sucrose

Modified sucrose (methyl-glucose)

Ordinary sucrose and a modified form of sucrose, methyl-glucose, synthesized by laboratory techniques. Fischer produced a number of sucroselike molecules, which he then tested as potential substrates with invertase.

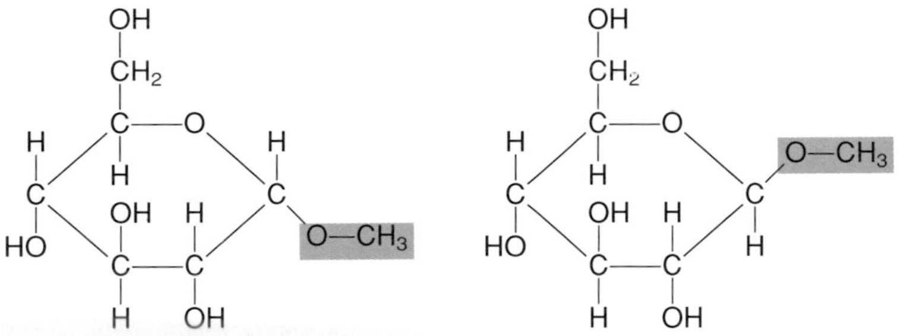

Methyl-glucose
(α configuration)

Methyl-glucose
(β configuration)

Two forms of methyl-glucose that differ only in the orientation of the methyl group. In one case, the methyl group projects "below" the "plane" of the central sugar ring (the α form). In the second case, the methyl group projects "above" the ring (the β form). Both molecules are virtually identical in chemical properties, but only one is a substrate for invertase.

A less astute scientist might have been perfectly satisfied with this new result. However, Fischer went on to probe the same reaction further, and he was surprised to be led to a somewhat contradictory and even more important conclusion—that even though enzymes work on localized areas within molecules, they can nonetheless distinguish between potential substrates with a high degree of specificity. For these experiments, Fischer prepared *pairs* of methyl-glucose molecules, with the methyl group attached to the same place but pointing either "up" or "down," as shown in the figure on this page. In terms of their *physical and chemical* properties, the two molecules were virtually identical. The striking finding, however, was that the enzyme was able to work on only one member of the pair. The α form of methyl-glucose (in which the methyl group points down) was rapidly cleaved into glucose and methanol by the enzyme, but the β form remained unaffected. This was a crucial experiment, for it led Fischer to the discovery that the enzyme could in fact be astonishingly demanding in recognizing the overall structure of the substrate.

Based on this study, Fischer drew the conclusion that, even today, is the best known general statement about how enzymes work:

In order to be able to act chemically on one another, an enzyme and its substrate must fit together like a lock and key.

Fischer had reached a conclusion with broad chemical implications. Because the substrate molecule had a precise shape that the enzyme was able to recognize, the enzyme itself must have an equally precise and complementary shape. Moreover, enzymes would probably be *large* three-dimensional structures, so that they can surround and thereby recognize the unique contour of their target molecules.

To Fischer, the classical chemist, the ability of enzymes to discriminate between molecules composed of essentially identical *chemical* groups (such as α- and β-methyl-glucose) came as something of a surprise:

Only factual observation could ever have convinced anyone that the geometrical structure of a molecule—even in the case of mirror image pairs—exerts such a large influence on its chemical reactivity.

Indeed, the *specificity* shown by invertase stands in marked contrast to the way methyl-glucose decomposes

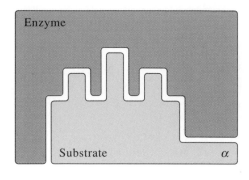

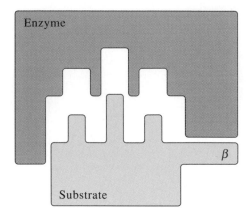

The complementarity in shape between an enzyme and its substrate. *Left:* The enzyme interacts with the substrate to promote reaction; often it can also work on a similarly shaped molecule (as in the case of sucrose and α-methyl-glucose). *Right:* The enzyme cannot interact with a substrate analog whose structure differs significantly from that of the natural substrate (as in the case of sucrose and β-methyl-glucose).

into methanol and glucose at high temperature or in an acid solution. In either of these cases, *both* the α and β forms of methyl-glucose are effectively cleaved. Invertase, however, only promotes the cleavage of α-methyl-glucose.

The important point, from the standpoint of the cell, is that either high temperature or a change in pH would increase the reactivity of *many* molecules. *But the enzyme, through the specificity provided by the lock-and-key mechanism, restricts its power to one particular molecule.*

When we extrapolate the results of Fischer's experiment with invertase to embrace all enzymes, we obtain an important insight into cellular physiology. We picture the cell, as did Traube and Hofmeister, as populated by a large number of different enzymes. Each enzyme, like invertase, shows selectivity with respect to the compound it works on. This extension of the lock-and-key hypothesis immediately provides an explanation for how enzymes manage the affairs of the cell. The number of enzymes in a cell is roughly equal to the number of different intracellular compounds, and, as a general rule, each enzyme works on only one cellular molecule. The result is that the cell's ensemble of enzymes moves molecules along well-defined pathways (not unlike a railroad system with molecular stations), ultimately generating a specific set of biomolecules with physiologically useful properties. Since uncatalyzed reactions take place only very slowly, the vast majority of potential chemical transformations never occur at a physiologically significant level.

In sum, the importance of lock-and-key specificity in enzyme–substrate interactions is that it allows biological systems to choose the chemical reactions they wish to promote, leaving other possible reactions aside. Different types of cells are thus able to develop their own unique characteristics by producing specific sets of enzymes to catalyze different chemical reactions.

An Early Proof for the Enzyme–Substrate Complex

From the ability of enzymes to recognize precise details in the architecture of their substrates, Fischer inferred that the enzyme and its substrate interact like a lock and key. However, because enzymes had not been isolated as pure chemical species in Fischer's time, he had no way of dem-

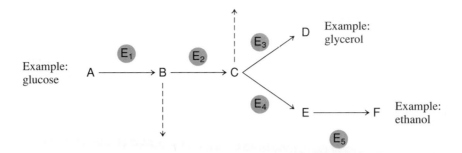

In living systems, specific enzymes move molecules along well-defined "metabolic pathways" to generate specific end products. This generalization is compromised only by the fact that some compounds (such as C above) are recognized by two or more enzymes, each carrying the molecule along a different pathway—that is, modifying it in a different way to form a different product. Even though alternative chemical transformations of the intermediate compounds will occur spontaneously (dashed arrows), because they are not catalyzed by enzymes, they proceed so slowly as to be of no physiological consequence.

onstrating this interaction directly. His enzyme was an invisible agent of change in an undefined preparation of material from living cells. How, then, could the lock-and-key concept be verified? A convincing proof was provided by the parallel experiments of two scientists, one near the end of his career—Adrian Brown of the British School of Malting and Brewing at the University of Birmingham— and the other at the beginning of his—Victor Henri (1872– 1940), an early collaborator of Alfred Binet, developer of the IQ test and, later, a leading French biochemist.

In their experiments, Brown and Henri analyzed the speed at which invertase catalyzed the breakdown of sucrose under conditions in which the amount of enzyme was held constant while the amount of substrate was varied. The experiment sounds abstract, as experiments often do, but the results and their interpretation were concrete. It was found that, as the amount of sucrose available to the enzyme was increased, the reaction velocity at first accelerated (the rising part of the curve in the graph on the next page). But then this increase in rate began to fall off, and eventually the addition of more substrate was without further effect in increasing the speed of the reaction.

This type of curve (a rectangular hyperbola asymptotically approaching a certain maximum limit, V_{max}) has been found for all enzymes examined, and was correctly

Adrian Brown's paper, in which he deduced the existence of the enzyme–substrate complex, is among the clearest and most insightful in biochemistry. The details of his life, however, have been obscured by time.

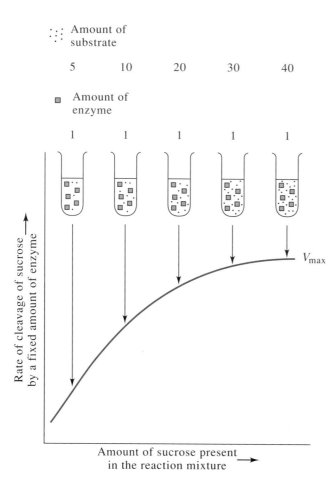

The experiment of Brown and Henri on the velocity of enzymatic reactions. A number of reaction mixtures were prepared, each containing the same amount of invertase but increasing amounts of sucrose. For each mixture, the speed of the reaction was measured by determining the rate at which sucrose was cleaved. The experiment shows that the enzyme has a finite capacity to process substrate, and eventually becomes "saturated," consistent with the formation of an enzyme–substrate complex in which the chemical reaction occurs over a finite period of time.

interpreted by Brown and Henri as reflecting the fact that *the enzyme works by forming a complex (like a lock and key) with the substrate and acting on it for a finite period of time.* Thus, beyond a certain concentration of substrate, the enzyme is fully occupied, or *saturated,* and cannot

absorb and process any more substrate molecules per unit time. A new substrate molecule will not be able to form a complex with the enzyme until the already bound substrate molecule has completed its reaction and departed. In effect, the enzyme behaves like a workshop with a limited capacity.

The chemical equation summarizing one cycle of enzyme action is

$$\text{E} + \text{S} \xrightarrow{\text{Binding}} \text{E·S}$$

(Enzyme + substrate)   (Enzyme–substrate complex)

$$\xrightarrow{\text{Reaction}} \text{E·P} \xrightarrow{\text{Dissociation}} \text{E} + \text{P}$$

(Enzyme–product complex)   (Enzyme + product)

This is quite different from a "traditional" chemical reaction of the form: $\text{A} + \text{B} \rightarrow \text{C} + \text{D}$, in which A and B are two reacting molecules. In that case, an interaction occurs, but it is not an enzyme–substrate interaction. Rather, A and B, at the moment of collision, form a *transient collision complex,* which leads to a reaction when the two molecules collide in the proper orientation and have sufficient kinetic energy upon contact. Such productive encounters are, of course, more frequent if the concentration of either of the reactants is increased—and, for this reason, ordinary chemical reactions can be accelerated by adding a greater amount of A *or* B. But in a reaction in which B is an enzyme, such as invertase, the situation is quite different. As we have seen, increasing the concentration of the substrate, A, eventually has no effect in accelerating the reaction. The point is that the enzyme does not participate in the chemical transformation as an ordinary reactant. Instead of relying on an energy transfer during a simple collision, it functions in some other way that requires time after the collision has occurred. Evidently, the enzyme uses this time to increase the intrinsic reactivity of the substrate. Thus, the enzyme–substrate complex represents a different type of strategy for promoting a reaction that we must now begin to unravel.

Today we recognize Fischer, Brown, and Henri as having provided some of the most fundamental statements about how enzymes work—insights that helped to crystallize and give impetus to the field of biochemistry. Their work revealed (1) that enzymes are specific for the struc-

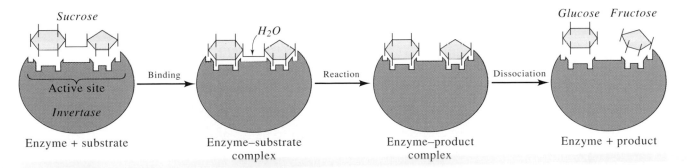

An enzymatically catalyzed reaction begins when a substrate molecule (in this case sucrose) binds to the active site of the enzyme (in this case invertase). This lock-and-key interaction is based on geometrical complementarity. In the enzyme–substrate complex, the force of the enzyme is exerted on the substrate, greatly increasing its inherent reactivity. After the reaction is completed, the product molecules (here glucose and fructose) remain associated with the enzyme (the enzyme–product complex) for a short period of time. Their dissociation from the enzyme (far right) frees the active site so that another substrate molecule can bind and undergo reaction.

ture of the molecules they work on, in contrast to such catalysts as acid; (2) that each enzyme has a structure that enables it to bind with a particular molecule to form an enzyme–substrate complex (the lock and key); and (3) that the chemical reaction occurs within the lock-and-key complex and requires a definite amount of time.

Our enzyme may still be a black box, but we are beginning to establish certain of its defining characteristics. In the next chapter, we begin to break open the black box. We will explore the chemical nature of enzymes and the way in which they promote reactions once the enzyme–substrate complex has been formed.

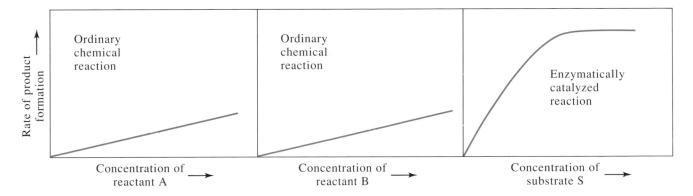

The rate for a standard chemical reaction can be continually increased by raising the concentration of either reactant A or B. This is not true for enzymatically catalyzed reactions, where increasing the concentration of the substrate eventually has no further effect because the active site of the enzyme has become saturated.

3

The Chemical Identification
of Enzymes

E ver since the work of Buchner, it had been clear that cells contain substances (not forces) that promote the conversion of one material to another. But although these substances had been given a name—*enzymes*—they thus far had no physical reality. Nonetheless, if enzymes were not a "vital force," then they must be some kind of molecule that actually exists in the cell. The next goal of biochemistry, therefore, was to break open cells, sort out the various types of molecules inside, and determine which ones were the reaction-promoting enzymes. This was no easy task, and it took many decades to identify the structures and functions of the different classes of cellular molecules.

The interior of the cell contains a bewildering variety of molecules and complex structures, all of which can be presumed to have their own special function. Chemically deciphering the living cell, therefore, presents a difficult problem. A straightforward approach to exploring the contents of the cell is to expose a sample of biological tissue to strong acid or base and then examine the molecules released. This was in fact the technique used to analyze the first simple biological substances. For instance, the predominant component of the sap of the sugarcane, sucrose, was found by acid treatment to be composed of two simple sugars, glucose and fructose (Chapter 2). However, the cell is vastly more complex.

Crystals of the enzyme chymotrypsin.

When entire cells or chunks of tissue were subjected to the same type of treatment with acid or base, a collection of several hundred molecules was recovered. As can be seen from the few representative examples shown on the facing page, these molecules range across a broad spectrum of chemical compositions and structures. Moreover, none of their structures gave any clue to the molecules' function.

It was, of course, inconceivable that such a complex organization as a cell would be composed merely of a collection of small molecules floating free in solution. Rather, it seemed evident that the acid or base treatment had decomposed the complex structures out of which the cell was actually made, just as sucrose had been broken apart into glucose and fructose under similar conditions.

Realizing this, physiological chemists began to examine biological material under less harsh conditions. They then succeeded in recovering larger structures in which the previously observed small compounds apparently retained many of the associations they had had in the intact cell. From this work it gradually became clear that the small compounds in fact serve as building blocks in the construction of four major classes of molecules that exist inside the cell—*proteins, carbohydrates, nucleic acids,* and *lipids*.

In this chapter, we will explore the various types of cellular molecules and discover that the proteins are the major class endowed with enzymatic activity.

Proteins

The first class of cellular molecules to come into focus were the proteins. Many students of the cell in the nineteenth century attached great importance to a group of substances that were collectively termed *albuminoids*. These materials were identified when they arose in such familiar natural processes as the coagulation of egg white by heat, the curdling of milk with acid, or the spontaneous clotting of blood. All these treatments caused a solid material to appear from an initially fluid substance. A widely known experiment performed in the 1800s demonstrated

the physiological importance of the albuminoids. Dogs fed a diet consisting only of sugar and fat would not survive—albuminoid substances were also essential for sustaining life.

The study of the albuminoid substances was guided by the prevailing chemical theory of the day. By 1820, the leading chemists had accepted John Dalton's atomic theory, which explained the chemical structures of common substances in terms of elemental building blocks. Specifically, molecules came to be viewed as arising from the "combining power" of their individual atoms—for instance, hydrogen could be thought of as associating with one other atom, oxygen with two others, and carbon with four. *The net effect of this theory was to define molecules in terms of their constituent atoms and to derive structural formulas to represent specific compounds.* Hence water, to the chemist, became H_2O and methane became CH_4.

$$H—O—H \qquad H—\overset{\displaystyle H}{\underset{\displaystyle H}{\overset{|}{\underset{|}{C}}}}—H$$

Water   *Methane*

More complex molecules, such as glycerol and glucose, also proved amenable to this type of analysis, giving rise to the formulas $C_3H_8O_3$ and $C_6H_{12}O_6$, respectively.

Mulder's Albuminoid Protein

It was only natural to extend this approach from simple, well-defined molecules to the study of albuminoid substances. Chemists thus became interested in determining the proportion of each element in the albuminoids so as to arrive at a molecular formula. The first systematic attack on the problem was that of the Dutch scientist Gerardus Mulder (1802–1880).

In his analysis of the albuminoids derived from eggs and blood, Mulder noted that these substances had a high content of nitrogen—in addition to the carbon, hydrogen, and oxygen atoms that always occur in the molecules of living cells. Moreover, he reported that his albuminoids

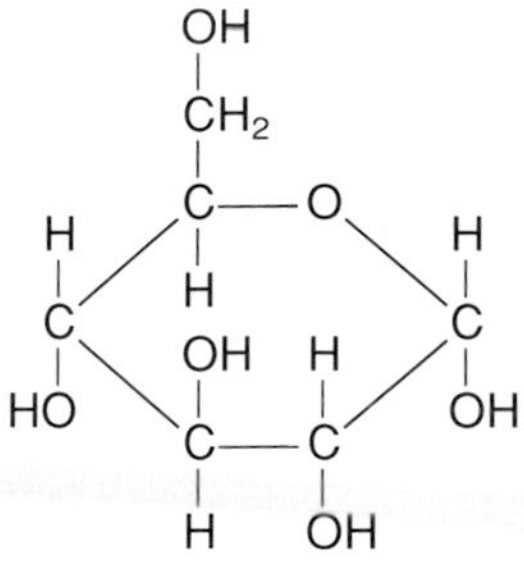

Glucose 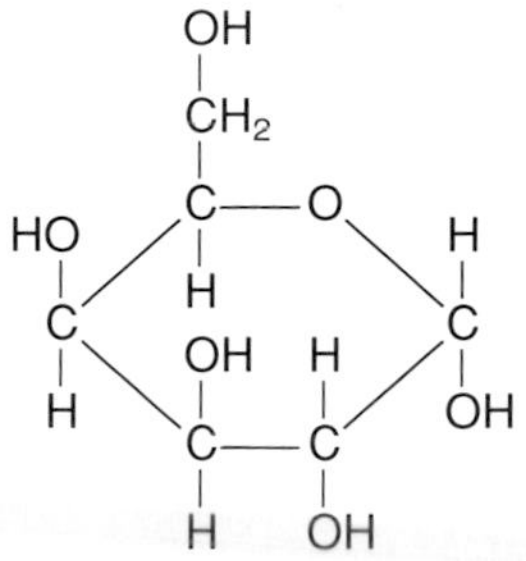 *Galactose* 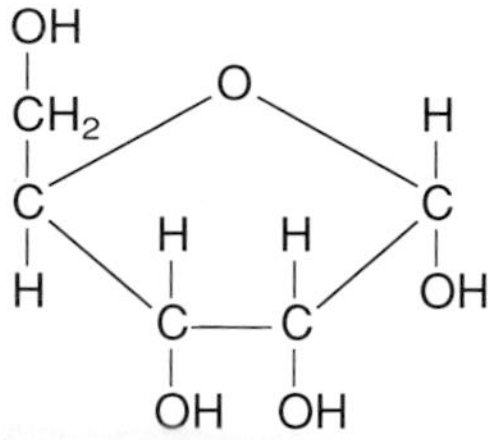 *Ribose*

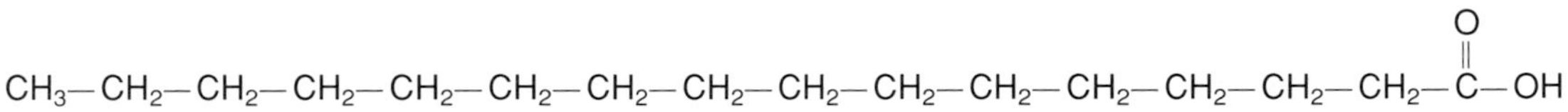

$$CH_3-CH_2-CH_2-CH_2-CH_2-CH_2-CH_2-CH_2-CH_2-CH_2-CH_2-CH_2-CH_2-CH_2-CH_2-\overset{\overset{\textstyle O}{\|}}{C}-OH$$

A fatty acid

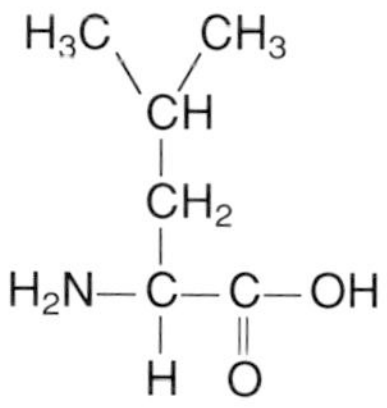

Leucine 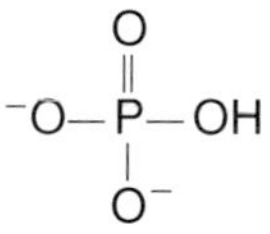 *Phosphate* 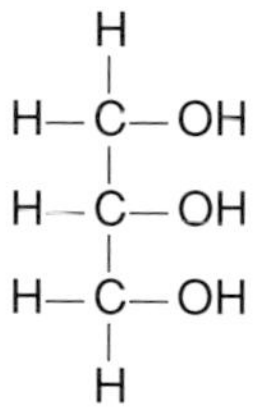 *Glycerol*

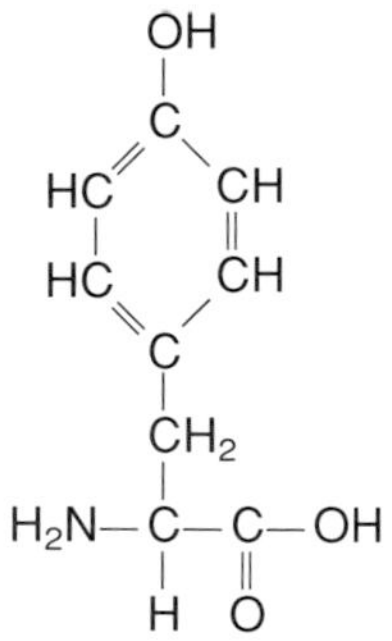

Tyrosine 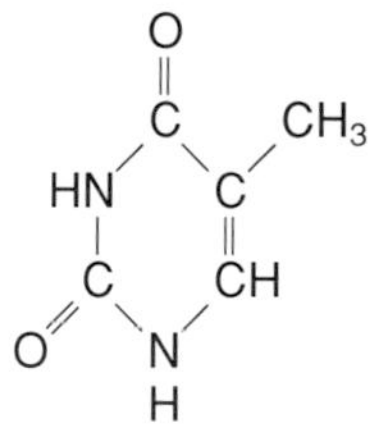 *Thymine* 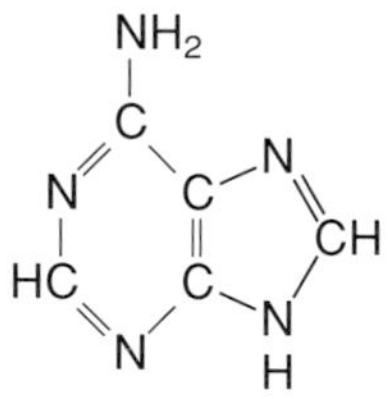 *Adenine*

A representative sample of the small molecules released from living cells after treatment with strong acid or base.

all appeared to have the same relative proportion of carbon, hydrogen, nitrogen, and oxygen, corresponding to the molecular formula $C_{40}H_{62}N_{10}O_{12}$—and that this basic unit was, in many cases, combined with an atom of sulfur.

Mulder's original claim, in 1838, roused considerable interest. Among those who were intrigued was the renowned Swedish biochemist Berzelius, who suggested that Mulder give the albuminoids another, more special name:

> Because albuminoids appear to be a principal substance of animal nutrition, I propose to you the word *protein,* which I derive from the Greek *proteios,* meaning "of primary importance."

But along with interest came controversy. When other biochemists sought to confirm Mulder's work, arguments soon developed over the specific composition of the albuminoid materials, or *proteins.* Typical of these battles were disagreements about the presence or absence of sulfur and about the exact proportions of carbon, hydrogen, oxygen, and nitrogen. Much of the experimental work was carried out in the laboratory of Justus von Liebig, the august German biochemist we encountered in Chapter 1. Liebig's albuminoids did not conform to Mulder's formula, and he was not slow to complain.

> We cannot isolate the particular substance described by Mr. Mulder. And so, it is a source of despair, after so much has been prattled and written about "protein," to have to say that there is no such thing.

In response, Mulder published a new series of studies and changed his "formula" for protein to $C_{36}H_{54}N_8O_{12}$. But his new results merely fueled the dispute. Eventually, the traditional chemists simply disengaged from the albuminoid studies. Confident in their highly precise knowledge of the atomic composition of small molecules, they were aghast at the conflicting claims made about the albuminoids. In the view of one exasperated researcher,

> the albuminoid substances do not constitute, properly speaking, a chemical species at all. They are organs, or

debris of organs, whose history should belong to biology rather than to chemistry.

Fortunately, not all chemists abandoned the study of albuminoids, and the next hundred years saw an evolution of Mulder's primitive and much too simple characterization of these substances into our modern understanding of proteins.

The Discovery of the Amino Acids

It was not that Mulder's approach was unsound, but rather that the complexity of protein molecules could not be approached by the limited technology of the day. Their large size meant that proteins were much more complex than such familiar compounds as CO_2, H_2O, and $C_6H_{12}O_6$, and it was not possible to derive accurate molecular formulas for them, owing to the errors inherent in obtaining precise measurements.

The observation that began to unravel the structure of proteins was, ironically, made by Liebig during his analysis of the problem. He found that acid treatment of the solid albuminoid substance derived from milk (the protein we would call *casein* today) released two small molecules that could then be isolated in pure form and analyzed. One was *leucine,* which had first been observed a quarter century earlier among the small molecules recovered from cells, and the other was a new compound, which he named *tyrosine.* It was possible to obtain accurate chemical formulas for such small molecules ($C_6H_{13}O_2N$ and $C_9H_{11}O_3N$) and eventually their complete chemical structures were determined:

Leucine   *Tyrosine*

As can be seen from these structures, the two molecules are somewhat similar. Both have a central carbon atom to which are attached a hydrogen (H), an amino group (NH₂), and a carboxylic acid group (COOH). The two compounds derive their individuality from the fourth component attached to the central carbon—their *side chain*.

In subsequent experiments, leucine and tyrosine were found to be released from many proteins upon treatment with acid, and their presence soon became a diagnostic test for protein. Even more important, it became apparent that other, related molecules were released from acid-treated proteins. *Thus, tyrosine and leucine became the forerunners of a set of 20 similar compounds that came to be known as amino acids because each contained an amino (NH₂) and an acid (COOH) component.* Two examples of the other 18 amino acids, the positively charged *lysine* and the negatively charged *aspartic acid,* are given to illustrate the chemical diversity of the protein building blocks.

The present list of amino acids was not completed until 1935—almost 75 years after Liebig's original work—but the essential point had been made much earlier: proteins are very large molecules—*macro*molecules—made up of various combinations of 20 well-defined smaller compounds, their amino acid building blocks.

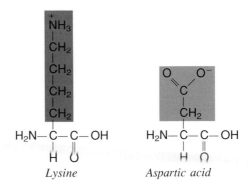

Lysine and aspartic acid, two examples of amino acids whose side chains carry an electric charge.

The Polypeptide Chain

Franz Hofmeister, along with Emil Fischer, formulated the first clear theory of protein structure. They proposed that the intact protein is a chain of amino acids joined together by bonds between the regularly repeating amino and carboxyl groups. The resulting chain, or *polymer* of linked amino acids, was termed a *polypeptide chain,* and an example is shown below.

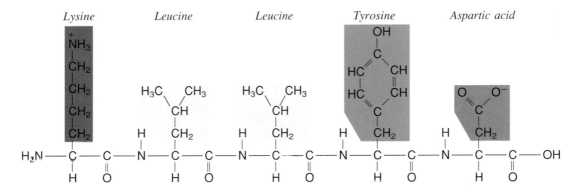

A short protein chain containing five subunits—two leucines, one tyrosine, one lysine, and one aspartic acid. Note that the chemically reactive components of water (H⁺ and OH⁻) are lost from the NH₂ and COOH groups of amino acids when they are linked together to form the protein chain.

By the beginning of the twentieth century, it had become evident that there was a multiplicity of different proteins in nature, rather than the one more-or-less universal albuminoid protein that had been envisioned by Mulder. Different proteins could be distinguished on the basis of their different amino acid compositions, among other properties. The subdivision of the old albuminoid category into individual protein species was easily accommodated by the new polypeptide-chain theory of protein structure. A polymer was a perfect structure for achieving variation. Given that a polypeptide chain could have any length, and that any one of the 20 amino acids could occur at a given position, the number of potential proteins was practically infinite. For example, even a very small protein with only 5 amino acid subunits could have any one of 20^5 (or 3,200,000) possible sequences. The number of possible variations for an average-size protein with 300 amino acids far exceeds the number of seconds the universe has existed (6×10^{17}). Each protein could potentially have a unique function.

Carbohydrates, Lipids, and Nucleic Acids

At the same time that biochemists were defining the structures of proteins, they were also discovering three other classes of cellular molecules, which, together with the proteins, are responsible for carrying out the physiological functions of the cell. These were the *carbohydrates,* the *nucleic acids,* and the *lipids*.

Structural studies of the four classes of cellular molecules led to an important generalization: they were all macromolecular polymers, constructed from simple subunits. In fact, the great variety of small compounds isolated from cells after harsh decomposition with acid—such as tyrosine, adenine, ribose, and glycerol—could be seen to fall into a few simple categories and to serve as building blocks for the construction of cellular macromolecules. Thus, just as amino acids were used to construct proteins, sugars served to make carbohydrates, nucleo-

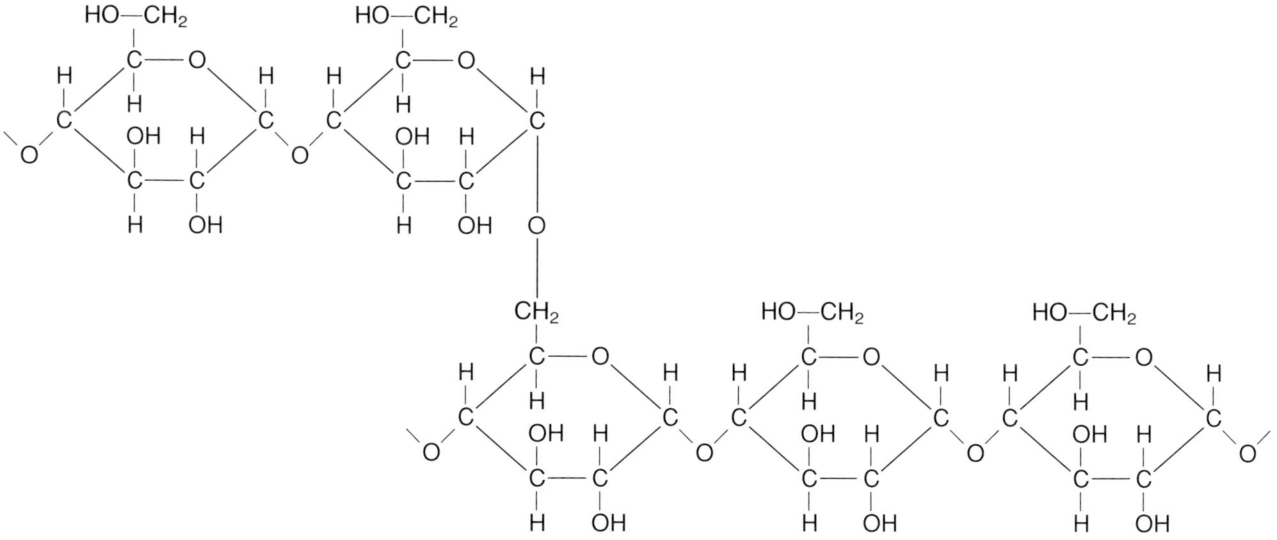

Glycogen

Carbohydrates range from simple sugars to long chains that contain thousands of connected sugar subunits. The wide variety of carbohydrates found in nature reflects the use of different sugars as building blocks, and their connection in different ways. A particularly simple carbohydrate is glycogen, which serves as a storage form of glucose.

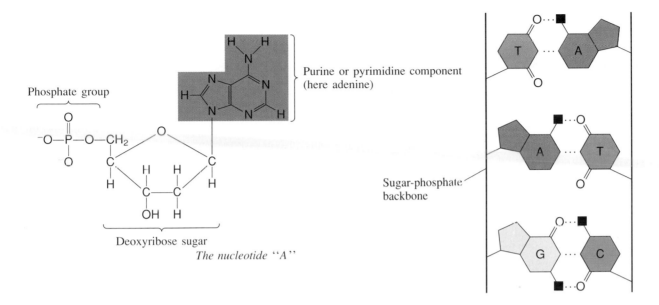

Phosphate group

Purine or pyrimidine component
(here adenine)

Sugar-phosphate
backbone

Deoxyribose sugar

The nucleotide "A"

Chains of nucleotide subunits are used to build the nucleic acids DNA and RNA, whose primary role is the storage and processing of genetic information. In DNA (shown above right), the sequences of A, T, C, and G nucleotides serve as a linear code that is used to align amino acids in their proper order for the production of a protein. The related nucleic acid RNA differs chemically only with respect to a minor modification of the sugar component in each nucleotide and a similarly small change in the "T" component.

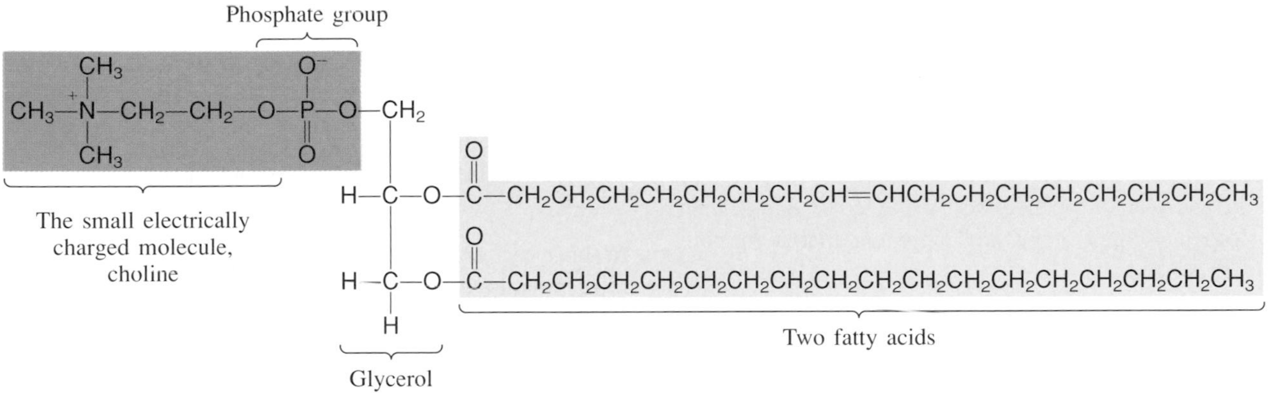

Phosphate group

The small electrically
charged molecule,
choline

Glycerol

Two fatty acids

The lipid membranes that surround cells, their nuclei, and their internal organelles, are made up of subunits such as the phospholipid shown above. These subunits spontaneously align themselves to form a bilayer in which the hydrocarbon fatty acids (yellow) are buried in the interior, while the electrically charged and OH-containing portions (blue) project outward to interact with the external environment of the cell on one side and the cell cytoplasm on the other. Embedded in the membrane are transport and other proteins.

tides to make nucleic acids, and fatty acids and glycerol to make lipids. In overview, carbohydrates were ultimately found to serve as energy stores and as structured elements, lipids as the building blocks of membranes, and nucleic acids as the carriers of the cell's inheritable genetic information. Even though the number and variability of the building blocks were limited, the potential complexity of the resulting macromolecules was not. In fact, this complexity began to mirror the diversity of life.

Proteins as Enzymes

The early biochemists could distinguish the four classes of cellular molecules only on the basis of their isolation properties and chemical composition. Proteins were nothing more or less than those substances built of amino acids, nucleic acids were the substances composed of nucleotides, and so forth. The functions of all of these classes of molecules were unknown.

But once the major components of the cell were in hand, it became possible to inquire into the chemical nature of enzymes. Naturally, the first question to be answered was whether enzymatic substances could be identified with one of the cell's known classes of molecules—carbohydrates, proteins, nucleic acids, or lipids. The alternative would be that enzymes are composite structures, or that some enzymes belong to one macromolecular class and some to another.

Most of the early workers—Fischer, for example—took it for granted that enzymes were proteins. But as the other classes of cellular molecules became better defined and the attention of biochemists broadened to include their physiological roles, it became apparent that a rigorous proof was needed.

Cell Fractionation

The initial attempts to provide this proof relied upon a procedure known as *cell fractionation*. In essence, this technique entails the physical separation of the cell into different molecular components, followed by their further analysis for function. The first methods of cell fractionation were based on the twin principles of *precipitation* and *extraction*. Cells were broken open (or a biological fluid, such as blood or milk, was used instead), and groups of components were selectively removed from the complex starting material by altering the character of the surrounding solution. For instance, the pH or salt concentration of the original solution could be changed, causing one or another group of molecules to precipitate. Or a new solvent could be added, drawing certain cellular species into itself and leaving other materials behind.

As fractionation procedures were applied to cell extracts, it was found that all of the 15-or-so enzymes that had been observed by the beginning of the twentieth century purified with the protein fraction. This observation suggested that enzymes were proteins, but it did not constitute a convincing proof. No matter how pure a "protein fraction" was, it was always possible that it contained a small (and undetectable) amount of some other cellular material. One could only say that the fraction that was most highly enriched for protein was equivalently enriched for a particular enzymatic activity. To the critical scientist, however, it would not be until enzymes were available in an absolutely pure state that the chemical nature of these catalysts could be established beyond doubt. But obtaining pure enzyme samples was considered a near impossibility at the time. Under the best experimental conditions available, enzymes were notoriously unstable, usually dying within a few hours or days of their isolation from living tissue, bringing the analysis to an end.

Willstätter, Colloids, and Chaos

In this treacherous state of affairs, the argument about the protein nature of enzymes took a serious wrong turn in the 1920s. The eminent German organic chemist Richard Willstätter (1872–1942) reported that he had succeeded in keeping the enzyme invertase alive through a series of purification steps and had obtained an enzyme preparation with a high level of activity. But, he announced, the active enzyme preparation did not contain even a trace of protein. Upon chemical analysis no amino acids could be detected. Willstätter was one of the first Nobel laureates in chemistry, and his views on enzymes were taken very seriously.

> It is not yet possible to obtain enzymes in an absolutely pure state. However, it has been possible to increase the degree of purity of some enzymes to such an extent that the basic questions regarding their chemical composition can be decided. Accordingly, our results show that enzymes cannot be included among the proteins or carbohydrates—and that, in fact, they do not belong to any of the known groups of complicated cellular molecules.

Willstätter's inability to detect any of the known classes of macromolecules in his enzymatically active invertase preparation led him to propose that enzymes were small organic molecules of unknown structure that were embedded in carrier substances—semidissolved complexes of protoplasmic material that went by the generic name of "colloids":

> If we attempt to sketch a picture of the chemical nature of an enzyme, I believe that the enzyme consists of a small active group that functions in a purely chemical way, associated with carrier material of a "colloidal" nature. Thus far it has not proved possible to permanently separate the chemically active group, that is, the enzyme molecule proper, from some form of supporting colloid.

Willstätter believed that he had succeeded in purifying the enzyme by absorbing it from its original cellular colloid carrier into an artificial colloid material that he had provided during the purification. To Willstätter, if enzymes purified like proteins during cell fractionation experiments, it was because the colloids in which they were embedded had proteinlike properties of adsorption, solubility, and precipitation. But as far as he was concerned, enzymes themselves could not be made of protein.

In retrospect we know that Willstätter was wrong, and therefore his data must have been misleading. But how had he managed to obtain an enzymatically active

Richard Martin Willstätter was a successor of Liebig as director of the prestigious Chemical Laboratory of the Bavarian Academy of Sciences in Munich. Winner of the 1915 Nobel prize in chemistry for his pioneering work on plant pigments, including chlorophyll, his analysis of the structure of enzymes proved less successful.

preparation that "contained no protein"? The answer reminds us of an important point about enzymes. It will be recalled that even the early studies had shown that enzymes, when first isolated, were highly active. For example, the diastase preparation of Payen and Persoz could convert more than ten times its own weight of starch into free glucose in a matter of minutes; and the enzyme could continue to function as a catalyst indefinitely—at least in principle. In other words, small amounts of an enzyme can be very active biologically. Willstätter's preparation must have contained predominantly colloids—inert and unrelated to catalysis—that were derived from the chemical materials used during purification. A trace amount of the enzyme he was studying made the preparation as a whole active even though the amount was below the level of chemical detectability. But this was not evident at the time, and for twenty years Willstätter's views appeared to have dealt a death blow to the protein theory of enzymes.

Sumner and the Crystallization of Enzymes

In 1927, Willstätter arrived at Cornell University as a Distinguished Visiting Professor to deliver a major set of lectures. He summarized the problems and methods involved in studying enzymes and discussed his own purification of invertase and its lack of any detectable protein component. He reiterated his conclusion that enzymes were small molecules with catalytic power, embedded in a colloid carrier. The function of the colloid was either to protect the enzyme or to "organize" it, but the catalytic force lay with the small molecule.

In the audience, and largely ignored by Willstätter, was a young assistant professor who the year before had published a relatively unknown paper on enzymes. The man was James Sumner (1887–1955), and his paper and the chain of discoveries it generated would eventually eclipse Willstätter's colloid theory and lead to the award of a Nobel prize for determining the general chemical nature of enzymes.

Sumner realized that, although the purification of enzymes had made considerable progress, especially with

James Batcheller Sumner (right) performed the first important study establishing that enzymes are proteins. Sumner had lost his left arm in a shooting accident at the age of 17 (he was left-handed). However, this did not prevent him from obtaining a doctoral degree and entering an academic career, making a major scientific discovery, and sharing in the award of a Nobel prize.

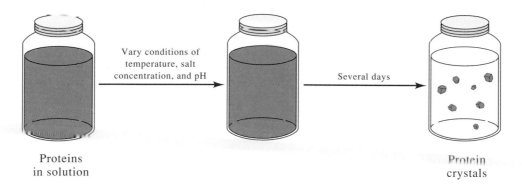

Crystallization of proteins. Whenever it is possible to obtain biological material in this way, a tremendous advantage accrues because the crystalline material is generally very pure.

regard to keeping the enzyme "alive," all attempts to isolate a pure enzyme and so determine its chemical nature directly had been unsuccessful. Even Willstätter's efforts resulted in an enzyme preparation that, while it was biologically active, could not be *chemically* defined. A new approach was in order.

The approach Sumner chose was based on the science and art of *crystallography*. It had been known for many years that small molecules (sodium chloride, glucose, tyrosine, and so forth) could often be prepared in a pure, crystalline form. The process of crystallization would position the atoms in a repeating arrangement, and the regularity of this crystal lattice would effectively exclude other dissimilar substances. The general procedure for preparing crystals was to begin with a solution containing a high concentration of the molecule of interest and attempt to "grow" the crystals in the surrounding solvent, generally after the addition of some concentrated salt solution to reduce the compound's solubility. It had long been known that many compounds readily formed crystals under given conditions in a particular solvent, leaving other substances behind in solution. Large molecules, such as proteins, proved to be much harder to crystallize than small molecules, but this too had more recently been accomplished for hemoglobin, the major protein of red blood cells. *Sum-ner's achievement was to crystallize—and thereby purify—the first substance with enzymatic activity.*

The enzyme Sumner chose to study, *urease*, carries out an important reaction in nitrogen metabolism. One molecule of urea is converted into two molecules of ammonia and one molecule of carbon dioxide:

$$H_2N-\overset{\overset{O}{\|}}{C}-NH_2 + H_2O \longrightarrow 2NH_3 + CO_2$$

$$\textit{Urea} \qquad\qquad \textit{Ammonia} \quad \textit{Carbon}$$
$$\textit{dioxide}$$

This decomposition reaction allows plants to unlock the nitrogen present in animal waste for subsequent use in their own metabolism.

Sumner set out to purify urease to a much higher level than had previously been achieved, taking advantage of the new approach of crystallization. He began with a finely ground preparation of plant material whose high biological activity suggested that it contained a correspondingly high level of urease enzyme.

The method which I use to obtain the crystals is extremely simple. A 32 percent solution of acetone in water is mixed with 100 grams of finely ground cellular material containing urease. The slurry is then poured onto a filter, and allowed to settle by gravity overnight in the cold. By the next morning it is observed that crystals have developed in the liquid that passed through the filter, and these may be collected by centrifugation. A second crystallization treatment—achieved by redissolving the original crystals in water and then adding fresh 32 percent acetone—gives a homogeneous collection of octahedral crystals. At any stage the crystals may be stored in the mother liquor [the solution in which the crystals develop], or they can be redissolved in water at room temperature and analyzed for enzyme activity.

The crystalline material could have been any of a number of substances in the cell extract. Fortunately, it appeared to be urease, since the redissolved crystals gave a solution with a greatly enhanced ability to break down urea.

The tremendous advantage of Sumner's success with crystallization was that it yielded very pure, enzymatically active material that offered the opportunity for a close chemical analysis. When he examined the crystalline material for the presence of the various types of cellular molecules—for example, by hydrolyzing the material with acid and analyzing the small molecules released—he found amino acids and nothing else. The crystalline urease was overwhelmingly protein.

Nonetheless, Sumner's original report was met with much skepticism—mostly because his conclusions ran counter to those of the more famous Willstätter. It was agreed that the crystallization of proteins, and especially their repeated recrystallization, was a valuable means of purification, since it yielded material that was much purer than that prepared by any of the traditional methods of cell fractionation. However, many investigators expressed doubt that the mere fact that Sumner's urease preparation was crystalline was sufficient to warrant the important conclusion that the enzyme was a protein. This was not an

Urease crystals.

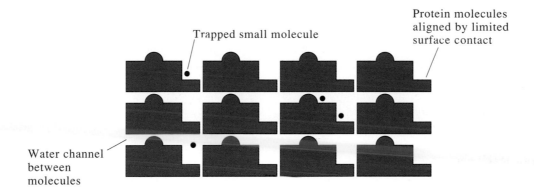

A schematic diagram of a protein crystal.

unreasonable concern because, as more was learned about protein crystals, it became clear that a significant fraction of their volume consists of water, in which small molecules might reside. Thus, the possibility existed that, although Sumner's material was much purer than any previous enzyme preparation, its catalytic activity resided not in the protein crystals themselves but in some small molecule trapped within the crystals. Although Sumner had made a compelling first case that enzymes are proteins, a significant element of doubt remained.

Northrop and Kunitz and the Proof That Enzymes Are Proteins

Sumner's conclusions gained full acceptance only about ten years later, through the strong support provided by John Northrop (1891–1987) and Moses Kunitz (1887–1978) of the Rockefeller Institute. Encouraged by Sumner's earlier work, Northrop and Kunitz took up the approach of protein crystallization and obtained crystals of a number of digestive enzymes—*chymotrypsin, trypsin,* and *pepsin.* The preparation of these enzymes was made easier by the fact that they were present in relatively high

concentration in digestive fluids, such as the fluid secreted by the pancreas into the small intestine. Like urease, the crystals of the new enzymes chymotrypsin, trypsin, and pepsin were found to be predominantly proteins.

Crystals of chymotrypsin. Digestive enzymes like chymotrypsin degrade proteins by severing the bonds between their constituent amino acids.

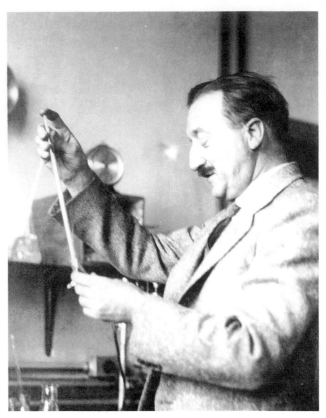

Moses Kunitz (left) and John Howard Northrop (right) convincingly showed that the catalytic activity of several enzymes was embodied in their protein component. Northrop shared the 1946 Nobel prize with James Sumner.

Northrop and Kunitz turned directly to the problem that had plagued Sumner—determining whether the catalytic activity of crystalline enzymes was an intrinsic property of their predominant protein component. They sought to disprove the alternative that, as proposed by supporters of Willstätter, the catalytic activity was due to small molecules that had accidentally been trapped in the protein crystals. The two men reasoned that, if such were the case, there would be two chemical species present in the crystals: one the protein material and the other a small, enzymatic substance. Because the two entities should have very different physical properties, it should be possible to redissolve the crystals and, through further analysis, separate them from one another. If it proved impossible to separate the hypothetical catalytic molecule from the bulk crystalline protein material, then it would become more and more difficult to maintain the argument that the secondary substance existed. Following this strategy, Northrop and Kunitz performed several types of experiments, of which we will consider only three.

1. A major technique for separating two different molecular species takes advantage of differences in their electric charge. It will be recalled that several of the amino acid subunits of proteins have side chains that

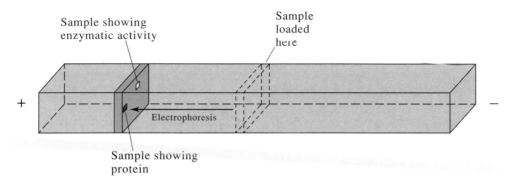

A schematic drawing of an electrophoresis experiment in which a sample of redissolved crystalline enzyme is subjected to the pull of an electric field. Both the enzymatic activity and the protein material migrate in the same direction and at the same rate.

carry either a positive or a negative charge. The total of such subunits confers a net electrical charge on the overall protein. When placed in an electric field, a protein will therefore migrate toward either the positive or the negative electrode, according to its charge. During such *electrophoresis*, each molecular species moves with its own characteristic speed. Northrop and Kunitz found that, when crystals of their digestive enzymes were redissolved and subjected to electrophoresis, the protein material (detected by the presence of amino acids) moved in a particular direction and at a particular speed. By testing the ability of samples recovered from the electrophoresis to work on test substrates, they could also monitor enzymatic activity. It was found that the protein material and the enzymatic activity had both migrated in the same direction and at the same speed. The catalytic material did not separate from the bulk protein.

2. Another powerful technique for separating two different molecular species makes use of *centrifugation*. Different molecules move more or less rapidly in a centrifugal field depending on their mass and shape, and each moves with a characteristic speed. If the enzyme were a small molecule merely trapped in a protein crystal, there is no reason why it should sediment at the same rate as a large protein. Thus, North-

rop and Kunitz redissolved their crystalline digestive enzymes and subjected them to centrifugation to see whether the catalytic activity sedimented together with the protein. Once again the physical material and the biological activity proved to be inseparable.

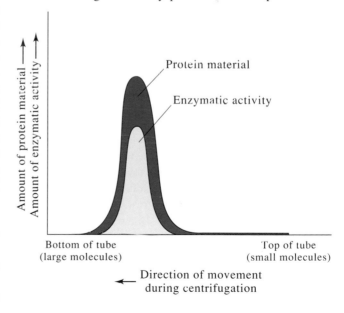

When a crystalline enzyme is redissolved and subjected to centrifugation, both the protein material and the enzymatic activity sediment at the same rate.

RNA ENZYMES

In the 50 years that followed the work of Sumner, Northrop, and Kunitz, more than a thousand enzymes were isolated and studied. All of these enzymes proved to be proteins. Organized into teams, they could be seen to carry out all of the physiological processes known to occur in living cells. Thus individual enzymes catalyzed a wide range of chemical transformations, including the metabolism of glucose and related molecules to generate the numerous building blocks of the cell, the production of energy in the mitochondria, the linkage of various sugars in different patterns to form complex carbohydrates, and the joining of nucleotides during the copying of the DNA molecules of genes.

It was not until the 1980s that the first exception to the rule that "enzymes are proteins" was found. Two research groups, one at Yale headed by Sidney Altman and one at the University of Colorado led by Thomas Cech, independently made the surprising—and initially not believed—finding that the chemical transformations under study in their laboratories were promoted by RNA molecules, not proteins. Both groups were characterizing reactions in which RNA molecules (chains of RNA nucleotides) were "processed." The essence of the processing reaction was the excision of material from the RNA chain to give a shorter, modified RNA molecule, which was then prepared to carry out its physiological function. In one case, a stretch of nucleotides was removed from the end of the RNA molecule; in the other case, an interior piece of the RNA molecule was excised and the flanking segments then spliced together. Based on the previous half century of experience, both Altman and Cech quite naturally expected that cellular proteins would catalyze these processing reactions, and the two research groups set out to isolate the enzymes involved. But when the reactions worked equally well in "control" experiments that contained RNA molecules, but no added proteins, the experimenters were led to the conclusion that the RNA molecules were catalyzing their own processing reactions.

In gaining acceptance for their work, both scientists were haunted by the ghost of Willstätter. It was widely argued that the processing reactions were carried out by a trace amount of a *protein* enzyme that had been isolated from cells along with the preparations of substrate RNA. Only the use of RNA molecules synthesized in test-tube reactions—molecules that necessarily had never come into contact with cellular proteins—put this concern to rest. The two original RNA enzymes, and the several related catalysts that have subsequently been found, have been given the name *ribozymes*. They are basically similar in that they carry out one general type of reaction: the breakage or formation of a covalent RNA linkage between two nucleotides.

The finding of ribozymes does not overthrow the protein theory of enzymes, but it does put boundaries on it. The fact remains that almost all chemical reactions in living systems are catalyzed by protein enzymes. The overwhelming preponderance of protein enzymes is attributed to the fact that they are built of 20 subunits with varied chemical properties, allowing a great potential for chemical diversity, whereas RNA enzymes, constructed of only four similar subunits, may have a more limited flexibility.

The deeper meaning of the finding of RNA enzymes does not lie in whether future research may identify many such enzymes, possibly catalyzing a larger variety of reactions. The object is not to win a numerical contest with protein enzymes. Rather, the importance of ribozymes lies in their demonstration that RNA molecules have the chemical sophistication necessary to function as catalysts. This means one can imagine that there might exist somewhere in the universe a world run by RNA enzymes. Clearly that world is not the Earth, 4.5 billion years into its evolution, but the intriguing possibility exists that the Earth was once such a world. Perhaps when living systems first evolved, the earliest enzymes were RNA enzymes. There is no evidence, but the idea is appealing because nucleic acids have a second, extremely useful property: they are designed to be inherently capable of replication. If the first enzymes, being made of RNA, could also replicate themselves, a major problem in understanding how life arose may become clear. Still, this idea is speculative, and it remains for future research to determine whether the existence of today's RNA enzymes reflects the origin of life.

3. Finally, we consider a completely different way of distinguishing between molecules that depends on their different sensitivities to various chemical treatments. Northrop and Kunitz took advantage of the fact that exposure to acid damages molecules at different rates. They incubated their redissolved enzyme preparations with various concentrations of acid, then returned the preparations to an appropriate pH and assayed them for enzymatic activity. During the treatment, the acid began to randomly break apart the amino acid connections in the intact protein, as shown by the falling curve in the graph on this page. At an exactly corresponding rate, the enzymatic activity was destroyed.

In all of these experiments, the fact that the physical material moves or is inactivated in the same way as the biological activity is the key to the conclusion that the catalytic power is an intrinsic property of the protein.

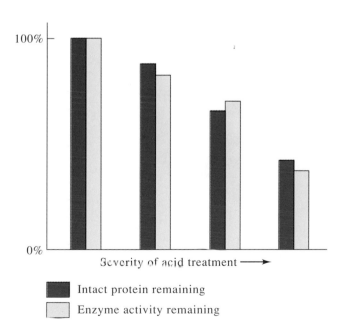

Intact protein remaining

Enzyme activity remaining

Upon exposure to acid, there is a correspondence between the destruction of intact protein and the loss of catalytic activity.

Our experiments have yielded no evidence that protein crystals contain a mixture of enzymatically active and inactive substances. It is reasonable to conclude that the crystalline material is either a pure substance or is composed of two very closely related substances. But if it were composed of two substances, it must be further assumed that these substances have the same electrophoretic and sedimentation properties. It must also be assumed that both substances are changed by acid at the same rate and to the same extent. This could hardly be true with the possible exception of two closely related molecular species. But [since the bulk of the crystalline enzyme preparation is clearly a protein], it would follow that the other component, the enzyme, itself must also be a protein and this, after all, is the main point.

Thus the small hypothetical catalyst of Richard Willstätter was finally laid to rest, dispatched by Occam's razor (the method of argumentation adopted by William of Occam, consisting of eliminating all unnecessary facts or assumptions from the question under analysis).

Looking back on the controversy, Northrop wrote:

The history of biochemistry is a chronicle of a series of controversies. These controversies exhibit a common pattern. There is a complicated hypothesis, which usually entails an element of mystery and several unnecessary assumptions. This is opposed by a more simple explanation, which contains no unnecessary assumptions. The complicated one is always the popular one at first, but the simpler one, as a rule, eventually is found to be correct. This process frequently requires ten to twenty years. The reason for this long time lag was explained by Max Planck. He remarked that scientists never changed their minds, but eventually they die.

The work of Sumner, Northrop, and Kunitz had produced conclusive evidence that the remarkable catalytic force of enzymes was somehow locked up in the structure of those molecules called proteins—presumably in their content and organization of amino acid subunits. In fact, the thousand or more enzymes that have been purified since then have, almost without exception, turned out to be proteins.

4

The Anatomy of an Enzyme

In the world of cellular chemical reactions, enzymes are the molecules of primary significance. They can promote a wide variety of reactions, under physiologically acceptable conditions of pH and temperature, that otherwise would occur so slowly as to be of no benefit to the cell. We now take up the study of the molecular architecture of enzymes, searching for those aspects of protein structure that can generate the ability to catalyze specific chemical reactions.

In order to gain a detailed understanding of the structure and function of an enzyme, we will concentrate our attention on one particular enzyme—*chymotrypsin*. This digestive enzyme is analogous to the one observed by Schwann in the 1830s that played such an important role in the discovery of enzymes. Ever since Northrup's and Kunitz's crystallization of chymotrypsin a century later, this protein has been under continuous study, and today more is known about its structure and mode of catalysis than is the case for almost any other enzyme.

Chymotrypsin is produced in the vertebrate digestive system by certain cells of the pancreas (the *acinar cells*). It is released into the pancreatic duct in an inactive form, and then activated when it reaches the small intestine. Through the action of chymotrypsin, large proteins, taken in as food, are broken into moderate-sized pieces, which are then further degraded by other protein-digesting enzymes *(proteases)*. The breakdown

A reconstruction of the enzyme chymotrypsin, generated on a computer from X-ray crystallographic data.

THE SYNTHESIS AND SECRETION OF CHYMOTRYPSIN BY THE PANCREAS

Much of the pancreas is organized into spherical aggregates of about a dozen cells, called *acini*. Each *acinus* has a central cavity (or lumen) into which the cells secrete the digestive enzymes they synthesize. The lumen is connected to a small duct, which merges with the ducts from other acini, eventually forming the large pancreatic duct that leads to the small intestine.

Not shown here is a second group of pancreatic cells. These form the *islets of Langerhans* and produce hormones such as insulin, which are secreted directly into the blood.

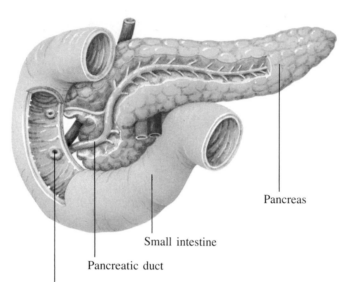

Pancreas

Small intestine

Pancreatic duct

Opening of pancreatic duct

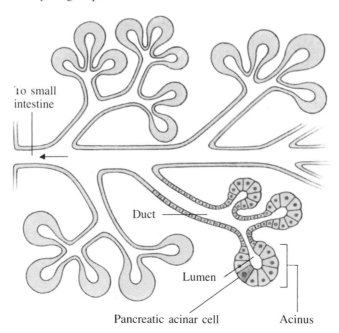

To small intestine

Duct

Lumen

Pancreatic acinar cell

Acinus

The electron micrograph to the lower left on the facing page shows a 5000-fold magnification of a cross section through a pancreatic *acinus*. The cells are arranged radially around a central lumen, with each wedge-shaped cell having a portion of its surface (the apical membrane) bordering on the lumen. The cell nuclei are located away from the apical surface, toward the base of the cells. The large black spots are storage vesicles containing chymotrypsin and the other digestive enzymes produced by the cell.

Central duct   Secretory vesicles

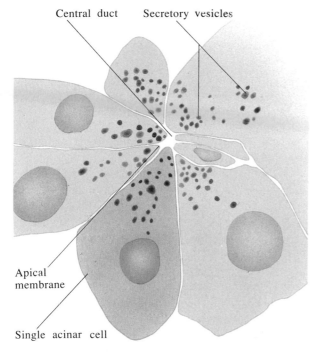

Apical
membrane

Single acinar cell

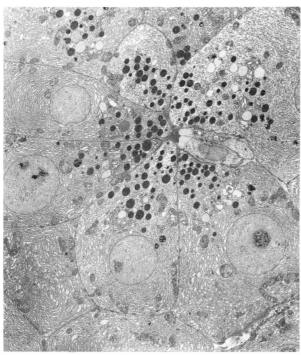

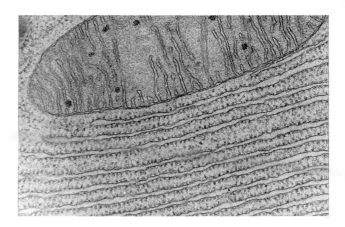

The 100,000-fold magnification of a cross section through an acinar cell above illustrates many of the intracellular structures essential for the synthesis, packaging, and release of pancreatic digestive enzymes. Much of the energy needed for protein synthesis and secretion is provided by mitochondria. Part of a mitochondrion appears as the large structure taking up the upper half of the photograph. The numerous small dense dots are the ribosomes—the cellular factories where amino acids are actually assembled into proteins.

Many of the ribosomes are attached to membranes, which form an internal network of interconnected tubes composing a cytoplasmic transport system, called the endoplasmic reticulum. When acinar cells produce chymotrypsin for export, the newly synthesized enzyme molecules pass from the ribosomes into the endoplasmic reticulum. Portions of the endoplasmic reticulum then bud off as small spherical transport vesicles, each containing millions of chymotrypsin molecules. These vesicles pass through another cellular organelle, the Golgi apparatus, where packaging is completed. The resulting enzyme storage vesicles come to lie at the apical end of the cell adjacent to the lumen of the acinus.

In the vesicles, the digestive enzymes are stored in an inactive form until hormones induce the movement of the storage vesicles to the lumen-facing surface of the acinar cell, where they fuse with the apical cell membrane and release their contents. The chymotrypsin molecules are free to flow down the pancreatic duct toward the small intestine, the site of their activation and participation in the process of digestion.

of ingested proteins yields free amino acids that can be assimilated by the cells of the organism for use in the synthesis of their own proteins. Enzymes are absolutely essential for this molecular recycling; in the case of digestion they allow the completion in a few hours of a process that, if it depended on uncatalyzed decomposition, would require centuries.

The Amino Acid Subunits of Chymotrypsin

Because chymotrypsin is easily crystallized after its recovery from the pancreatic fluid, biochemists have long had a simple method for its purification. Thus, *the one basic prerequisite for structural analysis is readily achieved— the enzyme is available in a pure form.* This purity is essential, for a biochemist could no more analyze the structure of a *mixture* of proteins than a psychiatrist could provide a profile of a particular individual based on data derived from an entire population.

Our analysis of the structure of chymotrypsin begins with a simple experiment. If crystals of the enzyme are dissolved in water and treated with concentrated acid for a prolonged time and at high temperature, about 20 different species of small organic molecules are released. These are the amino acids—a group of basic building blocks of such great chemical diversity that nature has found it possible to use them in varying combinations to construct virtually all the catalysts of all the organisms on earth.

The 20 amino acid building blocks are shown on the facing page. At first glance, the structures of these molecules may seem formidable, but in fact they are all variations on a single theme and are not really as complicated as they appear. The theme, expressed in chemical terms, is depicted below and in the box on pages 96–97. Each amino acid is composed of a central carbon atom (the α carbon), which is bonded to four distinctive groups:

- an amino group (NH_2)

- a carboxylic acid group ($\overset{\displaystyle O}{\overset{\displaystyle \|}{C}}$—OH, or COOH)

- a hydrogen atom (H)

- a variable group, called the side chain, or R

The 20 different side chains account for the 20 different amino acids, and each has a special role to play in the structure and function of the enzyme.

Generally, we will represent amino acids in the rectangular planar display shown in panel A of the box because this format is easiest to comprehend. Occasionally, however, we will want to consider the detailed three-dimensional structure of these molecules. In anticipation of this, panels B and C show the more complex "ball-and-stick" and "space-filling" models biochemists use to represent amino acids.

Like most proteins, chymotrypsin contains all 20 amino acid subunits. They occur in the amounts shown in the table on page 98. There is nothing very unusual about chymotrypsin's amino acid composition. The same amino acids are used, in more or less the same proportions, in all other proteins. Thus, our problem is to learn how this collection of otherwise inert building blocks can be organized so that the aggregate is an enzyme capable of promoting a specific chemical reaction.

In anticipation of the answer we will ultimately reach in this chapter, we can look forward to the model on page 139, which shows the structure of chymotrypsin. We see a constellation of 4000 carbon, hydrogen, oxygen, nitrogen, and sulfur atoms, organized into 241 amino acids, and held together in a very specific three-dimensional configuration. The result is a structure with surface features precise enough to accommodate Fischer's lock-and-key hypothesis.

Even the most casual glance at the structure shows that it is too complex to be appreciated immediately. Thus our goal in this chapter will be to build the structure step by step so that its underlying organizational themes are clear and stand out amid the array of chemical detail. Then, in the next chapter, with the structure established, we will be in a position to achieve our major goal: to see how this three-dimensional assembly of amino acids gives rise to an enzyme with the ability to promote a specific chemical transformation.

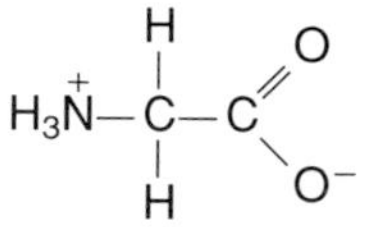

Glycine (Gly)

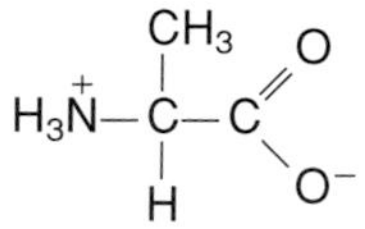

Alanine (Ala)

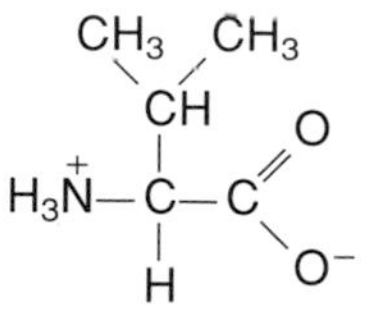

Valine (Val)

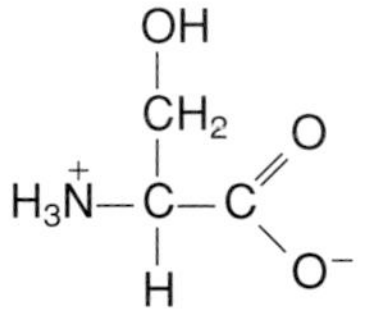

Serine (Ser)

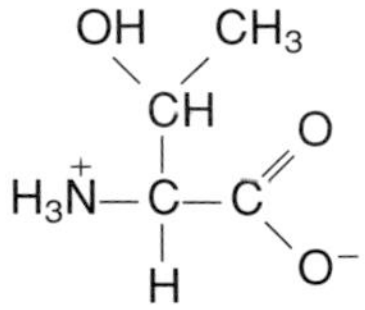

Threonine (Thr)

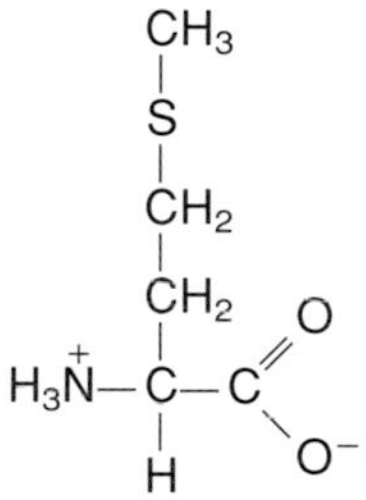

Methionine (Met)

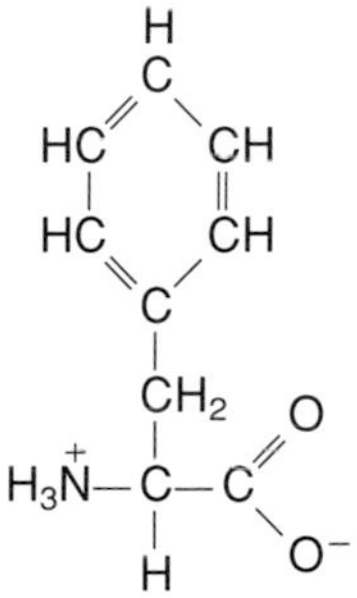

Phenylalanine (Phe)

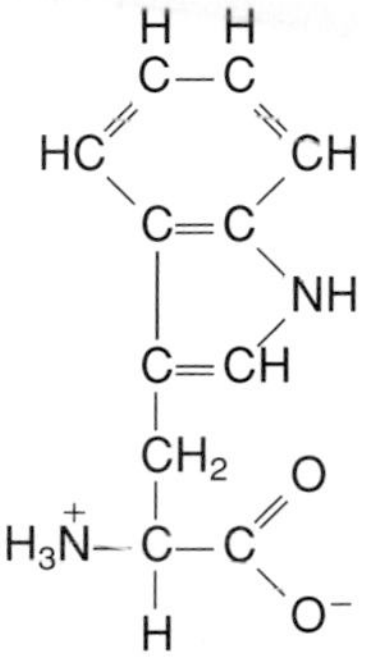

Tryptophan (Trp)

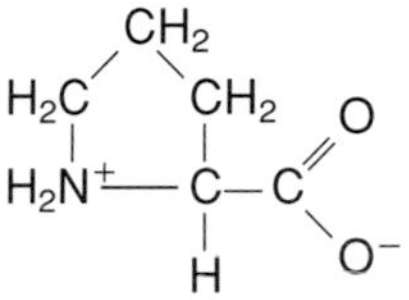

Proline (Pro)

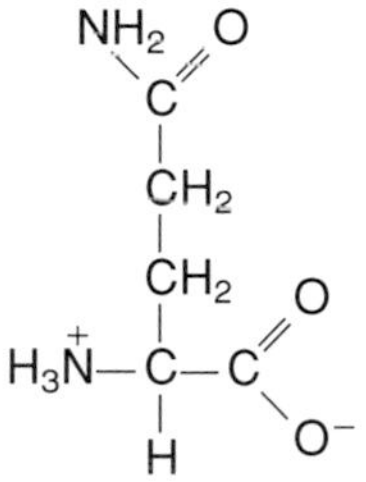

Glutamine (Gln)

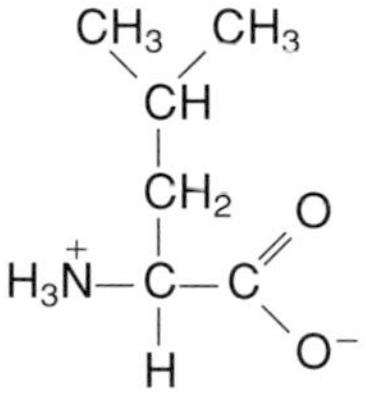

Leucine (Leu)

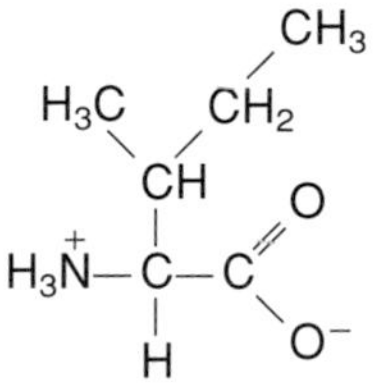

Isoleucine (Ile)

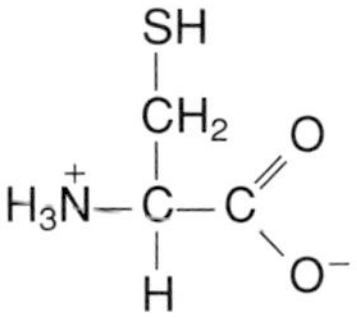

Cysteine (Cys)

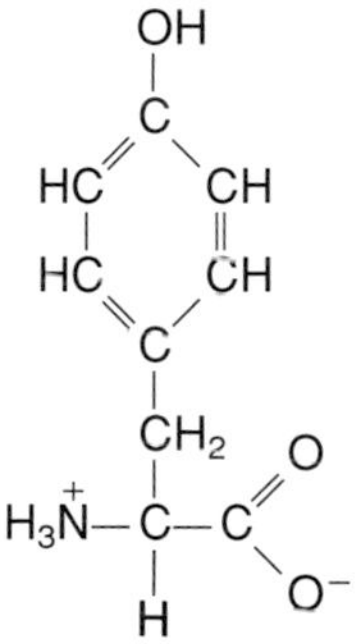

Tyrosine (Tyr)

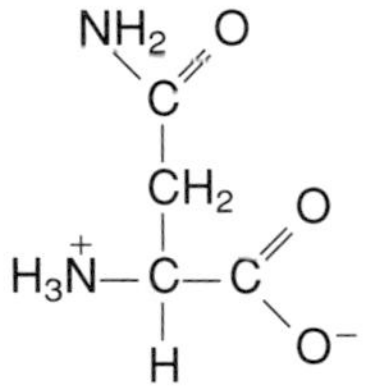

Asparagine (Asn)

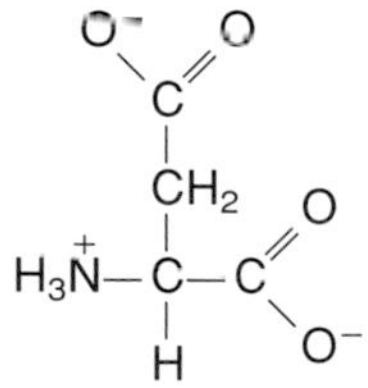

Aspartic acid (Asp)

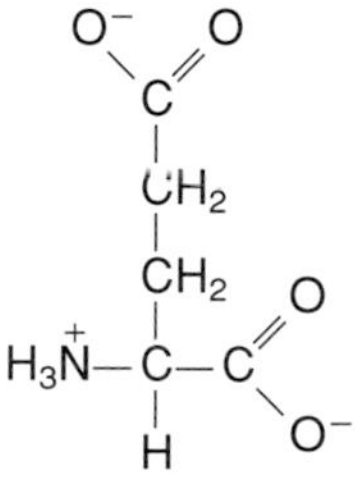

Glutamic acid (Glu)

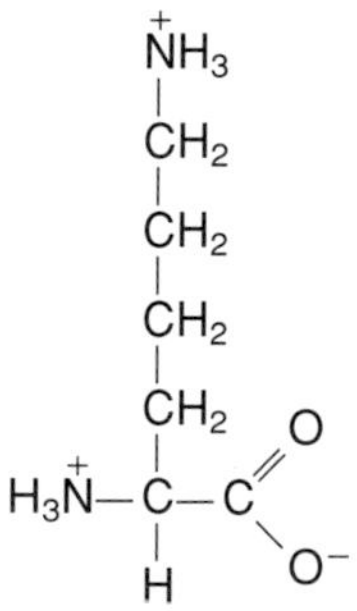

Lysine (Lys)

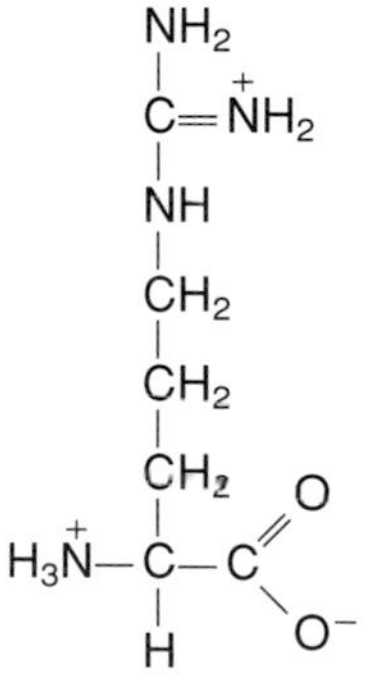

Arginine (Arg)

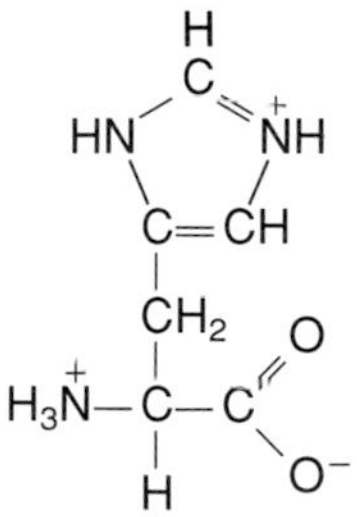

Histidine (His)

THREE WAYS
CHEMISTS REPRESENT THE MOLECULAR
STRUCTURE OF AMINO ACIDS

Panel A: The simplest method for representing amino acids is the *Fischer projection*. The molecule is drawn in a planar, rectangular format with the four groups attached to the central carbon atom projecting outward at right angles to each other. This method of diagramming chemical structures was introduced a century ago by Emil Fischer.

An important property of amino acids is shown below at the right. In the neutral pH environment of a cell (pH 7), the amino and carboxylic acid groups are both *ionized*—that is, they carry an electric charge. In the case of the carboxylic acid group, the charge arises from the dissociation of H^+ ($COOH \rightarrow COO^- + H^+$). Only in a very acidic environment (for example, pH 2) where there is an excess of H^+ in solution, does the carboxylic acid group retain its hydrogen (COOH). At neutral pH, the nitrogen atom of the amino group also becomes ionized: it uses its unbonded pair of electrons ($\dot{N}$) to gain H^+ and become positively charged ($\ddot{N}H_2 + H^+ \rightarrow NH_3^+$). This would fail to occur only in an OH-rich (alkaline) environment of pH 9.5 or higher.

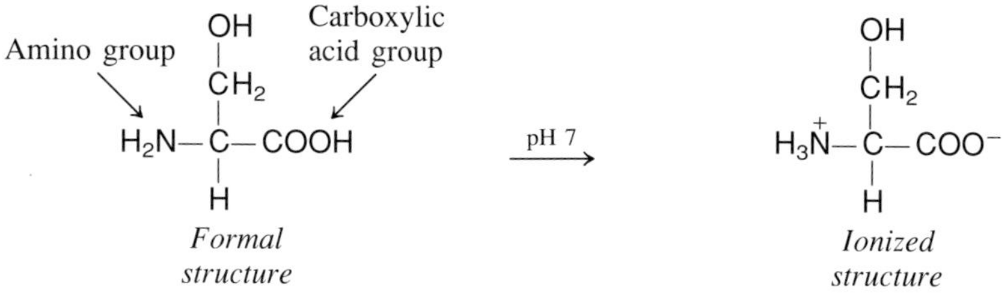

Amino group
OH   Carboxylic
| acid group
CH₂

$H_2N-C-COOH$

H

*Formal
structure*

pH 7 →

OH
|
CH₂

$H_3\overset{+}{N}-C-COO^-$

H

*Ionized
structure*

A

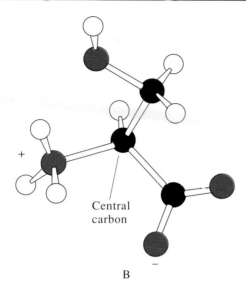

Central
carbon

B

−

C

Panel B: The *ball-and-stick model* is a more accurate representation of the three-dimensional structure of an amino acid. The four groups attached to the central carbon atom are shown radiating outward to the corners of a tetrahedron; one of the four groups (H) projects backward from the plane of the paper, away from the reader, while the remaining three groups (NH_3^+, H, and COO^-) project forward from the plane of the paper, toward the reader. In addition, the ball-and-stick representation attempts to convey the relative sizes and positions of the atoms, the lengths of the bonds between them, and the angles between the bonds. This relatively open structure gives the clearest impression of the three-dimensional structure of the amino acid building blocks. The specific amino acid shown is *serine,* whose side chain is CH_2OH. (By convention, carbon atoms are shown in black, hydrogen in white, oxygen in red, nitrogen in blue, and sulfur, when present, in yellow.)

Panel C: The most accurate representation for the structure of an amino acid is obtained from a *space-filling model*. This type of model takes into account the fact that molecules do not actually contain open space between the atoms. Thus, in this representation atoms (nuclei with their surrounding electron clouds) are shown closely juxtaposed to one another. We see not only that all of the internal space of the molecule is filled, but also (as in the case of the ball-and-stick model) that the external surface of each amino acid presents a number of well-defined contours, which can serve in lock-and-key interactions.

The Amino Acids of Chymotrypsin

Amino acid	Abbreviations		Number
Alanine	Ala	A	22
Arginine	Arg	R	3
Asparagine	Asn	N	13
Aspartic acid	Asp	D	9
Cysteine	Cys	C	10
Glutamine	Gln	Q	10
Glutamic acid	Glu	E	5
Glycine	Gly	G	23
Histidine	His	H	2
Isoleucine	Ile	I	10
Leucine	Leu	L	19
Lysine	Lys	K	14
Methionine	Met	M	2
Phenylalanine	Phe	F	6
Proline	Pro	P	9
Serine	Ser	S	27
Threonine	Thr	T	22
Tryptophan	Trp	W	8
Tyrosine	Tyr	Y	4
Valine	Val	V	23
Total			241

The Amino Acid Building Blocks

Since the amino acids are all alike except for their side chains (R groups), the place to begin an analysis of protein structure is with a careful study of the side chains and their properties. Both the structure of the enzyme and its mechanism of catalysis will ultimately depend on the specific biochemical properties of the amino acid side chains. In the following discussion, the most important point to note is the way in which the 20 kinds of side chains vary in size, shape, electrical charge, bonding capacity, and chemical reactivity. This variation generates a great reservoir of chemical flexibility which the cell draws upon to form a biologically active enzyme from a collection of inert building blocks.

Hydrophobic Amino Acids

The simplest amino acid is *glycine,* whose side chain consists of only a single hydrogen atom. This minimal side chain gives glycine a more or less neutral character—it generally serves as a flexible spacer in the finished protein molecule where no other amino acid would fit, but otherwise contributes no distinctive properties.

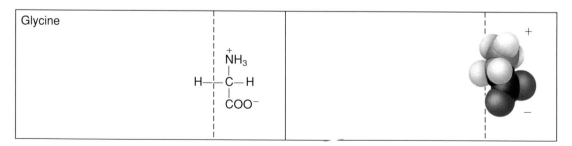

Glycine, the smallest amino acid, whose side chain is a simple hydrogen atom. The diagram shows both a space-filling model (on the right) and a modified Fischer projection (on the left). In both representations, the "common" portion of the amino acid is shown to the right of the dashed line, while the side chain projects to the left.

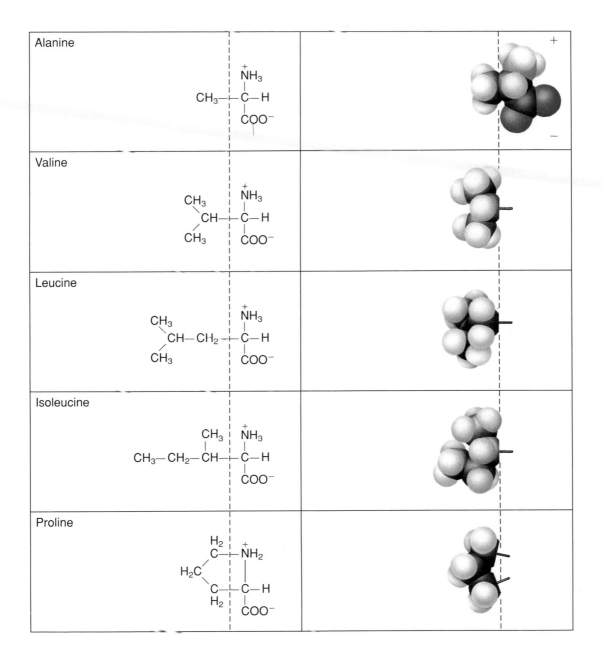

The five amino acids with pure hydrocarbon side chains: alanine, valine, leucine, isoleucine, and proline. Proline differs from the other amino acids in that the end of its side chain is bonded back to the amino group, resulting in an overall cyclic structure.

The next most complex amino acid is *alanine,* which has a methyl group (—CH₃) as its side chain. Alanine is the first of a set of five amino acids having hydrocarbon side chains (that is, side chains composed only of carbon and hydrogen atoms). The other amino acids in this group are *valine, leucine, isoleucine,* and *proline.*

From a chemical point of view, the important property of these amino acids is that their hydrocarbon side chains are *hydrophobic*—literally, "water fearing." As a result, these amino acids prefer to position themselves so that their side chains interact with each other rather than remain exposed to water molecules in the surrounding solution. The more massive the hydrocarbon component of the side chain, the more pronounced is the hydrophobic effect (see the table below). Thus the amino acids in this group tend to be clustered together, and they are often turned inward toward the interior of the protein. Here they serve as space-filling elements of various sizes and shapes. As will be discussed later, the tendency of hydrophobic amino acids to associate with each other gives rise to the hydrophobic bond, the first of four types of chemical interactions that hold the amino acid subunits in position with respect to each other and determine the overall structure of the enzyme.

Amino Acids with Hydroxyl Groups

Two more amino acids, *serine* and *threonine,* also contain short hydrocarbon side chains, but their side chains are augmented by the addition of a hydroxyl group. These hydroxyl groups (—OH) are important because they confer a special chemical capability upon the side chain. Specifically, OH groups are able to participate in the formation of *hydrogen bonds*—interactions in which a positive hydrogen is shared by two relatively negative atoms such as nitrogen and oxygen. Nitrogen has one ($\dot{N}$) and oxygen has two (:O:) unshared (or "lone") pairs of electrons, represented by the dots, that are not involved in forming covalent bonds with other atoms. These electron pairs are thus available for the formation of hydrogen bonds. An example of a hydrogen bond formed between two serine side chains is shown below:

Relative Hydrophobic Character of the Amino Acid
Side Chains

	Amino acid	Hydrophobic character*
Most hydrophobic	Isoleucine	1.83
	Leucine	1.80
	Phenylalanine	1.69
	Tryptophan	1.35
	Valine	1.32
	Methionine	1.10
	Proline	0.84
	Cysteine	0.76
	Tyrosine	0.39
	Alanine	0.35
Standard	Glycine	0
	Threonine	−0.27
	Serine	−0.63
	Histidine	−0.65
	Glutamine	−0.93
	Asparagine	−0.99
	Arginine	−1.50
	Lysine	−1.54
Least hydrophobic	Glutamic acid	−1.95
	Aspartic acid	−2.15

*High positive values indicate very hydrophobic amino acids. Negative values are assigned to amino acids whose side chains are hydrophilic— that is, water-seeking.

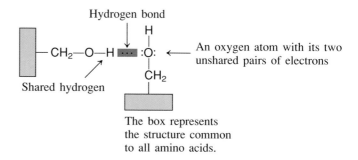

Hydrogen bonds are the second of the four types of chemical interactions that hold proteins together in their proper three-dimensional structure.

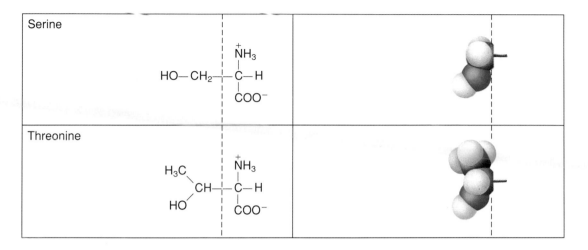

The amino acids with a hydroxyl (—OH) group on their side chain: serine and threonine.

Amino Acids with Flat Hydrocarbon Rings

There are three amino acids whose side chains are planar hydrocarbon rings: *phenylalanine, tyrosine,* and *tryptophan* (see the top of the following page). Like the other amino acids considered so far, these are basically hydrophobic. However, because the side chains of phenylalanine, tyrosine, and tryptophan are relatively flat, they are capable of forming geometrical arrangements that are somewhat different from those generated by branched side chains (such as those of leucine, valine, and isoleucine). In addition, tyrosine and tryptophan are further distinguished because they contain OH and NH groups that can form hydrogen bonds.

Amino Acids with a Positive or Negative Charge

The side chains of the 11 amino acids considered so far are uncharged at physiological pH. Five amino acids, however, are fundamentally different because their side chains normally carry a positive or negative charge. Such charges result from the gain or loss of H^+ by a side-chain amino (NH_2) or carboxylic acid (COOH) group. The presence of these electrical charges renders these amino acids *hydrophilic* (water seeking). The charges also endow these amino acids with the ability to form *ionic bonds*—a potential third type of link holding proteins together, which results from the attraction of the opposite charges, as shown below.

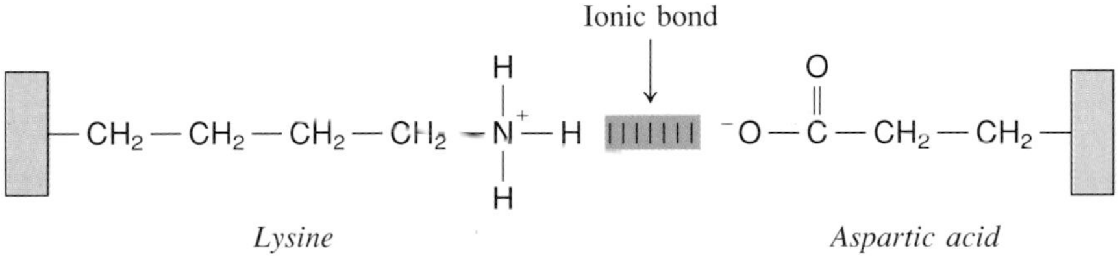

An ionic bond formed between two amino acids with opposite charges on their side chains.

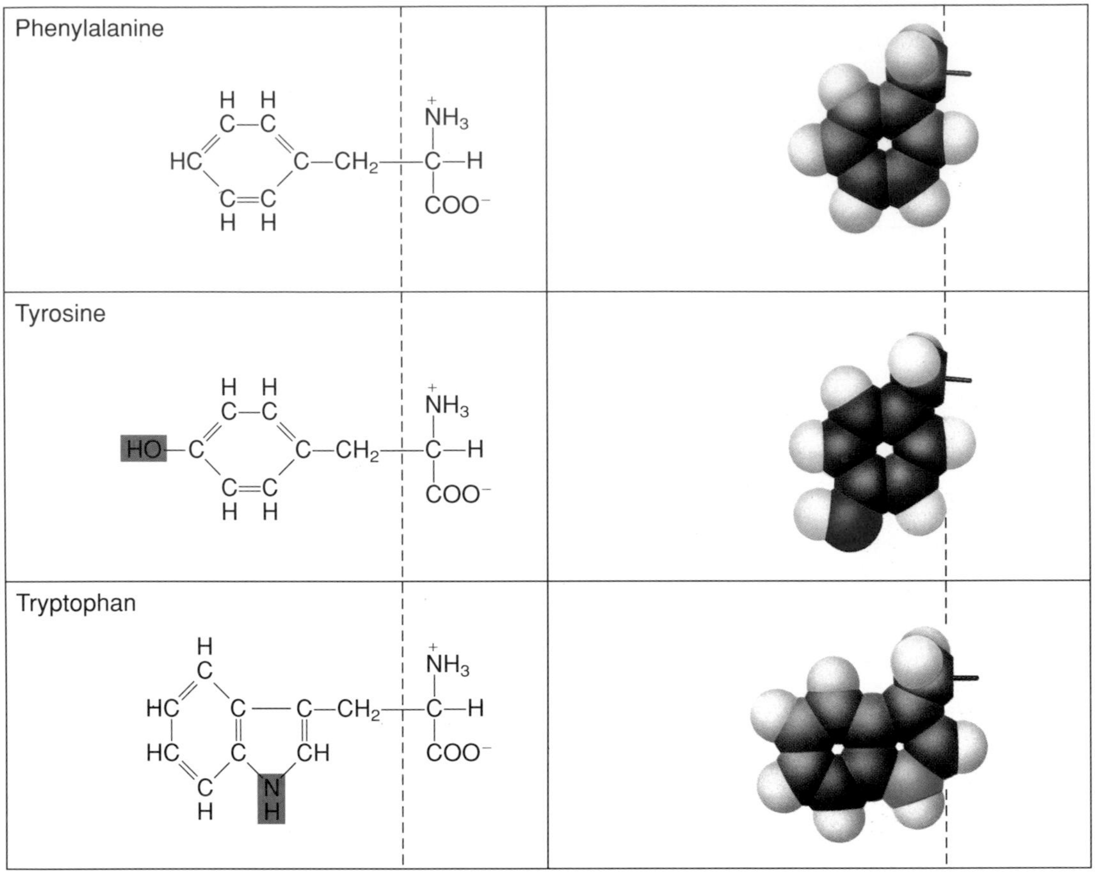

Phenylalanine

Tyrosine

Tryptophan

The planar hydrophobic amino acids: phenylalanine, tyrosine, and tryptophan. The hydrogen-bonding —OH and —NH groups of tyrosine and tryptophan are shown in red.

Lysine and *arginine* are positively charged amino acids. In both cases a nitrogen-containing group ($=\dot{N}H$ or $-\dot{N}H_2$) in the hydrocarbon side chain has used its un- shared pair of electrons to acquire a hydrogen ion (H^+) from the surrounding water solution and become $=NH_2^+$ or $-NH_3^+$ at physiological pH, as shown below:

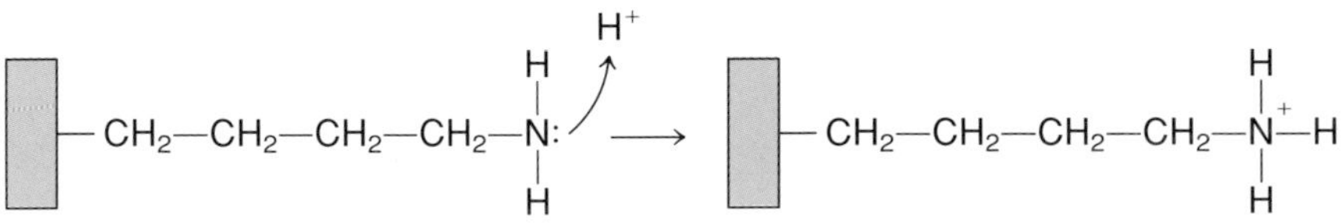

Lysine

Histidine is also capable of acquiring a hydrogen ion and becoming positively charged. However, the histidine side chain maintains a relatively weak hold on this hydrogen, easily losing and regaining the H^+ at neutral pH. Thus, histidine can be either neutral or positively charged and, in fact, shuttles back and forth between these two states. This chemical flexibility is of particular usefulness, as we will later see.

The amino acids with negatively charged side chains are *aspartic acid* and *glutamic acid* (see the following page). They differ only in the length of the hydrocarbon chain leading up to the negative charge. Like the carboxylic acid group (—COOH) attached to the central carbon of every free amino acid, those at the end of the hydrocarbon side chain in aspartic acid and glutamic acid readily dissociate at physiological pH to produce the negative charge.

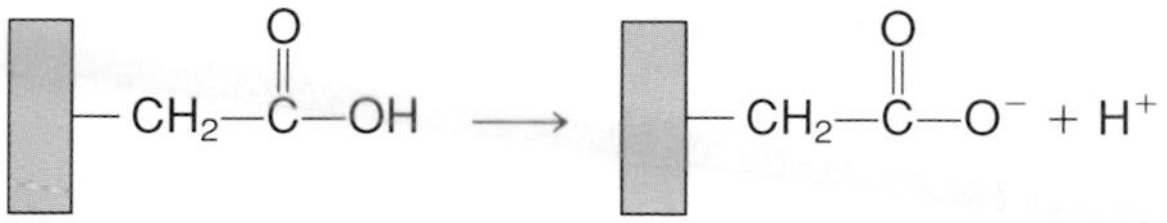

$$-CH_2-\overset{\overset{\textstyle O}{\|}}{C}-OH \longrightarrow -CH_2-\overset{\overset{\textstyle O}{\|}}{C}-O^- + H^+$$

Aspartic acid

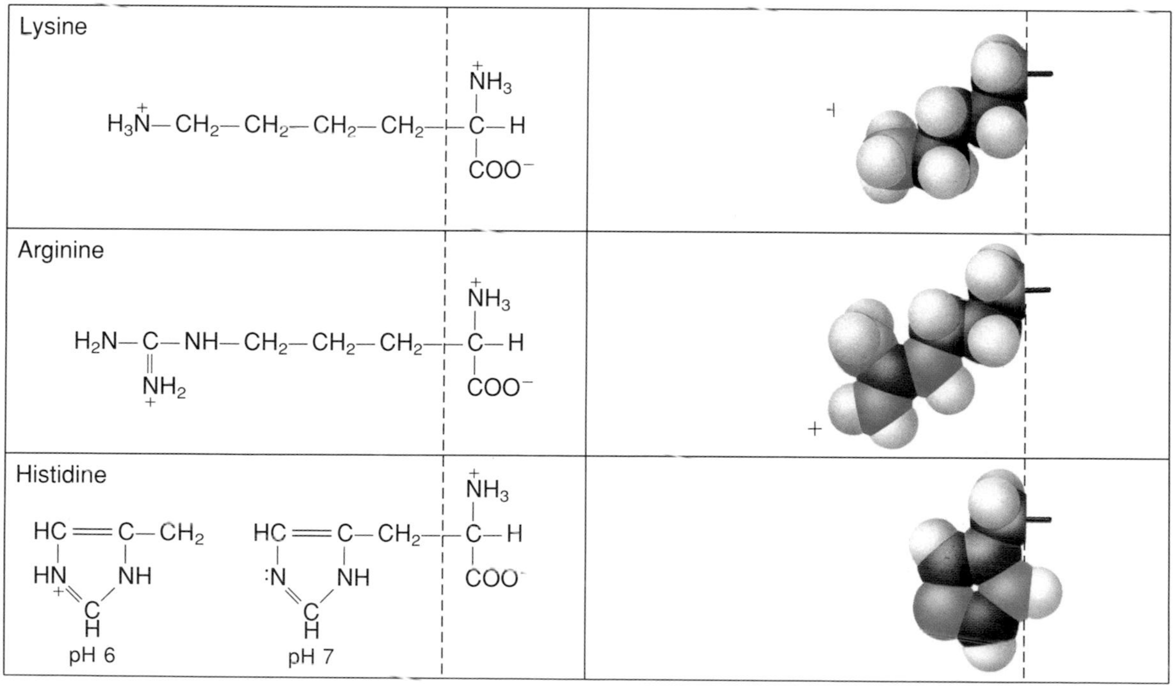

The positively charged amino acids: lysine, arginine, and, part of the time, histidine.

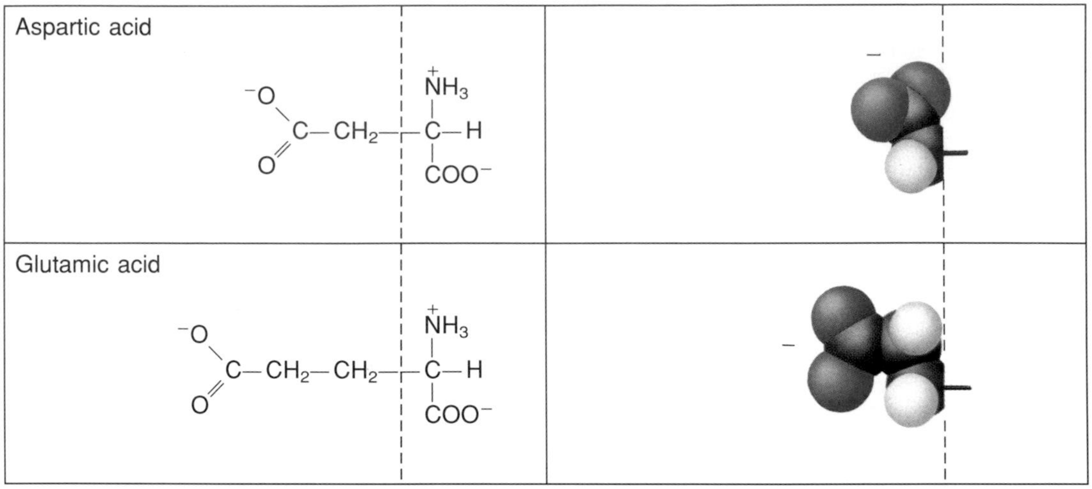

Aspartic acid	
^-O $\diagdown$ $C{-}CH_2$ $\diagup$ O — $\overset{+}{N}H_3$ $\mid$ $C{-}H$ $\mid$ COO^-	$^-$
Glutamic acid	
^-O $\diagdown$ $C{-}CH_2{-}CH_2$ $\diagup$ O — $\overset{+}{N}H_3$ $\mid$ $C{-}H$ $\mid$ COO^-	$^-$

The negatively charged amino acids: aspartic acid and glutamic acid. (Once the side chain COOH group has lost H^+, these amino acids are more properly referred to as aspartate and glutamate.)

Derivatives of Aspartic and Glutamic Acid

Modified forms of aspartic and glutamic acid are also found in proteins. These are *asparagine* and *glutamine,* in which a terminal amino group has been placed on the carboxylic acid component, removing its charge. Because they are not formally charged, these amino acids can no longer form ionic bonds. However, the N—H and C=O groups on their side chains can still participate in the formation of hydrogen bonds. The structure on the bottom of the facing page shows two of the many geometrical arrangements for hydrogen bonding by asparagine and glutamine. The strongest hydrogen bond is formed when the three participating atoms lie in a straight line, as on the left.

Sulfur-Containing Amino Acids

Finally, we come to the last two amino acids, *methionine* and *cysteine* (see top of page 106). These two amino acids are characterized by the presence of sulfur in their side chains. Methionine is essentially a hydrophobic amino acid and, except for its distinguishing, not very reactive, sulfur buried in a hydrocarbon chain, it might have been included in the group containing glycine, alanine, valine, leucine, isoleucine, and proline. Cysteine, however, is special: the *sulfhydryl* (—SH) group on its side chain is chemically reactive. As a consequence, two cysteine subunits can use their sulfhydryl groups to form a *covalent bond* with each other. The resulting *disulfide bond* then becomes an internal cross-link holding the two amino acids close together in the final protein structure.

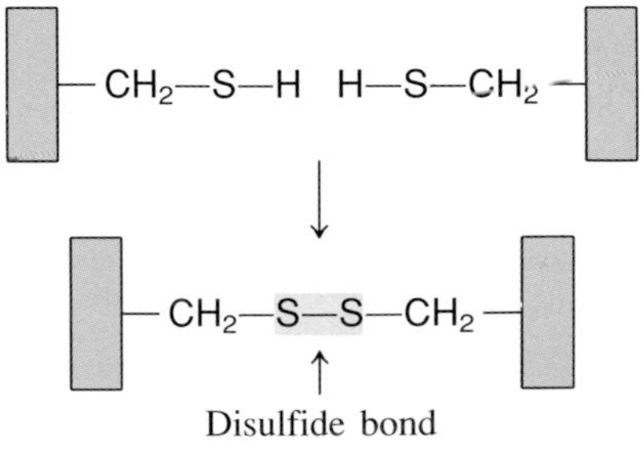

Disulfide bond

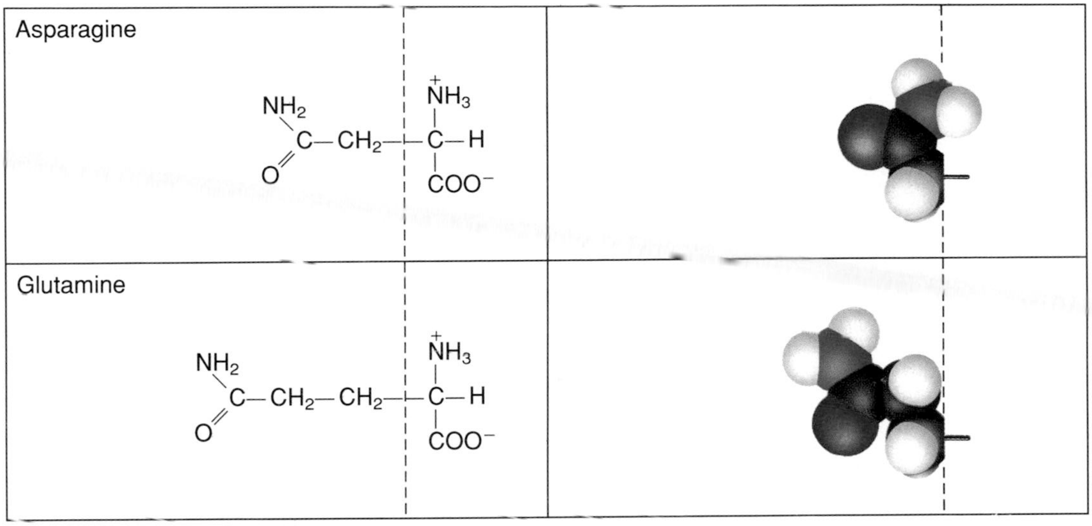

The uncharged derivatives of aspartic acid and glutamic acid: asparagine and glutamine.

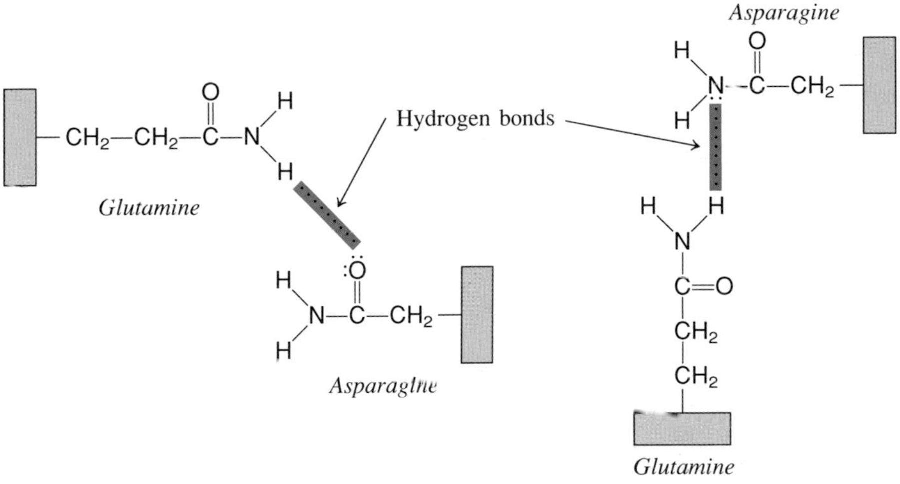

Two possible hydrogen bonds formed between glutamine and asparagine.

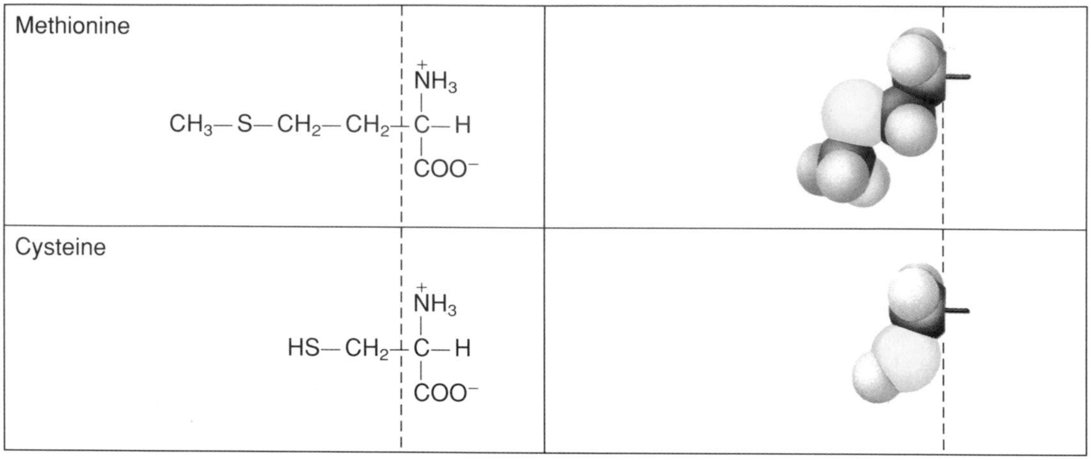

Methionine	
$$CH_3-S-CH_2-CH_2-\overset{\overset{+}{N}H_3}{\underset{COO^-}{C}}-H$$	
Cysteine	
$$HS-CH_2-\overset{\overset{+}{N}H_3}{\underset{COO^-}{C}}-H$$	

The sulfur-containing amino acids: methionine and cysteine. In methionine, the sulfur atom is buried in the interior of a hydrocarbon side chain, whereas in cysteine it is part of an exposed, chemically reactive —SH group.

These, then, are the amino acid subunits of proteins. Each of them makes a unique contribution to protein structure and function, a point that will become increasingly clear as we proceed to consider how these amino acids combine to form an enzyme such as chymotrypsin.

The Polypeptide Chain

While the side chains of the amino acids ultimately provide the protein with its unique properties, the two other substituents of each amino acid—the amino and carboxyl groups—serve the elementary function of tying the amino acids together. Their role became clear in experiments in which proteins were subjected to a brief acid treatment. Because the short exposure to acid could not completely decompose the protein into its individual amino acid subunits, such experiments yielded intact segments of the protein with two, three, or more amino acids still covalently joined together. Chemical analysis of, for instance, those fragments consisting of two amino acids showed that they contained only one free amino and one free carboxyl group, implying that the "missing" NH_2 and COOH groups had been used to link the amino acids together. From such results it could be deduced that the basic unit of organization in proteins is a linear polymer of amino acids connected to one another by covalent bonds formed between the carboxyl group of one subunit and the amino group of the adjacent subunit. Thus, the intact protein consists of a long chain of amino acids, connected to each other by characteristic linkages of the form:

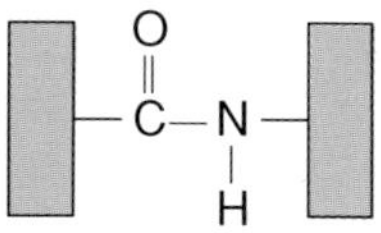

These linkages are termed *peptide bonds*, and the resulting *polypeptide chain* is a linear unbranched structure.

Although different proteins vary in length from less than 50 to more than 1000 amino acids, there is no corresponding increase in the complexity of amino acid connections as proteins become larger. *The peptide bond is the universal method of linking amino acids, and the polypep-*

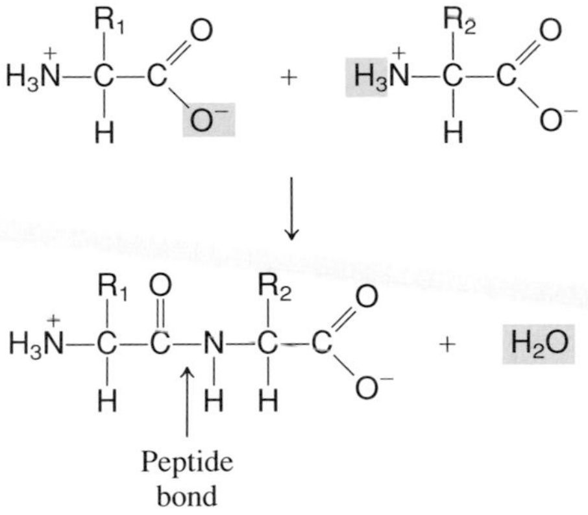

The amino and carboxyl groups of the amino acids hold the subunits of a protein together. Shown here is the formation of a peptide bond between two free amino acids. Note the loss of the elements of water during formation of the linkage.

mon agreement, the end with the free amino group is taken to be the beginning of the polypeptide chain. The sequence of amino acids in the protein is then written from the amino-terminal subunit (aa_1) to the carboxyl-terminal subunit (aa_n). Thus,

alanine-histidine-proline-serine-tyrosine

denotes a short protein chain (a peptide) five units long having alanine as the amino-terminal subunit and tyrosine as the carboxyl-terminal subunit. Naturally occurring proteins are, of course, much larger and, moreover, they may contain more than a single polypeptide chain—many have a number of identical chains, and still others are composed of several different chains. In the case of chymotrypsin, the 241 amino acids of the finished enzyme are distributed in three distinct polypeptide chains 13, 131, and 97 subunits long.

tide chain is the first or primary level of organization in protein structure.

Because the amino acid building blocks are themselves asymmetric structures, the finished polypeptide chain has an inherent chemical directionality. This "polarity" arises naturally from the difference between the groups present at either end. After the amino acids have been linked, only one amino and one carboxyl group remain unlinked, at opposite ends of the polypeptide chain. All the other similar groups have been used to form internal peptide bonds, as shown in the figure below. By com-

The Sequence of Amino Acids in Chymotrypsin

We have seen that a polypeptide chain consists of two parts: a regularly repeating *backbone* held together by peptide bonds and a variable part formed by the projecting amino acid side chains. Yet no protein exists in nature as a simple, flexible chain of amino acids. Rather, its biologically useful properties arise from the folding of the chain into a precise three-dimensional conformation with each

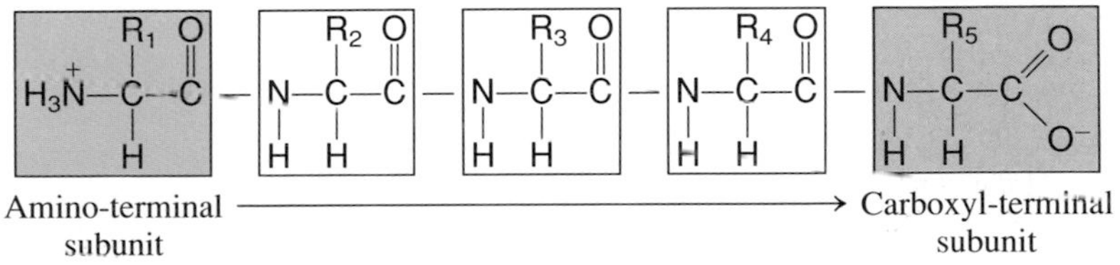

A short polypeptide chain containing five amino acids.

amino acid in a specific place. Importantly, it is the order of amino acids in the polypeptide chain—the *primary structure* of the protein—that determines precisely how it folds up. Thus, an essential step in our analysis of the structure of chymotrypsin is to determine the sequence of its amino acids.

The methods for determining an amino acid sequence were developed by Frederick Sanger in Cambridge, England, and Pehr Edman in Australia. By the middle of the 1950s, Sanger's ground-breaking work had made it possible to describe the amino acid sequence of a few small proteins, and Edman's subsequent improvements allowed protein sequencing to become a standard procedure used in laboratories throughout the world.

Preparing a Protein for Sequencing

In the simplest case, a protein consists of a single polypeptide chain, and its amino acid sequence can be analyzed directly. But if, like chymotrypsin, the protein has two or more polypeptide chains, a direct analysis would yield an uninterpretable mixture of several sequences—one from each chain. For this reason it is essential to first separate the chains from each other so that they can be studied individually.

When chymotrypsin is produced in the cells of the pancreas, it is first assembled as a single polypeptide chain. In the intestine, however, it is activated by cleavage in two places so that the mature enzyme is, in fact, composed of three associated polypeptide chains:

A chain: 13 amino acids long
B chain: 131 amino acids long
C chain: 97 amino acids long

To separate these chains, it is first necessary to break any cysteine-to-cysteine disulfide bonds holding the chains together. The protein is incubated with a chemical reagent that regenerates independent cysteine side chains (—SH), which are then modified with a reactive blocking agent (I-CH$_2$COOH) to yield inert —S—CH$_2$COOH groups. The three polypeptide chains remain held together only by noncovalent bonds (ionic bonds, hydrogen bonds, and

Frederick Sanger, who, together with Pehr Edman, devised the procedures for determining the amino acid sequence of proteins. For his early work, Sanger was awarded the first of his two Nobel prizes, in 1958.

hydrophobic interactions), and various chemical treatments can then be used to dissociate them. For example, urea and guanidine interfere with noncovalent interactions between the amino acid subunits so that proper folding of the protein chains can no longer be maintained. The un-

Urea and guanidine, two small organic molecules that unfold or *denature* proteins by interfering with the internal interactions between the amino acid subunits. Note the hydrogen-bonding potential of the NH$_2$ and CO groups in these molecules.

folded, or *denatured,* chains can then be separated from each other by means of treatments taking advantage of differences in their size, their net electrical charge, or their content of hydrophobic amino acids. The three polypeptide chains of chymotrypsin differ markedly in these properties and are thus readily separated by such procedures as electrophoresis and sedimentation.

	Length	Net charge at pH 7
A chain	13	0
B chain	131	0
C chain	97	+5

The Chemistry of Amino Acid Sequencing

Sanger's and Edman's strategy for sequencing proteins takes advantage of the polypeptide chain's free N-terminal amino group ($-NH_3^+$). Because of the influence of the nearby backbone C=O component, this amino group has a special chemical property: it tends to lose its extra H^+ more readily than other amino groups in the protein (those on the side chains of lysine and arginine subunits). At a pH of about 9, this group changes freely between two states:

$$-NH_3^+ \longleftrightarrow -\ddot{N}H_2 + H^+$$

Once in the $-\ddot{N}H_2$ state, the terminal amino group can react with organic molecules that are subject to attack by its unshared pair of electrons. The result is the formation of a covalent bond permanently attaching the organic molecule to the N-terminal amino group. In this way, the first amino acid in the protein chain becomes marked.

The preferred end-labeling reagent was developed by Pehr Edman, who found an effective marker in the chemical phenyl isothiocyanate. Its reaction with the N-terminus of a polypeptide chain is simple to understand and is shown on the following page. In essence, the unshared pair of electrons on the terminal amino group reaches out and forms a covalent linkage with the carbon atom of phenyl isothiocyanate. Since carbon can accommodate

only four bonds, one pair of electrons previously associated with this carbon (in the C=N double bond) withdraws entirely to the nitrogen. These electrons subsequently become stabilized there when they establish a new linkage with an H^+ obtained from solution.

Once phenyl isothiocyanate has been added to the N-terminal amino group, the polypeptide chain acquires an important new chemical property. The peptide bond between the first and second amino acids becomes destabilized. It is then subject to breakage through an attack that can be thought of as coming from the unshared pair of electrons on the nitrogen atom of the attached phenyl isothiocyanate (see page 111)—a reaction that can be carried out under relatively mild conditions that do not result in the hydrolysis of any other peptide bonds. Specifically, incubation of the marked polypeptide chain in a moderately acidic solution (pH 2) releases a modified cyclic form of the N-terminal amino acid, which can then be recovered and identified. When the Edman reagent is applied to the A chain of chymotrypsin, a modified form of cysteine is released, identifying cysteine as the N-terminal amino acid.

Since the Edman degradation cleaves only the aa_1–aa_2 peptide bond, the rest of the protein chain ($aa_2 \rightarrow aa_n$) remains intact, although shorter by one unit. The shortened polypeptide chain can be recovered and returned to a pH 9 environment, where once again it can be subjected to the Edman reagent. A second round of degradation and analysis reveals the aa_2 amino acid in the original chymotrypsin A chain: glycine. Sequential rounds of degradation give the complete sequence of the A chain:

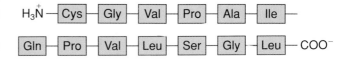

Subdividing Large Polypeptide Chains

Polypeptide chains up to 40 amino acids long can be readily sequenced by applying the Edman reagent repeatedly.

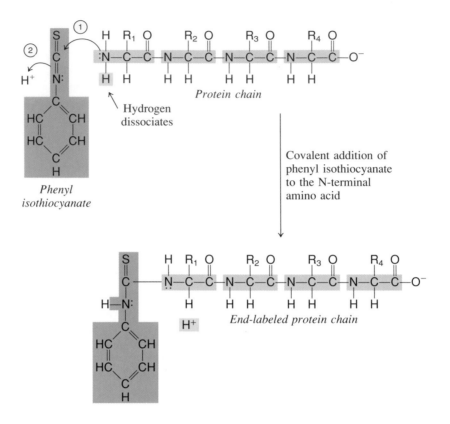

The reaction in which the Edman reagent (phenyl isothiocyanate) is used to tag the amino group at the N-terminus of a polypeptide chain. Where a nitrogen atom gains or loses a hydrogen, its unshared pair of electrons is explicitly shown.

However, there is an inevitable small loss of material after each step, since it is not possible to recover 100 percent of the protein after the cleavage reaction. Nor can perfect synchrony be maintained among the protein chains being sequenced, because of the failure of phenyl isothiocyanate to react with all possible target N-terminal amino groups at each stage.

Thus, although the Edman reagent allows the chymotrypsin A chain to be sequenced directly, the B and C chains are too large to be analyzed using this approach alone. Therefore the Sanger-Edman sequencing procedure calls for a "divide-and-conquer" strategy. The B and C chains are dissected into smaller, more analyzable units,

and the sequential degradation reaction is applied to each piece.

The most effective way to subdivide a large polypeptide chain is first to denature the target protein, exposing the backbone of the amino acid polymer, and then to treat it with a proteolytic (protein-digesting) enzyme. Trypsin—and chymotrypsin itself, the very enzyme we are studying—seem almost designed by nature to help in this regard.

The useful characteristic of trypsin and chymotrypsin is that they cleave protein chains after particular amino acids. In essence, they exhibit the specificity that is expected on the basis of Fischer's lock-and-key concept.

The attack displaces a pair of electrons that
were previously bonded to the C=O carbon,
forcing their withdrawal to the adjacent nitrogen
atom and breaking the protein chain. The electrons
eventually acquire H⁺ from solution.

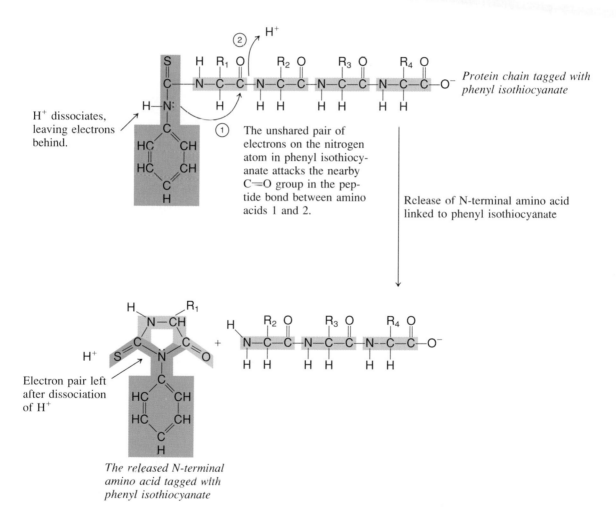

H⁺ dissociates,
leaving electrons
behind.

*Protein chain tagged with
phenyl isothiocyanate*

① The unshared pair of
electrons on the nitrogen
atom in phenyl isothiocy-
anate attacks the nearby
C=O group in the pep-
tide bond between amino
acids 1 and 2.

Release of N-terminal amino acid
linked to phenyl isothiocyanate

Electron pair left
after dissociation
of H⁺

*The released N-terminal
amino acid tagged with
phenyl isothiocyanate*

A simplified representation of the cyclization reaction in which the N-terminal amino acid of a polypeptide
chain is released by the Edman reagent.

Trypsin, for example, cleaves polypeptide chains only after the positively charged amino acids arginine and lysine. Each of the resulting *tryptic peptides* thus ends with either arginine or lysine (except, of course, for the carboxyl-terminal peptide of the protein).

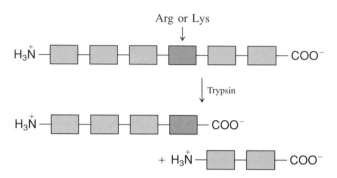

Arg or Lys

Trypsin

The application of trypsin to the denatured chymotrypsin B chain subdivides the 131-unit polypeptide chain into the set of 10 peptides shown in the table on this page. Initially, of course, we do not know the precise sequence of amino acids in these peptides, but this information is easily obtained. The peptides can be separated from each other on the basis of differences in their size and charge, and they are small enough to be individually sequenced by the Edman procedure.

Like trypsin, chymotrypsin also reduces a protein chain to a set of specific fragments. This protease cleaves the exposed backbone of the polypeptide chain adjacent to amino acids with planar, ring-shaped hydrocarbon side chains (phenylalanine, tyrosine, and tryptophan). For example:

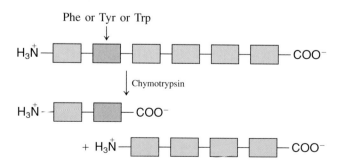

Phe or Tyr or Trp

Chymotrypsin

The Peptides Released from the Chymotrypsin B Chain by Digestion with Trypsin

Peptide	Amino acid sequence
T-1	Ile-Val-Asn-Gly-Glu-Glu-Ala-Val-Pro-Gly-Ser-Trp-Pro-Trp-Gln-Val-Ser-Leu-Gln-Asp-Lys
T-2	Thr-Gly-Phe-His-Phe-Cys-Gly-Gly-Ser-Leu-Ile-Asn-Glu-Asn-Trp-Val-Val-Thr-Ala-Ala-His-Cys-Gly-Val-Thr-Thr-Ser-Asp-Val-Val-Val-Ala-Gly-Glu-Phe-Asp-Gln-Gly-Ser-Ser-Ser-Glu-Lys
T-3	Ile-Gln-Lys
T-4	Leu-Lys
T-5	Ile-Ala-Lys
T-6	Val-Phe-Lys
T-7	Asn-Ser-Lys
T-8	Tyr-Asn-Ser-Leu-Thr-Ile-Asn-Asn-Asp-Ile-Thr-Leu-Leu-Lys
T-9	Leu-Ser-Thr-Ala-Ala-Ser-Phe-Ser-Gln-Thr-Val-Ser-Ala-Val-Cys-Leu-Pro-Ser-Ala-Ser-Asp-Asp-Phe-Ala-Ala-Gly-Thr-Thr-Cys-Val-Thr-Thr-Gly-Trp-Gly-Leu-Thr-Arg
T-10	Tyr

The *chymotryptic peptides* derived from the B chain of chymotrypsin, and their sequences, are shown in the table on the facing page.

Ordering of the Peptides

At this point in our analysis, the chymotrypsin B chain has been subdivided, and the amino acid sequences of the released peptides have been determined. But the *order* of the peptides in the intact polypeptide chain is not yet known. The information needed to find the order is obtained from what are known as *overlap peptides*. If trypsin has been used to derive one set of peptides and chymotrypsin another set, numerous peptides will overlap for part of their sequence so that one can establish their relative order. The procedure is analogous to reconstructing the word *biochemistry* from the fragments *biochemist* and *chemistry*.

The Peptides Released from the Chymotrypsin B Chain
by Digestion with Chymotrypsin

Peptide	Amino acid sequence
C-1	Ile-Val-Asn-Gly-Glu-Glu-Ala-Val-Pro-Gly-Ser-Trp
C-2	Pro-Trp
C-3	Gln-Val-Ser-Leu-Gln-Asp-Lys-Thr-Gly-Phe
C-4	His-Phe
C-5	Cys-Gly-Gly-Ser-Leu-Ile-Asn-Glu-Asn-Trp
C-6	Val-Val-Thr-Ala-Ala-His-Cys-Gly-Val-Thr-Thr-Ser-Asp-Val-Val-Val-Ala-Gly-Glu-Phe
C-7	Asp-Gln-Gly-Ser-Ser-Ser-Glu-Lys-Ile-Gln-Lys-Leu-Lys-Ile-Ala-Lys-Val-Phe
C-8	Lys-Asn-Ser-Lys-Tyr
C-9	Asn-Ser-Leu-Thr-Ile-Asn-Asn-Asp-Ile-Thr-Leu-Leu-Lys-Leu-Ser-Thr-Ala-Ala-Ser-Phe
C-10	Ser-Gln-Thr-Val-Ser-Ala-Val-Cys-Leu-Pro-Ser-Ala-Ser-Asp-Asp-Phe
C-11	Ala-Ala-Gly-Thr-Thr-Cys-Val-Thr-Thr-Gly-Trp
C-12	Gly-Leu-Thr-Arg-Tyr

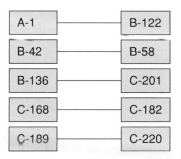

These assignments have been made by treating the target protein with trypsin (or chymotrypsin) *before* breaking the disulfide bonds. One obtains several fragments with the following general structure:

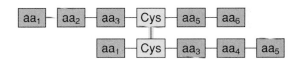

These composite fragments can be isolated and their cysteine-to-cysteine bonds broken at this stage. Sequence analysis of the resulting short peptides then establishes the location of the disulfide bridges in the intact protein.

Folding the Polypeptide Chain into a Biologically Active Protein

The amino acid sequence of chymotrypsin and the location of its disulfide bonds give only a two-dimensional representation of the enzyme. In contrast, the enzymatically active protein fills a three-dimensional space. Our next goal, therefore, is to understand the mechanism by which a flexible polypeptide chain folds into a geometrically specific structure.

There are two basic aspects to protein folding, which give rise to the *secondary* and *tertiary* structure of the molecule. In the first, nearby areas of the *backbone* of the

As an example, consider the tryptic peptide called T-1; it includes the amino acids contained in the chymotryptic peptides called C-1 and C-2, with seven amino acids left over at the C-terminal end (Gln-Val-Ser-Leu-Gln-Asp-Lys, shown in red in the table on page 112). These extra amino acids indicate which chymotryptic peptide follows C-2 in the B chain. Only one begins with the appropriate amino acids: *Gln-Val-Ser-Leu-Gln-Asp-Lys*-Thr-Gly-Phe. Hence, it is identified as C-3.

Through the divide-and-conquer strategy, the complete amino acid sequences of the chymotrypsin B and C chains can be determined. When this information is added to the sequence of the A chain, we obtain the complete primary structure for chymotrypsin as shown on page 114.

The figure also shows cysteine-to-cysteine disulfide bonds at five positions in the chymotrypsin molecule:

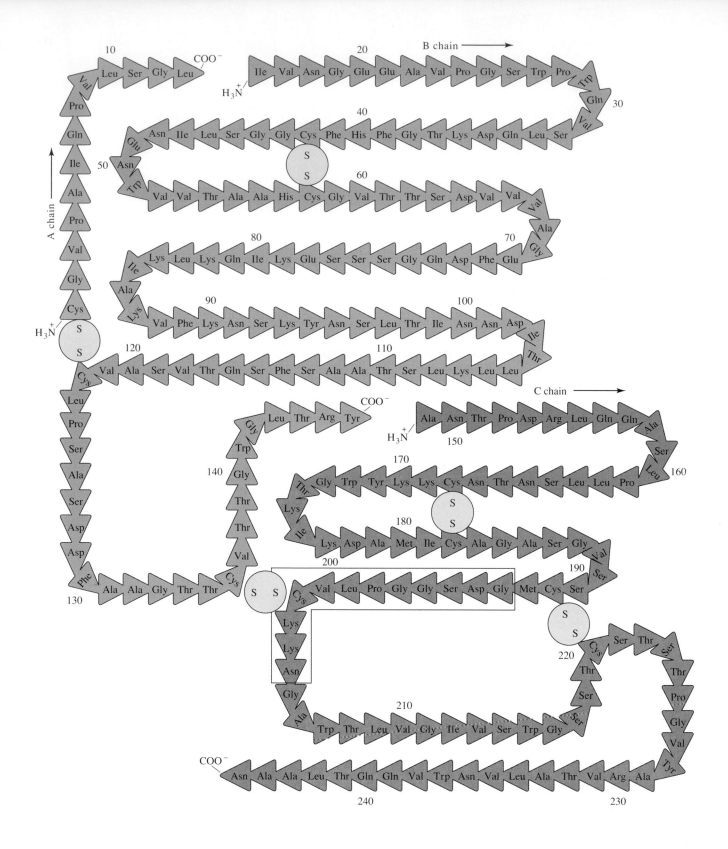

The order of the amino acid subunits (the primary structure) of chymotrypsin. The three distinct polypeptide chains (A, B, and C) are laid out to emphasize the five disulfide bonds covalently connecting the chains. The missing amino acids at positions 14 and 15 and 147 and 148 are removed when the enzyme is activated in the intestine. In this diagram, amino acids 193 to 204 have been boxed. At the end of the following section, we will consider how the chemistry of their side chains influences the three-dimensional folding of the protein into a biologically active enzyme.

polypeptide chain interact in a regular manner to create local regions of folding. In the second, interactions between specific *amino acid side chains* bring distant areas of the polypeptide chain together, leading to a solid structure with unique surface cavities and projections. Both mechanisms of folding are important and will be considered in turn.

Folding the Backbone:
The α-Helix and the β-Sheet

The first insights into protein folding were obtained in the 1940s by Linus Pauling and Robert Corey at the California Institute of Technology. They analyzed the structures of various "model" compounds—crystallized amino acids and very short protein chains—using the technique of X-ray diffraction. In this procedure, the precise arrangement of the atoms in a crystal is derived by observing the diffraction (or redirection) of X rays as they pass through the repeating layers of the crystal. Pauling's and Corey's aim was to determine the complete set of distances and bond angles between the atoms of the small model compounds and then to use this information to predict the conformation of larger protein structures. The figure on this page shows an example of the type of data they obtained.

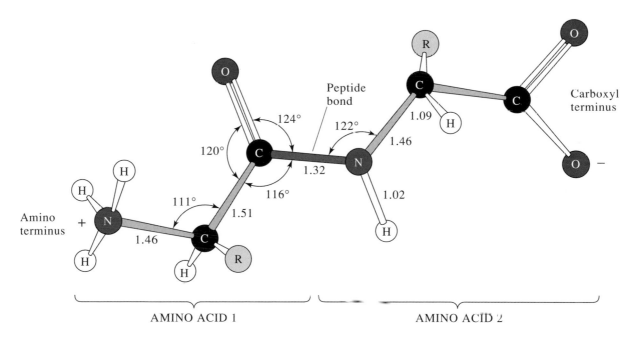

All of the interatomic distances and bond angles are defined in this structure of a small model protein containing two amino acids. Taken from the X-ray crystallographic study of Pauling and Corey.

Their results led Pauling and Corey to several conclusions about protein structure, the most important of which can be summarized as follows:

1. In a *peptide bond,* the connection between two amino acids, the hydrogen of the amino group is nearly always *trans* to the oxygen of the carbonyl group. That is, the oxygen and the hydrogen project in opposite directions rather than in the same direction (*cis*):

C—N single bond would allow free rotation.

Partial double bond prevents free rotation.

C=N double bond would prevent free rotation.

Three possible arrangements for the bonding electrons in a peptide linkage.

2. The immediate area of the peptide bond is flat—the atoms of the N—H and C=O groups all lie in a single plane. Evidently there is no rotational freedom at the covalent bond between the C=O group and the N—H group, even though at first sight the peptide bond appears to be an ordinary, single bond, which should allow free rotation of the participating atoms. This lack of flexibility will ultimately be important for protein folding, but what is its cause? The answer became apparent through a careful analysis of the distances between the various atoms. The length of the bond from the C=O to the N—H group was 1.32 Å, and this length is intermediate between the length of an ordinary C—N single bond (1.49 Å) and a C=N double bond (1.27 Å), as determined from the study of other simple molecules. Pauling and Corey proposed that the peptide C—N linkage actually has a structure intermediate between that of a single bond and double bond. This partial double-bond character results from a delocalization of the electrons in the region of the peptide bond, allowing them to move more freely among the participating atoms. The resulting composite arrangement of electron clouds imposes the same lack of rotational freedom that is characteristic of a pure double bond.

3. Although the peptide bond itself has a partial double-bond character, the adjacent bonds do not. The bonds between the central carbon atom and the carbonyl group on one side, and between the central carbon atom and the amino group on the other, are pure single bonds. Rotation at these positions is not restricted.

In sum, Pauling and Corey had found that certain areas of the backbone of the polypeptide chain are rigid and planar (the peptide bonds), whereas in the region of the central carbon and its attached side chain the component atoms are free to rotate with respect to their neighbors.

Pauling and Corey considered what shapes might be assumed by a polypeptide chain having a restricted degree of rotational flexibility. In 1951 they proposed two regularly repeating structures: the *α-helix* and the *β-sheet.* Both of these structures are attempts to explain how a polypeptide chain begins to adopt an ordered three-dimensional form through interactions between segments of the backbone.

The α-Helix

In the α-helix, the backbone of the polypeptide chain is coaxed into a helical conformation by the formation of hydrogen bonds between the N—H and C=O groups of the backbone peptide bonds. As can be seen in panels C and D of the figure on page 118, the C=O group in each amino acid uses its oxygen to attract the hydrogen of the N—H group of the amino acid situated four subunits further along in the polypeptide chain.

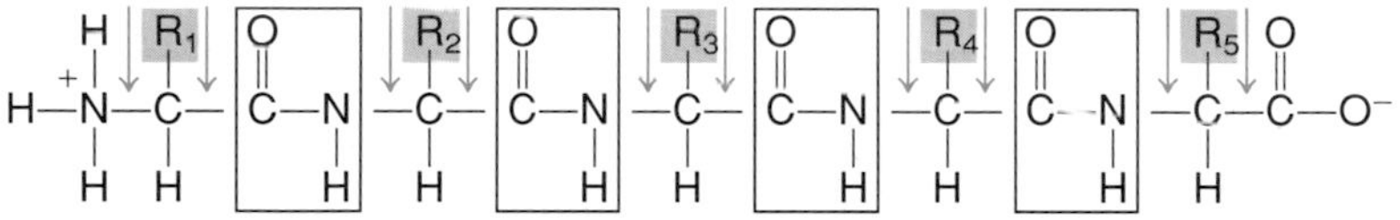

$$H-\overset{+}{\underset{H}{N}}-\overset{H}{\underset{H}{C}}-\overset{R_1}{\underset{}{C}}-\overset{O}{\underset{}{C}}-\overset{}{\underset{H}{N}}-\overset{R_2}{\underset{}{C}}-\overset{O}{\underset{}{C}}-\overset{}{\underset{H}{N}}-\overset{R_3}{\underset{}{C}}-\overset{O}{\underset{}{C}}-\overset{}{\underset{H}{N}}-\overset{R_4}{\underset{}{C}}-\overset{O}{\underset{}{C}}-\overset{}{\underset{H}{N}}-\overset{R_5}{\underset{}{C}}-\overset{O}{\underset{}{C}}-O^-$$

The limited degree of rotational freedom in the polypeptide chain. Boxes show the rigid, planar peptide bonds. Arrows mark the positions in the backbone, between the peptide bonds, at which the polypeptide chain is free to rotate. (In this and most other diagrams of the polypeptide chain, we will not represent the precise three-dimensional bonding angles, but instead use the convenient rectangular Fischer projections. A more precise representation, as in the figures on pages 122 and 125, would show the tetrahedral orientation of the groups radiating from each central carbon and the orientation of adjacent amino acid side chains in alternating directions.)

The stability of the α-helix results from the fact that when the protein chain is positioned in this spiral form, *all* the peptide bond C=O and N—H groups can participate in the formation of hydrogen bonds: amino acid 1 is bonded to amino acid 5, 2 is bonded to 6, and so forth. The position of each amino acid in the α-helix is related to the next by a single, repeated operation consisting of a *translation* and a *rotation*. That is, in moving from amino acid 1 to amino acid 2, one travels 1.5 Å along the axis of the helix and, in addition, turns 100 degrees.

The important point is that a protein chain takes on a precise structure when it assumes the α-helical conformation. A more-or-less solid cylinder is formed, with the polypeptide chain backbone forming a central core and the amino acid side chains projecting outward. An important characteristic of the α-helix is that amino acid side chains located three or four apart in the linear sequence become quite close to one another, projecting outward from the same side of the helix (for example, see amino acids 2, 6, 10, and 14 in the figure on page 119). These side chains can cooperate to form a stable facet inside the protein or on its surface. In contrast, the side chains of amino acids that are immediately adjacent in the linear sequence project outward at different angles from the helix—and thus are not in a position to make contact with each other to form a particular surface in the protein. They can, however, participate in the formation of other, different surfaces.

The α-helix is a dominant organizational theme in many proteins. These are generally proteins that serve in a

For his work on the nature of chemical bonds and their role in molecular structure, Linus Pauling was awarded the Nobel prize in 1954.

structural role, for the regularity of the α-helix is well suited to the formation of extended, highly ordered fibers. The "intermediate filaments," for example, are a class of α-helical, cablelike proteins that are used to construct biological structures ranging from the internal support filaments within cells to such external protective structures as skin and hair (see the box on pages 120–121). In en-

zymes, however, the role of the α-helix is more limited. The entire polypeptide chain does not fold up into an α-helix—only short segments of about eight to ten amino acids do, creating local regions of three-dimensional structure. As we will see in the case of chymotrypsin, the last 10 amino acids of the C chain exist in the form of an α-helix.

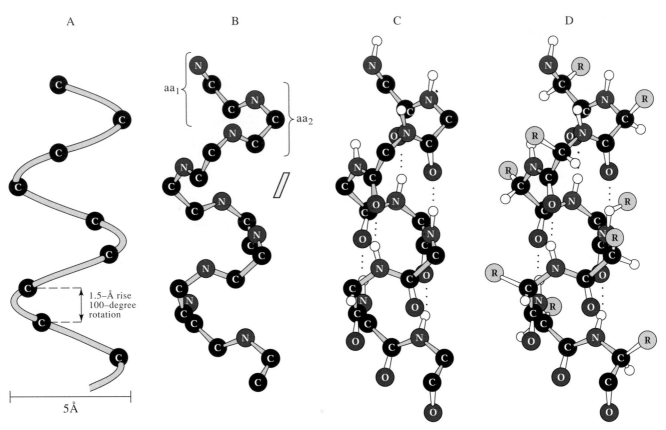

In the α-helix, the N—H and C=O groups of nearby peptide bonds interact with each other, drawing the polypeptide chain into a spiral with a central core (the backbone) and leaving the amino acid side chains to extend outward in a radial array. The diagram builds the structure schematically, step by step. (A) The path traced by the central carbon atoms of nine amino acids in an α-helix. (B) The backbone of the polypeptide chain showing the N—C—C atoms of all the amino acid subunits. (C) The oxygen and hydrogen atoms of the backbone have been added. Note how all the N—H groups point "up" and all the C=O groups point "down," allowing the formation of C=O···H—N hydrogen bonds. (D) The side chains have been added and are seen projecting outward from the cylindrical α-helix.

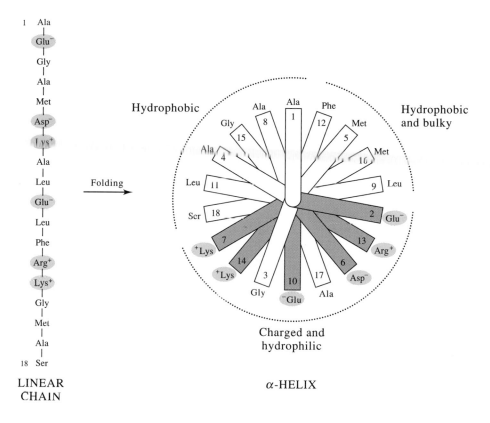

Left: A linear polypeptide chain with 18 amino acids. Scattered at various positions are five electrically charged amino acids (blue). *Right:* The same polypeptide chain drawn in the form of an α-helix shows that nonadjacent amino acids can project outward from the same side of the α-helix to form a chemically distinct surface.

The β-Sheet

The second structure proposed by Pauling and Corey for folding the backbone of a polypeptide chain is the β-sheet. It is a basically planar structure of potentially unlimited extent, in contrast to the defined solid cylinder of the α-helix. The polypeptide chains in the β-sheet are not spiraled but instead are almost fully extended. In this state, they interact with other polypeptide segments lying alongside, once again by forming hydrogen bonds between peptide N—H and C=O groups. The interacting backbones in a β-sheet can come from different polypeptide chains or from distant parts of the same chain folded back on itself. The chains can run in the same direction (a parallel β-sheet) or in opposite directions (an antiparallel β-sheet, see page 122). An important structural feature of the β-sheet is that successive amino acid side chains project alternately above and below the plane of the sheet. Thus, the two surfaces of the β-sheet can have quite different chemical properties.

The β-sheet is the major structural motif in several well-known proteins. A strand of silk, for example, consists almost entirely of stacks of antiparallel β-sheets. As we will see in our discussion of chymotrypsin, the β-sheet

INTERMEDIATE FILAMENTS

Intermediate filaments are an important group of proteins that are based on the α-helical motif. Occurring in all animal cells, they are used to construct much of the cell's cytoskeleton—a network of thin, strong structural support fibers. As can be seen in the accompanying micrograph, the intermediate filaments crisscross the cell, lending it structural integrity and shape and holding the nucleus and other organelles in place.

A variety of intermediate filament proteins are produced in different cells, but they all share a common, important property. In each case, the central portion of the polypeptide chain forms a long, almost perfect region of α-helix. In contrast, the amino-terminus and the carboxyl-terminus of the protein chain do not participate in the α-helix but project outward as independent spherical domains. It is their central α-helical region that underlies the use of these proteins as structural support elements in the cell. The finished intermediate filament is an assembly of thousands of such elongated units, in which the α-helical regions define the filament's basic structure and longitudinal axis.

As can be seen in the diagram, the assembly of the intermediate filament begins when two subunit proteins associate, forming a dimeric structure in which the two α-helices wind around each other like strands in a cable. This interwrapping is engineered by the inclusion of a small amino acid, like glycine, at every third position in the α-helix. The projection of these subunits from one side of the helix (page 119) forms a relatively flat surface that allows the two chains to approach each other very closely. Once formed, this dimeric structure associates with a similar dimer to give the basic tetrameric building block used to construct the finished intermediate filament. The tetramers link together through end-to-end noncovalent interactions between their terminal domains. In the last stage of assembly, eight groups of linked tetrameric units line up side-by-side to form a solid, tubular structure—the finished intermediate filament.

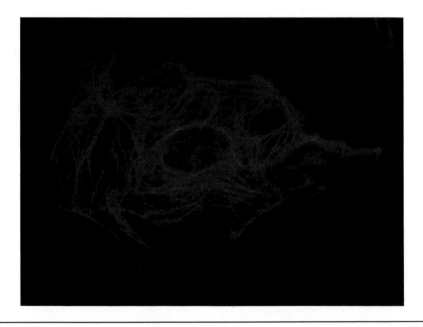

An epithelial cell has been processed by a technique that generates the molecular equivalent of an X-ray photograph. The cell has been stained with a reagent that binds specifically to intermediate filaments, thus highlighting a major part of the cell's internal "cytoskeleton."

Three examples of cells in which intermediate filaments play an especially prominent role are nerve cells, muscle cells, and epithelial cells. The intermediate filaments in nerve cells, called *neurofilaments,* occur throughout the length of the axon, where they resist the mechanical stresses caused by movement—stresses that would otherwise fracture these very long, thin cellular processes.

In muscle cells, the intermediate filaments, called *desmins,* run perpendicular to the cell axis and hold the contractile muscle fibers in alignment so that the fibers all pull the same way when the cell contracts.

In epithelial cells, which line the surfaces of the body, the organization of intermediate filaments is especially elaborate. Known as *keratin,* these intermediate filaments become anchored to the cell membrane at specialized docking regions called *desmosomes.* The desmosomes of adjacent cells are in direct contact and serve as connector elements to hold the cells together. Because the keratin filaments are attached to desmosomes, the filaments in each cell are connected to the filaments in neighboring cells, forming a continuous fiber network throughout the entire epithelium. It is this network that gives the epithelium its tensile strength and resistance to mechanical stress.

The most complex types of intermediate filaments are found in the epidermis, the multilayer epithelial covering of the body. The keratin intermediate filaments produced in the outermost layers eventually become covalently crosslinked to one another and to a variety of associated proteins (for example, by cysteine-cysteine disulfide bonds). As the cells in the most superficial layer of the epidermis die, these crosslinked keratins persist and are used to generate such protective structures as hair, nails, horns, beaks, and feathers. These structures differ, in large measure, because of the other proteins and mineral deposits aggregated with the intermediate filaments.

Though very different in appearance and biological function, the intermediate filament–based structures—not only those of the intracellular cytoskeleton, but also those of such extracellular structures as hair and nails—trace their molecular origin back to the α-helical proteins of which they are made.

α-Helical regions shared by all intermediate filaments

Monomer

Dimer

Tetramer

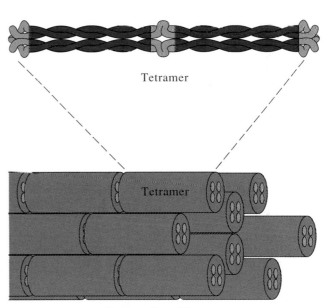

Tetramer

Intermediate filament
(assembly of 8 chains of tetramers,
forming a cylinder)

The assembly of α-helical protein units to form an intermediate filament.

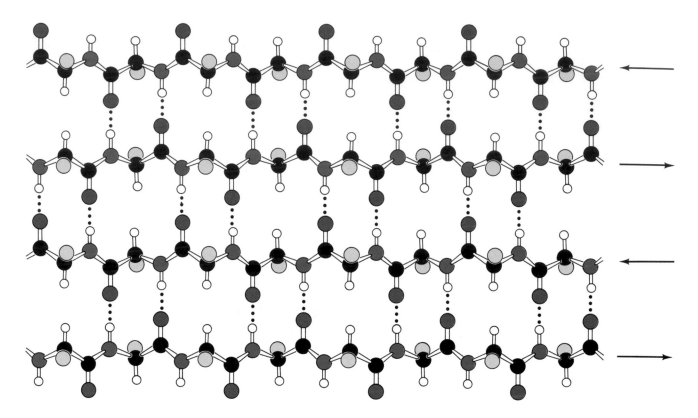

The side-by-side interactions of polypeptide chains in an antiparallel β-sheet. C=O and N—H groups from the backbone of one stretched-out chain interact regularly with the C=O and N—H groups from another chain. The result is a planar structure of potentially unlimited breadth. The amino acid side chains (shown in gold) project alternately above and below the plane of the β-sheet—in this diagram, into and out from the plane of the page. Note that instead of showing the polypeptide backbone in our usual simple planar format (page 117), this drawing more precisely depicts the tetrahedral arrangement of the chemical bonds around each central carbon atom.

also plays an important, though less extensive, role in enzymes by helping the polypeptide chains to fold back and forth across each other. Especially common are short folding units in which stretches of two to six amino acids in antiparallel segments touch each other. The first and last polypeptide chains of the β-sheet often interact with each other to close the structure into a solid barrel-like form. The individual polypeptide segments correspond to the staves of the barrel.

The β-Turn

Related to the α-helix and the β-sheet is the *β-turn*, a backbone interaction that helps a polypeptide chain reverse direction. The essence of this hairpin turn is that the N—H and C=O groups of subunit *n* of the protein chain form one, and sometimes two, hydrogen bonds with the corresponding groups of subunit *n* + 3. It is as if the type of bonding in a β-sheet were ''borrowed'' and used at a

SILK

The architectural possibilities of the β-sheet motif are illustrated by the protein fibroin, which is the principal component of the silk of spider webs, of the cocoon of the silkworm, and of the fine fabric.

A fiber of silk contains hundreds of thousands of fibroin chains, fully stretched out and aligned side by side. The antiparallel chains are linked by backbone N—H and C=O hydrogen bonds to form the characteristic arrange-

A caterpillar of the cecropia moth spinning a cocoon.

ment of polypeptide chains in a β-sheet. In the finished silk fiber, the large flat surfaces of numerous β-sheets interact with each other, an association engineered at the molecular level by a particular property of the fibroin amino acid sequence. The 3000-amino-acid fibroin protein has a primary structure dominated by the repeated amino acid sequence Ser-Gly-Ala-Gly-Ala-Gly. Every second subunit is glycine, which alternates with either alanine or serine, two amino acids with similar small side chains. Since, in the β-sheet, adjacent amino acids project alternately above and below the planar structure, two distinctive surfaces arise. On one side, the projecting amino acids are all glycine, and their minimal H side chains give rise to a relatively flat surface. On the other side, the projecting side chains of alanine and serine create a knobbed surface. This arrangement allows the β-sheets to interact with each other back to back and front to front. That is, the flat glycine surfaces contact each other, as do the knobbed surfaces, whose projecting alanine and serine side chains perfectly interdigitate. The resulting stack of β-sheets gives rise to a fiber that is both strong, because the individual β-sheets interact with each other over broad areas, and also relatively inextensible, because the component polypeptide chains are already extended to nearly their maximum possible length. Still, the fibers are very pliant because the interactions between adjacent polypeptide chains and adjacent β-sheets involve only weak, noncovalent interactions. The overall silk fiber is thus tough but flexible—a perfect building material.

When it is first produced in the silk gland, fibroin occurs in the form of random coils. However, the primary structure of fibroin is able to form a β-sheet structure under favorable conditions, which the silkworm or spider brings about in the course of spinning its web or cocoon. The animal first extrudes a highly viscous liquid solution of unorganized fibroin and then places mechanical stress on the fibroin chains by backing away. This stress, together with the evaporation of the surrounding solvent, promotes the conversion of the initial random coil structures into extended β-sheet assemblies.

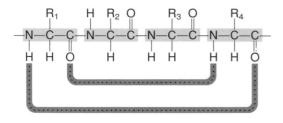

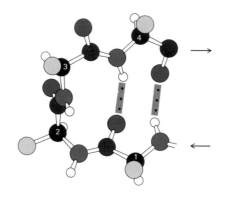

A *β*-turn, or *tight* turn, shown schematically at top and in three dimensions at bottom.

specific location to help a polypeptide chain fold back upon itself. By means of the *β*-turn, the polypeptide chain can undergo an abrupt but stable change in direction.

Protein Folding Induced by the Amino Acid Side Chains

When the folding of a protein chain is based mainly upon interactions between areas of the polypeptide backbone, the result is a highly ordered three-dimensional structure dominated by such motifs as the *α*-helix and the *β*-sheet. These two motifs, however, are most common in structural proteins, whose role is to provide a mechanical support, both within the cell and in the connective matrix between cells. For enzymatic proteins, like chymotrypsin, we must search more deeply for folding mechanisms that will create unique three-dimensional structures, more complex and less repetitive than those of structural pro-

teins. The amino acid side chains provide the basis for this aspect of folding.

To illustrate the ways in which interactions between side chains contribute to protein folding, we will focus our discussion on a ''model peptide''—a stretch of 12 amino acids that actually occurs in the chymotrypsin molecule:

Gly-Asp-Ser-Gly-Gly-Pro-Leu-Val-Cys-Lys-Lys-Asn

These are amino acids 193 to 204, found in the middle of the chymotrypsin C chain (seen in the figure on page 114). This sequence has been chosen for two reasons. First, it includes representative examples of all of the various types of amino acid side chains—those that form covalent bonds, ionic bonds, hydrogen bonds, and hydrophobic bonds. Collectively these interactions draw the flexible polypeptide chain into its final biologically active structure. Furthermore, this stretch of amino acids will prove to be of particular interest for chymotrypsin's mechanism of action, since it contains two of the key amino acids in the enzyme's catalytic center.

In the discussion that follows, the model peptide is represented in a planar format designed to show clearly the various bonds that can be formed by the amino acid side chains. Once we have obtained the complete three-dimensional structure of chymotrypsin, it will be possible to picture the much more complex geometrical arrangement that actually exists in the folded enzyme.

Cysteine and the Formation of Covalent Bonds

We begin our discussion of the role of side chain interactions in protein folding with the *cysteine* subunit in the model peptide.

Gly-Asp-Ser-Gly-Gly-Pro-Leu-Val-Cys-Lys-Lys-Asn

Cysteine plays a unique and well-defined role in protein folding because it is the only amino acid whose side chain is capable of spontaneously forming a covalent bond. As mentioned earlier, the sulfhydryl (—SH) groups on the side chains of two cysteine subunits can lose the

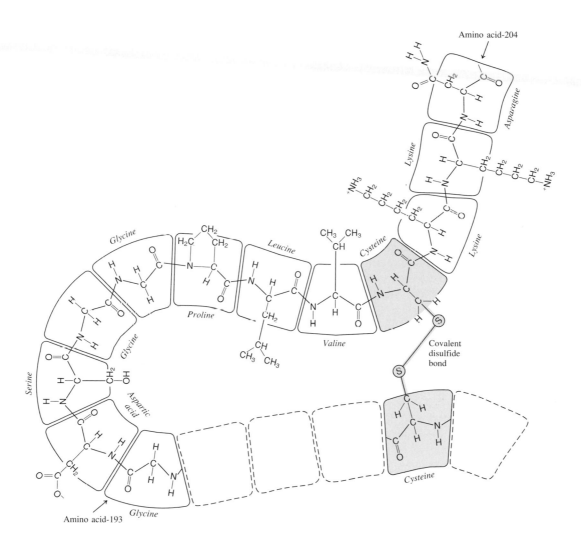

Two cysteines stabilize the polypeptide chain in a particular conformation by forming a disulfide bond.

COVALENT BONDS

The sharing of electrons between atoms underlies the covalent bond. In forming such bonds, the participating atoms obtain a complete set of electrons to fill their outer electron shells. Carbon, for example, has four electron vacancies in its outer shell and thus seeks to form four covalent bonds. In the same way, nitrogen lacks three electrons and bonds with three other atoms. Oxygen forms two covalent bonds and hydrogen one. These numerical relationships are evident in all of the chemical structures we will encounter.

Covalent bonds are by far the strongest bonds that hold molecules together. Evidence of their strength is provided by the considerable amount of energy (on the order of 100,000 calories per mole, or 400,000 joules per mole) required to separate two covalently bonded atoms. It is also difficult to distort the angles between the covalent bonds emanating from an atom. This is because the sharing of electrons depends on the merging of orbitals, which occurs favorably only in specific directions. Thus, we can picture covalent (including disulfide) bonds as holding groups of atoms together with linkages of a fixed length, disposed at a limited number of rigidly controlled angles.

A good example is seen in the ball-and-stick representation of the amino acid cysteine, which is shown below.

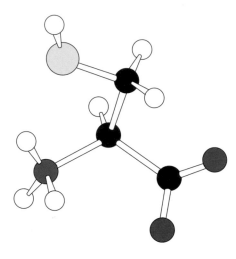

Covalent bonds hold atoms together at fixed distances; the rigidly controlled angles between them maintain the structure of the molecule.

---

elements of molecular hydrogen and become linked by a disulfide bond:

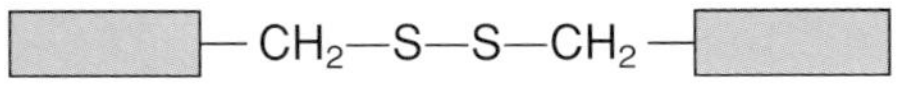

$$\boxed{}-CH_2-S-S-CH_2-\boxed{}$$

The bond forms a covalent bridge between two areas of the polypeptide chain that might have been quite distant in the linear amino acid sequence. We therefore expect that the cysteine subunit in our model peptide may be covalently cross-linked to another cysteine in the folded chymotrypsin molecule, thereby locking a certain degree of structure into the enzyme. A schematic example of a disulfide cross-link is shown in the figure on the preceding page.

Disulfide bridges are the only covalent bonds normally found in proteins other than those occurring in the amino acid subunits themselves. These linkages are thus equal participants in the stable network of covalent bonds that holds together the 4000 atoms of the chymotrypsin molecule.

Electrically Charged Amino Acids and Protein Folding

Covalent bonds may be the strongest, but they are by no means the only important bonds involved in protein struc-

IONIC BONDS

The ionic bond is an electrostatic interaction that results from the attraction of two oppositely charged species. This type of bond exhibits a greater degree of chemical flexibility than the covalent bond, which always has a fixed length and orientation. The ionic bond is not "broken" if the two charged groups are pulled further apart—instead its strength simply falls off with distance. Furthermore, an ionic bond can occur at virtually any angle, since electron orbitals do not have to be merged as in covalent bonds. For both reasons, the possible geometrical arrangements of ionic bonds are more varied than those of covalent bonds.

The strength of an ionic bond is given by the equation

$$F = \frac{q_1 q_2}{d^2 \epsilon}$$

This equation summarizes the experimental findings that the bond strength is directly proportional to the amount of charge on each molecular species (q), is inversely proportional to the distance (d) between the two charged species, and is also inversely proportional to the chemists' term ϵ, which is the "dielectric constant" of the solvent (a measure of the ability of the surrounding molecules to reorient themselves to neutralize the charge).

Because of the influence of ϵ, the strength of electrostatic interactions such as ionic bonds depends critically on the local environment. In the interior of a protein, charged amino acids would be surrounded by chemically dissimilar hydrocarbon side chains and thus forced to interact with each other. Here ϵ is low (for a methyl group, —CH_3, $\epsilon = 1$), and the conditions exist for a strong electrostatic attraction. Thus, if two charged amino acids were to assist protein structure by forming an ionic bond in the interior of the protein, that bond would be almost equal in strength to a covalent bond (about 50,000 calories per mole, or 200,000 joules per mole).

In contrast, when charged amino acids lie on the surface of the protein, they can form electrostatic interactions with shells of surrounding water molecules. Through their electrical asymmetry, the water molecules partially neutralize the charge on the amino acids. For water, $\epsilon = 85$, and there is a much reduced force driving oppositely charged amino acids to find each other and interact. However, the interactions of charged amino acids with water still contribute importantly to protein folding by directing certain parts of the polypeptide chain to the surface of the protein. These interactions also promote the protein's overall solubility in the surrounding aqueous environment.

ture. The vast majority of amino acid side chains contribute to protein folding by participating in a variety of weaker interactions.

To illustrate this point, let us return to our model peptide and consider its *aspartic acid* and two *lysine* subunits. These are, of course, representative of the 31 charged amino acids that occur in chymotrypsin.

Gly-Asp-Ser-Gly-Gly-Pro-Leu-Val-Cys-Lys-Lys-Asn

How will such amino acids play a role in determining the three-dimensional structure of the protein? On chemi-

cal principles, there are two quite different possibilities. One is that oppositely charged amino acids will interact in the interior of the protein to form *ionic bonds*. Such an electrostatic (charge-based) interaction is shown in the top of the figure on the following page, where a positively charged lysine in the model peptide is seen interacting with a negatively charged glutamic acid from elsewhere in the protein. This ionic interaction helps to hold together in position distant areas of the protein chain.

Although a pair of oppositely charged amino acids might join together in the interior of the protein to form an ionic bond, it is unlikely that a *single* charged amino acid

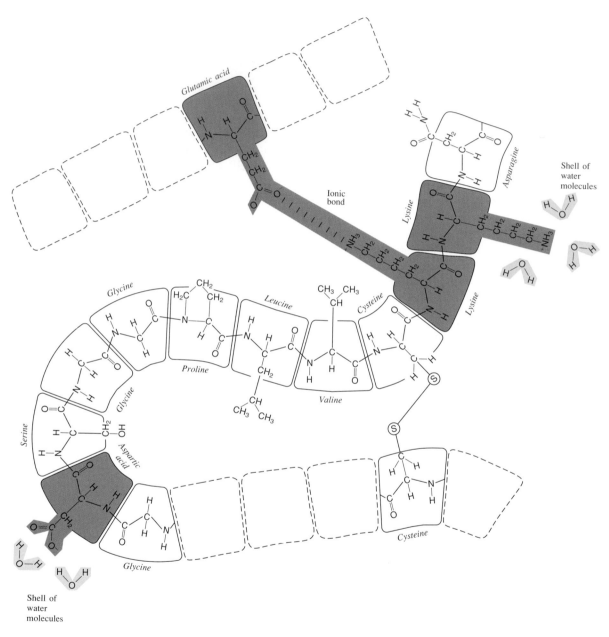

Examples of an ionic bond and other electrostatic interactions. *Top:* Two charged amino acids (lysine from the model peptide, and glutamic acid from elsewhere in the protein) are shown forming an ionic bond and folding the polypeptide chain. *Upper right:* A single, positively charged amino acid (lysine) at the surface of the protein is surrounded by a shell of oriented water molecules. These weak electrostatic interactions promote the solubility of the protein. *Lower left:* Another external amino acid (aspartic acid) has a negative charge and is surrounded by water molecules oriented in the opposite direction.

would tolerate this environment. Were it to become trapped in the interior of the protein, it would be unable to interact with the surrounding amino acid side chains, which are hydrophobic. Instead, charged amino acids are more likely to seek the surface of the protein, where their side chains can interact with the surrounding water molecules. The basis for this interaction stems from an important property of water. As shown below, water molecules are electrically asymmetric structures in which the more negatively charged atom (the oxygen) lies on one side while the two more positively charged atoms (the hydrogens) project away in the opposite direction:

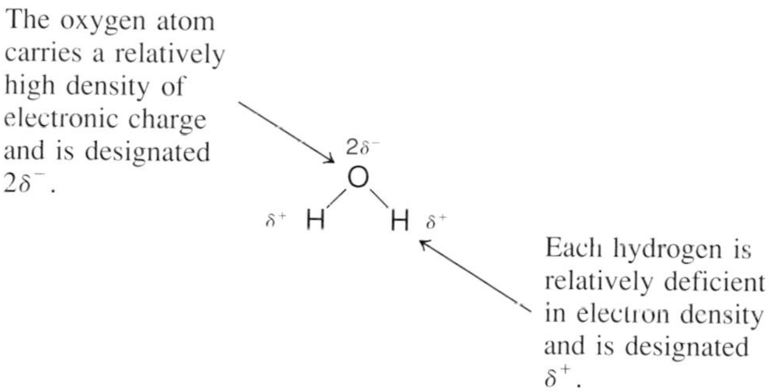

The oxygen atom carries a relatively high density of electronic charge and is designated $2\delta^-$.

$$2\delta^- \quad O$$
$$\delta^+ \; H \qquad H \; \delta^+$$

Each hydrogen is relatively deficient in electron density and is designated δ^+.

The consequence of this asymmetry is that positively or negatively charged amino acids projecting from the sur-

face of the protein will each be surrounded by a shell of water molecules oriented so that either their oxygens (δ^-) or their hydrogens (δ^+) face the side chain in such a way that opposite charges can interact. Such interactions allow the protein to associate with the surrounding solvent, which is essential if the protein is to be soluble and free-floating. Two examples of charged amino acids interacting with water molecules are shown in the upper right and lower left of the figure on the facing page.

In sum, charged amino acids are expected to contribute to protein structure either (1) by forming strong ionic bonds in the interior of the protein or (2) by first defining the external surface during the process of protein folding and later making the protein soluble in the aqueous environment of the cell.

Amino Acid Side Chains That Form Hydrogen Bonds

Looking again at our model peptide,

Gly-Asp-Ser-Gly-Gly-Pro-Leu-Val-Cys-Lys-Lys-Asn

we see two amino acids whose side chains can form hydrogen bonds. These are *serine* and *asparagine*. In fact, groups capable of participating in hydrogen bonding are

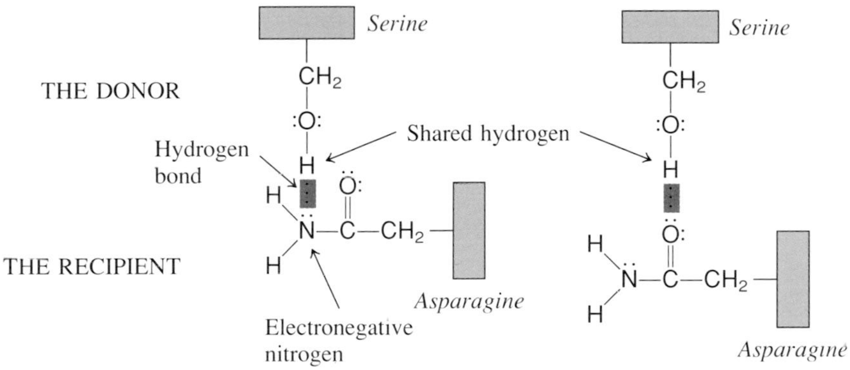

Hydrogen bonds formed between serine and asparagine. Serine serves as the hydrogen donor and asparagine as the recipient.

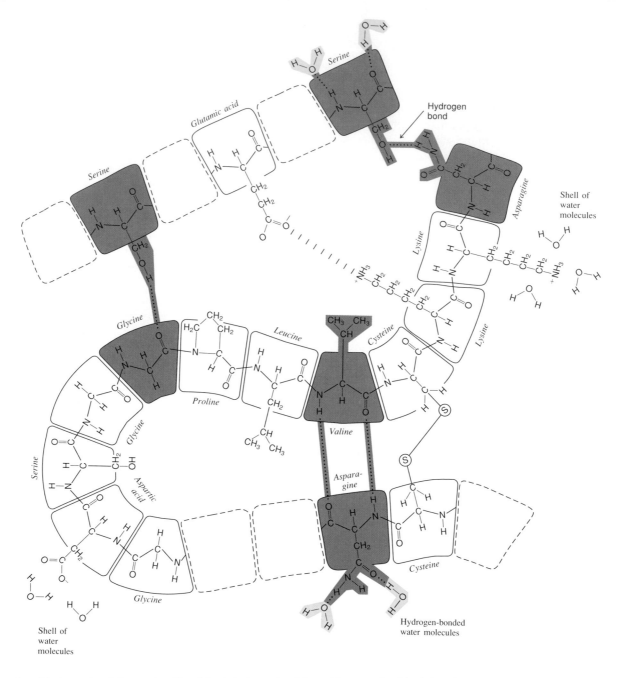

Examples of hydrogen bonds in proteins. *Top right:* A hydrogen bond formed between the side chains of serine and asparagine in the interior of the protein. *Top left:* A hydrogen bond formed between the side chain of
serine and the polypeptide backbone. *Center:* A pair of hydrogen bonds formed between the C=O and H—N
groups of two parts of the polypeptide-chain backbone—as would occur in a β-sheet folding unit. *Top and
bottom:* Two examples of amino acids on the protein surface, forming backbone and side-chain hydrogen
bonds with water molecules in the surrounding solvent.

=== HYDROGEN BONDS ===

Hydrogen bonds arise because of the way certain atoms share electrons unequally when they form a covalent bond. Consider, for example, the covalent bond between nitrogen and hydrogen. If this bond had been formed between two identical atoms such as two carbons, there would be an equal sharing of the electrons. However, in the case of an N—H bond, the nitrogen atom is inherently more *electronegative* (electron attracting). Its higher nuclear charge enables it to keep the shared electrons in its vicinity a greater percentage of the time. Hence, the electron cloud has a greater density in the region of the nitrogen atom. As a consequence of this asymmetric electron density, the N—H bond exists as a *permanent electric dipole*—the nitrogen atom has a slight excess of electrons and therefore carries a partial negative charge (denoted δ^-). The hydrogen atom is left with a slight electron deficit and therefore carries a partial positive charge (denoted δ^+). The N—H bond is thus represented as

$$\delta^+ \boxed{\text{H—N}} \delta^-$$

A C=O covalent bond is polarized in an exactly analogous way. In this case, the carbon atom carries the partial positive charge, and the oxygen carries the partial negative charge:

$$\delta^+ \boxed{\text{C=O}} \delta^-$$

The essential point is that, as electric dipoles with partial charges, both the N—H and C=O groups are capable of participating in electrostatic interactions. The interaction between a hydrogen-containing dipole (such as N—H) and another dipole (for example, C—O) gives rise to the hydrogen bond.

$$\delta^+ \boxed{\text{C=O}} \delta^- \cdots \delta^+ \boxed{\text{H—N}} \delta^-$$

When two amino acids form a hydrogen bond in the interior of a protein, the maximum potential strength of their interaction is about 5000 calories per mole (20,000 joules/mole)—some 20-fold weaker than a covalent bond, and 10-fold weaker than the strongest ionic bond. Moreover, for a strong hydrogen bond to form, the two amino acids must approach each other very closely, and they must be positioned at an angle that orients the donor hydrogen toward one of the unshared electron pairs on the recipient nitrogen or oxygen atom ($\dot{\text{N}}$ or $\ddot{\text{O}}$:), where the negative electron density is greatest. Despite these restrictions, the cumulative effect of a large number of hydrogen bonds is to provide a network of moderate-strength interactions that are important in determining protein structure.

abundant in proteins—12 of the 20 amino acids have side chains that can form this type of interaction, and every amino acid has backbone N—H and C=O groups capable of hydrogen bonding.

The essence of the hydrogen bond is that a hydrogen atom serves as a positively charged element that is shared by two relatively *electronegative atoms* (electron-attracting atoms)—generally oxygen or nitrogen. The hydrogen atom is covalently bound to one of the electronegative atoms; the corresponding amino acid is considered to be the *donor* in hydrogen-bond formation. The oxygen or nitrogen of a second amino acid shares the hydrogen in a looser association and is considered to be the hydrogen-bond *recipient*. In all cases, the hydrogen is oriented toward an unshared pair of electrons on the recipient oxygen or nitrogen atom. (See bottom of page 129.)

An example of a hydrogen bond is shown in the upper right of the figure on the facing page. Here the asparagine from the model peptide interacts with a serine subunit from a distant part of the protein. In this case, the asparagine is serving as the donor in hydrogen bond formation; the serine, through its unshared pair of electrons on the oxygen atom, serves as the recipient. The result of this hydrogen bond is that two amino acids interact in the interior of the protein to help anchor in position separate parts of the polypeptide chain.

In contrast to the covalent and ionic bonds we have previously discussed, the number of potential hydrogen bonds in proteins is enormous. This follows from three facts: (1) so many amino acid side chains can participate in hydrogen-bond formation; (2) the side chains are varied in size and shape; and (3) each can act as either a hydrogen donor or a recipient. For example, variation is possible even in the bond between the two amino acids we have just considered, asparagine and serine. If the roles of the side chains are reversed—so that the serine side chain acts as the hydrogen donor and the nitrogen of the asparagine serves as the recipient—a geometrically different hydrogen bond results. In fact, even greater structural diversity is possible, still using asparagine and serine: instead of interacting with the other side chain, the side chain of either amino acid could form a hydrogen bond with an N—H or C=O group in the *backbone* of the polypeptide chain. An example of such a side chain–backbone interaction is shown in the upper left of the figure on page 130: a serine OH group can be seen interacting with the backbone C=O group of a glycine subunit in the model peptide. In sum, virtually any N—H, C=O, or O—H group in the entire protein is potentially able to participate in the formation of a hydrogen bond. Which hydrogen bonds will actually form depends on the final folded structure of the protein, in which the number of various types of bonds is maximized.

Not all hydrogen-bonding amino acids are located in the interior of the protein. Many are found on the surface, where, like the charged amino acids we discussed earlier, they have a special role to play. Rather than forming hydrogen bonds with each other, they interact with the surrounding water molecules, which allows the protein to remain soluble and free-floating. The figure on page 130 shows two examples of amino acids at the protein surface participating in hydrogen-bond formation with water. In the first case, water molecules form hydrogen bonds with the exposed backbone C=O and N—H groups of serine;

in the second, the water interacts with the projecting side chain of asparagine.

Amino Acid Side Chains That Form Hydrophobic Bonds

Lastly, our model peptide contains six amino acids with hydrophobic side chains—*leucine, valine, proline,* and three *glycines.*

Gly-Asp-Ser-Gly-Gly-Pro-Leu-Val-Cys-Lys-Lys-Asn

The simple hydrogen side chains of the glycine subunits are essentially neutral and play little part in the bonding interactions that hold proteins together. But the larger hydrocarbon side chains of leucine, valine, and proline are able to influence protein folding by means of the *hydrophobic bond*. This is not quite a bond in the traditional sense, but rather a gathering of amino acids whose hydrogen- and carbon-containing side chains would prefer to interact with each other rather than remain separated and surrounded by water molecules. As a consequence, hydrophobic amino acids tend to cluster together and turn inward toward the interior of the protein, where they serve as space-filling elements of various shapes and sizes. A hydrophobic amino acid can associate with another, purely hydrophobic amino acid (such as isoleucine, phenylalanine, valine, proline, or tryptophan), or with the hydrocarbon components that are present in the side chains of various other amino acids (for example, the —$CH_2CH_2CH_2CH_2$— stem of lysine). Two examples of hydrophobic interactions involving leucine and valine, and valine and the stem of lysine, are shown in the center of the figure on the facing page.

The interaction between two hydrocarbon groups is much weaker than a covalent or ionic bond, and even somewhat weaker than a hydrogen bond. Nonetheless, because of their very large numbers, hydrophobic interactions play a major role in establishing and maintaining protein structure.

Van der Waals Interactions

Before leaving our theoretical discussion of the basis of protein folding, we will consider one further, very general

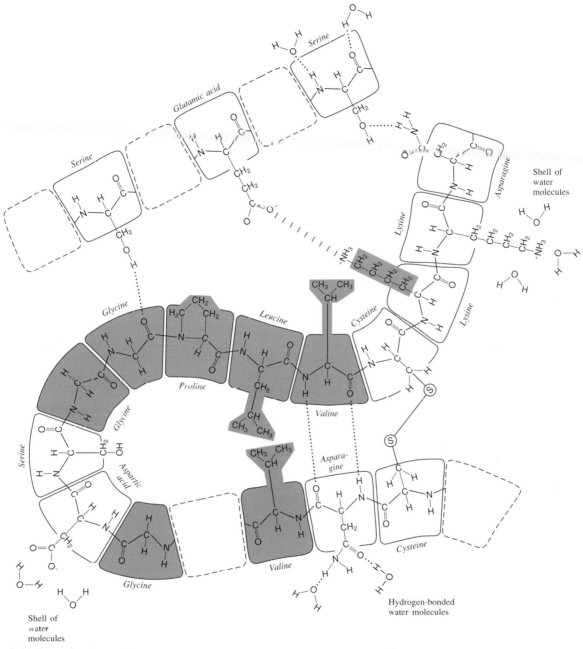

A hydrophobic interaction between leucine in the model peptide and valine from elsewhere in the protein, and a hydrophobic interaction between valine and the hydrocarbon stem of a nearby lysine subunit. The remaining subunits shown in color are glycines, which function as flexible space-filling elements but do not influence protein folding through side-chain interactions. However, their backbone C=O and N—H groups can be used to form hydrogen bonds.

HYDROPHOBIC INTERACTIONS

The theory underlying the hydrophobic bond involves an understanding of the lattice structure of liquid water. The symbol H_2O is an adequate representation of the structure of water only in the gas (water vapor) phase, where molecules exist individually. In an aqueous environment such as that of the cell, the liquid properties of water become important. Each O—H unit in a water molecule is a permanent electric dipole in which the oxygen atom has a relative electron surplus (δ^-) and the hydrogen atom a relative electron deficit (δ^+). This imbalance of charge enables water molecules to form hydrogen bonds with each other whenever the random thermal motion of molecules is low, as at physiological temperatures. Such hydrogen bonds are stable, and they dominate the behavior of water. The water molecules tend to interact with each other so as to maximize the number of energetically favorable hydrogen bonds. This leads to the formation of a lattice structure, where every water molecule is potentially bonded to four nearest neighbors.

The important point is that, in the temperature range of living systems, the structure of water is best represented as $(H_2O)_n$, where n has a value between 20 and 30. We can therefore picture water as a collection of aggregates of approximately this size continually gaining and losing individual H_2O molecules from the periphery of the lattice.

If we consider an unfolded polypeptide chain in the aqueous environment of the cell, we can predict what will happen to the hydrophobic amino acids it contains. Events will be governed by the fact that these side chains cannot participate in the hydrogen-bonded structure of water. On the contrary, each hydrophobic side chain will create a cavity, disrupting the water structure. Although other amino acids also interrupt the lattice structure, they can at least compensate in part for the energy lost by forming electrostatic interactions with the surrounding water molecules. But the water molecules in direct contact with hydrophobic side chains necessarily participate in fewer hydrogen bonds, and there is a net loss of hydrogen bonding

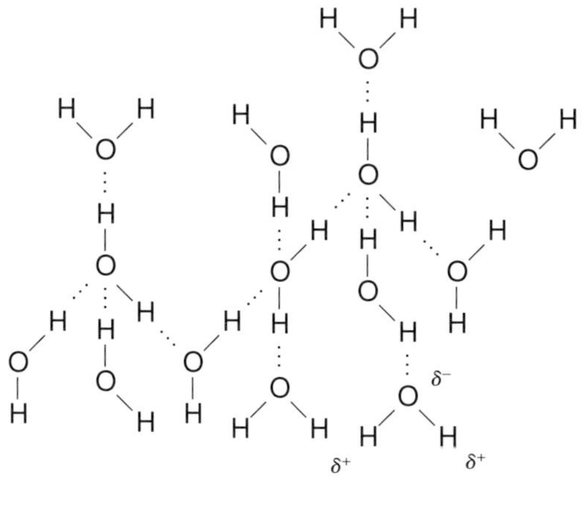

The lattice structure of water, which contains about 30 subunits at room temperature. Below 0°C, this structure becomes more extensive and more regular, forming ice.

in the system. Furthermore, these water molecules become immobilized and suffer an energetically unfavorable loss of freedom of movement (or *entropy*).

The number of incompletely bonded water molecules can be lowered, however, if the hydrophobic side chains were to interact with each other. As shown in the diagram at the bottom of this page, a smaller surface area is exposed to the surrounding solvent molecules, and the number of water molecules left unbonded around the hydrophobic material is lower. The water molecules that are released are then available to participate in energetically more favorable interactions by once again joining in the

hydrogen-bonded structure of water. They gain not only the ability to form additional hydrogen bonds but also increased *entropic freedom* (increased "disorder" due to enhanced movement), in as much as they no longer need to form a shell around the hydrocarbon side chains. The energy derived from the release of water molecules is on the order of 1000 calories per mole of hydrocarbon —CH_2— group. As a consequence of the drive toward entropic freedom and the formation of energetically favorable hydrogen bonds, the hydrophobic side chains of proteins will tend to position themselves in the interior of the molecule.

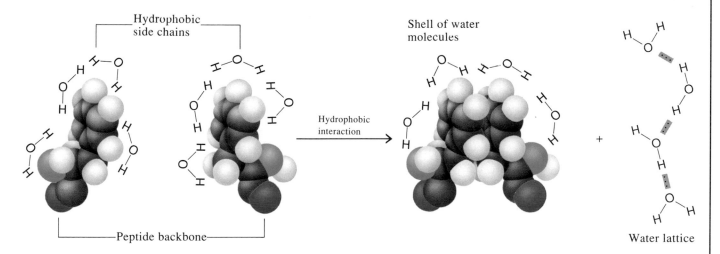

On the left, two separated hydrophobic amino acid side chains are surrounded by water molecules with which they cannot chemically interact. On the right, the side chains have been placed next to each other, allowing the release of several water molecules. The "energy" of the hydrophobic interaction is actually derived from the entropic and bonding potential gained by the released water molecules, which become free to interact with each other rather than form unfavorable contacts with the hydrocarbon side chains.

type of chemical interaction that applies to all of the amino acids. It is most commonly known as the *van der Waals force*.

The electron clouds that surround atoms are never still. There are continuous slight fluctuations as, in classical terms, the electrons travel their orbits. The result is the formation of *transient* electric dipoles—temporary asymmetries in the distribution of electron density. For example, the two methyl groups ($—CH_3$) at the end of a leucine side chain may come to differ in the density of their electron clouds:

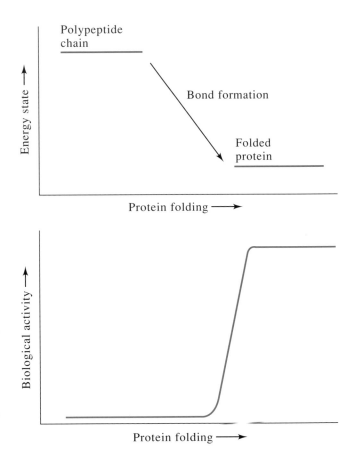

Leucine side chain   *Transient dipole form*

As they flicker in and out of existence, these transient dipoles can participate in electrostatic interactions, either with permanent electric dipoles (such as $C≡O$ and $N—H$) or with charged groups (such as NH_3^+ or COO^-). But—and this is the important point—they can also *induce* similar types of transient dipoles in neighboring chemical groups. The interactions between the transient dipoles and the dipoles they induce give rise to a significant attractive force between otherwise neutral atoms.

Individually, van der Waals interactions are tenuous connections, and they are further weakened if there are water molecules present. But in the interior of a protein, where water molecules are virtually excluded and the atoms pack tightly together, interactions between transient dipoles and induced dipoles attain binding energies of about the same strength as weak hydrogen bonds or hydrophobic interactions. Van der Waals attractions, though short-lived, become an important force in maintaining the tightly folded, three-dimensional structure of the protein because almost all of the thousands of atoms in the interior of the molecule are involved, not just those chemical groupings with the particular capacity for forming, say, ionic or hydrogen bonds.

Covalent bonds, ionic bonds, and hydrogen bonds—and even hydrophobic interactions—can be identified by

a direct inspection of the structure of a protein. It is easy to see, for example, when positive and negative charges are close together, or to identify the characteristic $N—H\cdots O≡C$ arrangement of atoms participating in a hydrogen bond. In contrast, it is not possible to identify specific interactions between transient dipoles through direct inspection. Van der Waals forces therefore tend to receive relatively less attention when the forces leading to protein folding are discussed. This, however, should not obscure the major role that these interactions play in protein structure.

The folding of a polypeptide chain with a random conformation into a biologically active protein is driven by the formation of covalent, ionic, hydrogen, hydrophobic, and van der Waals bonds.

To summarize our discussion of bonding, the important point is that the formation of each of the types of chemical interactions we have described—covalent, ionic, hydrogen, hydrophobic, and van der Waals—allows a polypeptide chain to assume a more favorable energy conformation than if it remained randomly coiled. The tendency of a molecule to find the state of least energy is the force driving the folding of the protein chain. *Indeed, the organization of the polypeptide chain to form the maximum number of chemical bonds can be viewed as a "reaction" leading to a lower, more favorable energy state.*

The Three-dimensional Structure of Chymotrypsin

Now that we know how protein folding comes about in theory, we can make the transition to an understanding of the precise three-dimensional structure of chymotrypsin.

It is apparent that millions of interactions are possible among the amino acids of a long polypeptide chain. For example, if we consider just the 10 cysteine residues in chymotrypsin, a simple mathematical analysis shows that there are more than 1000 different ways to form disulfide linkages. We find an even larger number of possible combinations of ionic, hydrogen, and hydrophobic bonds. Because of the immense potential for bonding, it is impossible to predict the exact three-dimensional structure of a protein solely from its amino acid sequence and general considerations about chemical bonding. Rather, this knowledge provides the theoretical basis for understanding the final folding pattern.

Fortunately there is a powerful, if difficult, method for determining the precise anatomy of an enzyme. It is a specific line of experimentation that allows one to define the exact position and orientation of every amino acid in the protein. The method is X-ray crystallography, which, over a period of 50 years, has grown from a technique for determining the atomic arrangements in simple salt crystals into a procedure that can be applied to large molecules such as proteins.

The fact that chymotrypsin and many other proteins form crystals indicates that they, like small molecules, possess highly ordered, well-defined structures. It is this property that allows billions of identical protein molecules to form similar interactions with their neighbors and build a regularly repeating structure (much like the repeating horsemen shown on page 54) that can be subjected to X-ray analysis. When a beam of X rays passes through a crystal, the electromagnetic waves are diffracted (redirected) as they make contact at regular intervals with the repeating patterns of atoms. The result is a series of parallel waves, which, when they reinforce each other, give rise to electronically detected spots on a photographic film placed behind the sample. (See the figure and the photograph on pages 138–139.) While small molecules produce only a few "diffraction spots," complex molecules such as proteins can give rise to diffraction patterns with 10,000 or more spots. The position and intensity of each spot provide information about the atoms in the molecule, and a sophisticated mathematical analysis allows one to reconstruct the three-dimensional structure of the protein. It is as if one constructed a precise portrait of a person from an immense set of anatomical measurements.

The X-ray crystallographic analysis of chymotrypsin was completed in the 1970s by David Blow and his colleagues Jens Birktoft, Richard Henderson, Brian Matthews, Paul Sigler, and Tom Steitz in Cambridge, England. After several years of work, the remarkable three-dimensional structure shown on the bottom of page 139 emerged. This "space-filling" model of chymotrypsin shows all 4000 atoms that make up the enzyme, in their exact positions relative to one another.

What can we learn from this structure? We shall see that, when it is examined carefully, it reveals all the basic principles that underlie the construction of enzymes.

At the outset, we see that the enzyme is an essentially solid structure: its three polypeptide chains fold back and forth through space until they occupy a more or less spherical area with a diameter of about 45 Å. Except for a few water molecules that are trapped at specific places within the folded molecule, the 241 amino acids of chymotrypsin entirely fill this space. For a moderate-sized protein like chymotrypsin, if one were to travel from one side to the other through the middle, one would encounter about 5 to 10 amino acids.

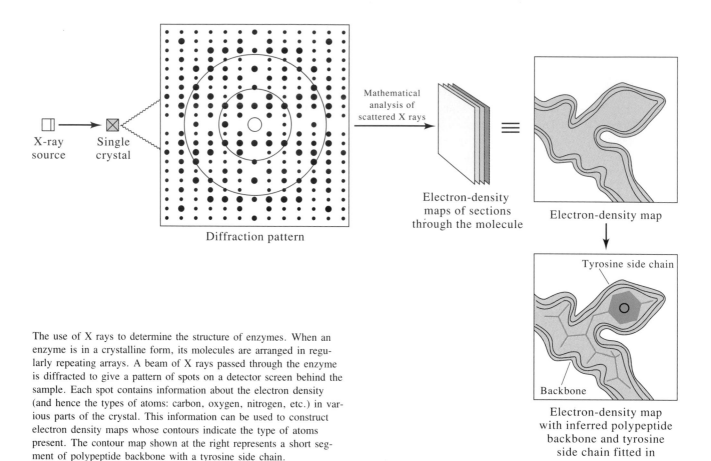

X-ray
source

Single
crystal

Diffraction pattern

Mathematical
analysis of
scattered X rays

Electron-density
maps of sections
through the molecule

Electron-density map

Tyrosine side chain

Backbone

Electron-density map
with inferred polypeptide
backbone and tyrosine
side chain fitted in

The use of X rays to determine the structure of enzymes. When an enzyme is in a crystalline form, its molecules are arranged in regularly repeating arrays. A beam of X rays passed through the enzyme is diffracted to give a pattern of spots on a detector screen behind the sample. Each spot contains information about the electron density (and hence the types of atoms: carbon, oxygen, nitrogen, etc.) in various parts of the crystal. This information can be used to construct electron density maps whose contours indicate the type of atoms present. The contour map shown at the right represents a short segment of polypeptide backbone with a tyrosine side chain.

The space-filling model is a highly accurate representation of chymotrypsin—indeed, the *most* accurate since it shows all the contacts between the various atoms. However, the interior of the molecule is hidden from our view. In order to get a clearer picture of the interior organization of the enzyme, we will develop a number of alternative representations that will simplify the relatively opaque space-filling model. We will rely on a set of dissections to reveal different aspects of the structure. The first such molecular dissection is shown on page 140. In this three-dimensional schematic diagram, the full structure of chymotrypsin has been simplified by shrinking each of its amino acids to a small circle. The side chain of each

amino acid, along with its amino and carboxyl group, has been omitted, leaving only the central carbon atom to represent the subunit. The result is that the flesh has been stripped away from the protein, leaving the polypeptide backbone clearly visible.

As can be seen, in chymotrypsin (as in other enzymes), the path followed by the backbone is neither regular nor predictable. This is evident in the behavior of our ''model peptide,'' Glycine-193 to Asparagine-204 (which is shown in red). Subunits 193 through 196 lie on the surface of the enzyme. Then, beginning with subunits 197 and 198, the model peptide bends toward the interior of the protein. Eventually subunits 199 to 204 emerge on the

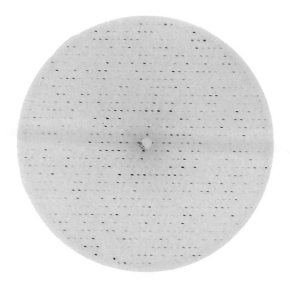

The X-ray diffraction pattern given by the enzyme chymotrypsin.

The three-dimensional structure of chymotrypsin. Carbon atoms are shown in black, nitrogen in blue, oxygen in red, hydrogen in white, and sulfur in yellow. The diameter of the enzyme is about 45 Å (somewhat less than a millionth of an inch). The hydrophilic side chain of Arginine-145 is clearly visible projecting outward from the right side of the molecule. The ridges and grooves on the surface of the chymotrypsin molecule are as unique as the mountains and craters of the moon, and herein lies the fulfillment of Fischer's lock-and-key hypothesis.

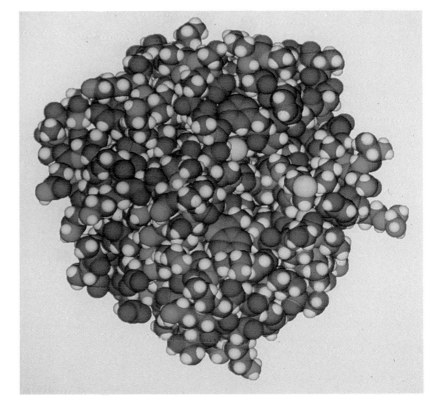

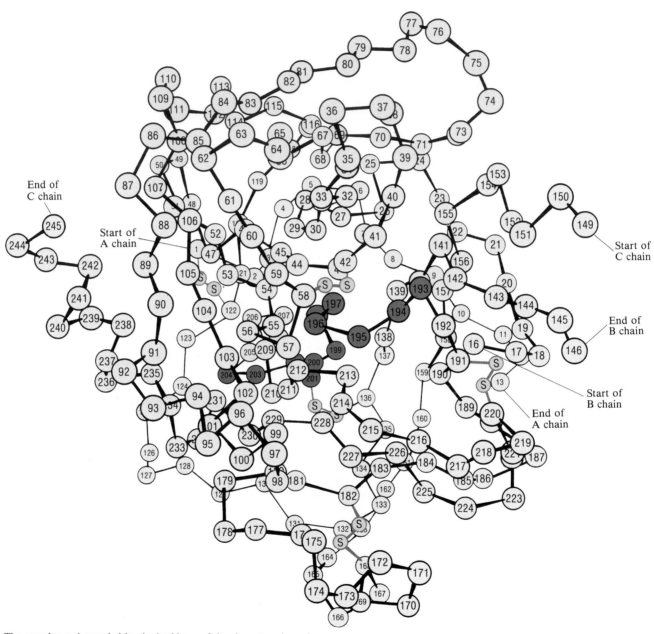

The complex path traveled by the backbone of the chymotrypsin molecule. The numbered circles trace the positions of the central carbon atoms of successive amino acids. The model peptide—amino acids 193 to 204—is shown in red. The orientation of chymotrypsin in this figure is exactly the same as in the space-filling model on the previous page.

opposite surface. The twisting path of the backbone in the model peptide illustrates that the protein does not have an internal geometry based on a simple repeating principle, and its three-dimensional folding pattern is therefore asymmetric. The irregular internal structure of chymotrypsin is in striking contrast to the highly regular backbone conformations found in proteins whose role is *structural* rather than *enzymatic* (recall the keratin protein in hair and the fibroin protein in silk).

Folding Involving Backbone Interactions

Although at first glance the internal structure of chymotrypsin appears random, a closer inspection shows that, within the apparent asymmetry of the overall construct, there are certain localized regions of structure. These regions contain the three types of backbone structures stabilized by hydrogen bonds that we have already discussed—the α-helix, the β-sheet, and the β-turn. These structural features become particularly evident in another molecular dissection of the original space-filling model—the "zigzag figure" on the following page. In this diagram no attempt is made to show the three-dimensionality of the path of the backbone; rather, that path has been untwisted and flattened out to obtain a planar figure. Then, applying the X-ray crystallographic data that give the exact positions of the atoms in the intact structure and the distances between them, those areas of the backbone that are connected by $C=O \cdots H—N$ hydrogen bonds are placed together in the diagram.

We immediately see two short regions (colored red) where the polypeptide chain is organized into an α-helix. The first is composed of amino acids 164 to 173 in the interior, and the other of amino acids 235 to 245 at the end of the protein. In the planar zigzag diagram, the α-helical regions are drawn as sawtooth areas, and the connecting hydrogen bonds between the participating backbone N—H and C=O groups are shown as short straight lines.

Although the α-helix is not used extensively to fold chymotrypsin into its final three-dimensional structure, the β-sheet motif plays a highly prominent role. The zig-

zag pattern shows that a large fraction of the protein backbone occurs as β-sheet, organized into two major β-folding units (shown in blue). The first includes most of the B chain from amino acids 29 to 112, and the second includes both the end of the B chain from amino acids 135 to 140 and most of the C chain from amino acids 156 to 230. Each of these folding units is composed of six polypeptide segments, aligned in an antiparallel fashion and linked by hydrogen bonding between backbone N—H and C=O groups. Hydrogen bonds also occur between the first and last elements of each β-sheet, rolling the six polypeptide segments into an irregular cylinder called a β-barrel. The interior of each cylinder is filled with hydrophobic amino acid side chains that project inward from one surface of the β-sheet. Portions of the antiparallel elements in the two β-folding units of chymotrypsin are highlighted in blue in the three-dimensional backbone diagram on page 144.

In structural proteins such as silk, β-sheets are extensive, highly regular structures. In enzymes, however, the β-sheet pattern is imperfect. As can be seen in both the three-dimensional and the planar backbone diagrams, the participating polypeptide chains are not perfectly aligned side by side but instead travel irregularly alongside each other, making only occasional contacts. Moreover, the interacting backbone segments are not always derived from immediately adjacent parts of the protein, as would be the case for the simplest zigzag β-sheet pattern. Rather, the folding units often occur in the conformation of a "Greek key," a mazelike structure in which segments from one area of the protein interdigitate with segments from other areas. The diagram on page 143 shows the Greek-key arrangement for the three elements of the first β-folding unit involving amino acids 52 to 54, 87 to 90, and 104 to 107. This portion of the β-folding unit can easily be identified in both the three-dimensional backbone and planar zigzag diagrams.

Lastly in our discussion of backbone interactions that contribute to protein structure, we come to the β-turn. As will be recalled, this is a hydrogen-bonded turnaround involving the N—H group of one amino acid and the C=O group of another amino acid three subunits away. Such hydrogen bonds help the polypeptide chain to reverse its direction abruptly and then run almost parallel to its origi-

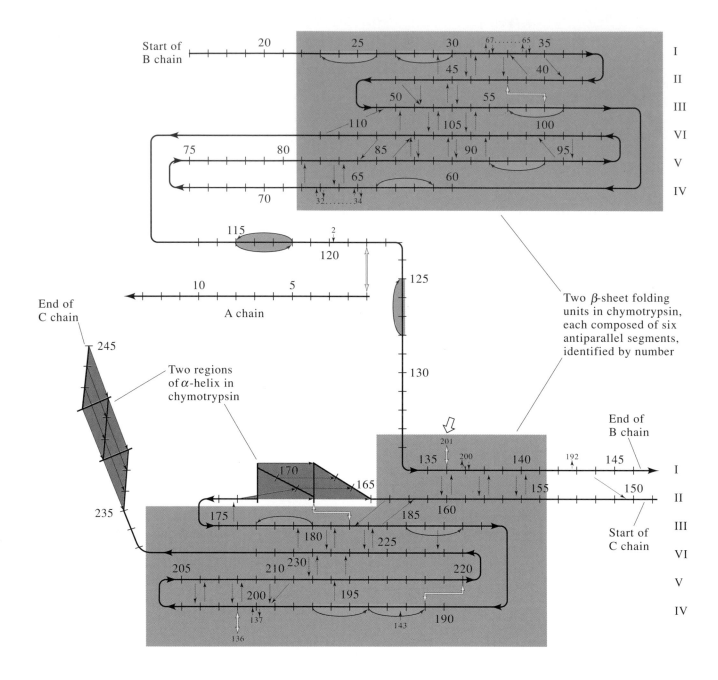

Start of B chain

20 25 30 67....65 35 I
45 40 II
50 55 III
110 105 100 VI
75 80 85 90 95 V
65 60 IV
70 32.....34

115 2
120

End of C chain

10 5
A chain

245

Two regions of α-helix in chymotrypsin

235

201
135 200 140 192 145 I
155 150 II
165 III
170 175 180 225 185 160 VI
205 210 230 220 V
200 195 190 IV
137
136 143

Two β-sheet folding units in chymotrypsin, each composed of six antiparallel segments, identified by number

End of B chain

Start of C chain

The chymotrypsin backbone has been redrawn in a planar form to emphasize regions of α-helix (red) and β-sheet (blue) and the occurrence of β-turns (green). The thin, straight arrows represent hydrogen bonds between N—H and C=O groups that are involved in the formation of α-helix or β-sheet structures. Generally, the two amino acids connected by such an arrow are opposite one another in the diagram. However, some arrows point off toward numbers, which refer to the "distant" amino acid subunits to which the arrows connect. For simplicity, only certain such hydrogen bonds are shown—for example, those connecting the first and last components of the first β-folding unit. A number of other backbone hydrogen bonds (about 20) are not shown; these pin the first and second β-folding units together. Curved arrows represent hydrogen bonds involved in β-turns, which stabilize turnarounds in the polypeptide backbone. Double-headed yellow arrows represent disulfide linkages. An attempt has been made to have equal distances between each amino acid. However, at certain positions, for example between amino acids 112 and 113, the distance has been exaggerated in order to facilitate the drawing of the planar diagram.

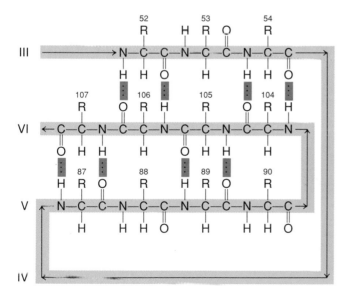

The Greek-key arrangement for three elements in the first β-folding unit of chymotrypsin. The roman numerals identify the individual elements of the β-folding unit, as in the figure on the facing page.

nal course. Hydrogen-bonded β-turns in chymotrypsin are highlighted in green in the planar zigzag diagram, and four examples are highlighted in green in the three-dimensional diagram of the enzyme's structure on page 144. These involve amino acids 23 to 26, 56 to 59, 61 to 64, and 91 to 94, and occur in the first β-folding unit.

In sum, major aspects of the structure of chymotrypsin arise from backbone interactions in which short stretches of amino acids are organized into α-helices, β-sheets, and β-turns. About 10 percent of the enzyme's amino acids are involved in forming α-helices, and another 75 percent are used to form β-sheets and β-turns. These backbone connections form the basic scaffold upon which the rest of the enzyme's structure is superimposed. In addition, these structural motifs allow the relatively hydrophilic (water-seeking) backbone of the polypeptide chain to be "neutralized" and accommodated in the interior of the protein. Indeed, almost half of the interior of the protein is occupied by backbone material.

The α-helix, the β-sheet, and the β-turn are all based on the hydrogen bond. In fact, this is the primary way this bond is used to achieve protein folding. Of the 200 hydrogen bonds that occur in chymotrypsin, 190 are found in α-helical, β-sheet, and β-turn structural elements.

Folding Involving Side-chain Interactions

While backbone interactions clearly play a major role in the folding of chymotrypsin, much of the precise three-dimensional structure of the enzyme is due to specific interactions between the amino acid side chains. These interactions are, of course, too numerous and varied to be described in a schematic diagram such as the planar zigzag drawing. Nevertheless, it is still possible to illustrate those aspects of the folding of chymotrypsin that are based on side-chain interactions by returning to our model peptide and analyzing the positions of its amino acids

Gly-Asp-Ser-Gly-Gly-Pro-Leu-Val-Cys-Lys-Lys-Asn

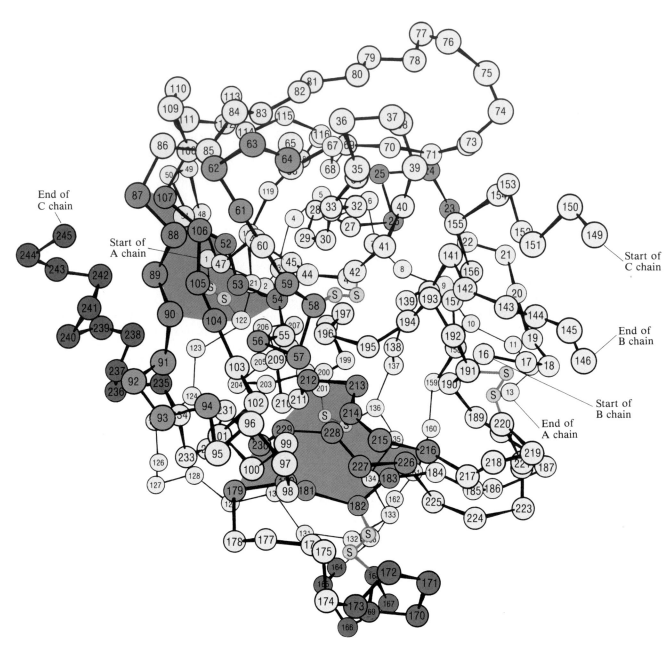

A three-dimensional representation of the path traveled by the backbone of the chymotrypsin molecule. The various colors in the diagram highlight structural motifs that help to organize the protein.
These include two α-helical regions (red), portions of two β-sheet
folding units (blue), and several β-turns (green).

The location of the model peptide in the final folded structure of chymotrypsin is indicated in red in the drawing on page 140. This sequence of 12 amino acids begins on the surface of the enzyme (193–197) and then turns inward toward the interior, finally emerging on the opposite side (198–204). The X-ray crystallographic analysis reveals how the side chains of these amino acids contribute to protein folding, and the data validate our earlier assumptions about the role of side-chain interactions in protein folding that initially were derived only from basic chemical principles.

The Cysteine Subunit

We noted earlier that the sulfhydryl (—SH) side chain of cysteine has the ability to form a covalent disulfide linkage with another cysteine. Indeed, the cysteine subunit of our model peptide (amino acid 201 of chymotrypsin) is involved in a disulfide linkage with a second cysteine from a distant part of the protein (Cysteine-136). This covalent interaction serves as a linchpin to secure the barrel structure of the enzyme's second β-folding unit (see the open arrow in the planar zigzag diagram on page 142).

In addition, Cysteine-201 also participates in two other bonds that are important for protein structure. Cysteine's backbone N—H and C=O groups are used to form a pair of hydrogen bonds with the corresponding C=O and N—H groups of the nearby amino acid Threonine-208. These hydrogen bonds contribute to the stability of the short β-sheet element containing amino acids 199 through 203 and 204 through 211 (see the diagram on page 142). Cysteine-201 is thus a particularly good example of how a single amino acid can participate in a network of bonding interactions—it holds three parts of the polypeptide chain in close proximity, thereby contributing in several ways to the overall structure of the enzyme.

Disulfide bonds are particularly prominent in chymotrypsin and a number of other enzymes that are secreted into the digestive tract. Cellular enzymes, in contrast, generally lack such bonds. The reason for this difference may lie in the strategy of digestion. In the stomach and the initial portion of the intestine, the pH is very acidic (as low as 2). This change from the normal cellular pH of 7 has

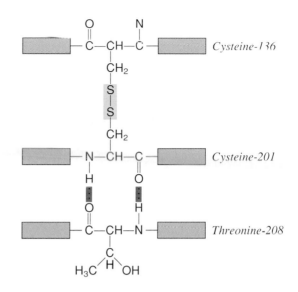

Cysteine-201 participates in (1) a disulfide linkage with Cysteine-136 and (2) two backbone C=O ⋯ N—H hydrogen bonds with Threonine-208.

evolved to place stress on ingested proteins. They begin to unfold and, in their unwound state, become subject to cleavage by such enzymes as chymotrypsin. (See top of page 146.) In contrast, the digestive enzymes themselves, fortified by their covalent S—S bonds, are less prone to unfolding. Because they are relatively resistant to digestion, they remain structurally intact and biologically active.

The Electrically Charged Amino Acids

The model peptide has three charged amino acids—two positively charged lysines immediately adjacent to Cysteine-201 and, several subunits away, a negatively charged aspartic acid. Our previous discussion of the chemical properties of electrically charged amino acids led to the conclusion that they could promote protein folding in two ways: (1) by forming ionic bonds in the interior of the protein, and (2) by interacting individually with the

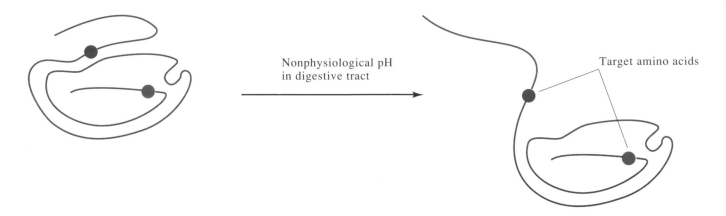

As compact proteins unfold in the hostile pH environment of the digestive tract, previously hidden target amino acids become accessible to digestive enzymes. Through the action of enzymes such as chymotrypsin (which cleaves after phenylalanine, tyrosine, and tryptophan) and trypsin (which cleaves after lysine and arginine), the ingested proteins are reduced to a number of peptide fragments. The breakdown process is then completed by other proteases that remove amino acids one at a time from the many ends thus created.

surrounding water solution at the surface of the protein. Indeed, charged amino acids serve both functions in chymotrypsin, as illustrated by the model peptide.

The X-ray data indicate that the negatively charged side chain of Aspartic acid-194 lies adjacent, in the final structure of the enzyme, to the positively charged —NH_3^+ group on Isoleucine-16 (see the diagrams on pages 144 and 206). This free amino group occurs as the N-terminus at the beginning of the enzyme's B chain. In the folded protein, the ionic bond between Aspartic acid-194 and Isoleucine-16 serves to anchor the end of the B chain. Indeed, the formation of this bond plays a key role in the activation of the enzyme (as will be discussed on page 206).

The two positively charged amino acids of the model peptide—both lysines—are not involved in forming ionic bonds. Instead, they are located on the surface of the protein, and their side chains project outward into the external medium. Each side chain is surrounded by a shell of water molecules, whose electronegative oxygen atoms are ori-

ented toward the positively charged —NH_3^+ group, providing a set of interactions that help the protein to float freely in the surrounding water solution.

Of the 17 positively charged amino acids in chymotrypsin (14 lysines and 3 arginines), all are on the surface of the enzyme. Similarly, 12 of the 14 negatively charged amino acids (glutamic and aspartic acid subunits) also lie on the surface. From this (and similar studies with other enzymes) it can be concluded that charged amino acids are only rarely used to achieve protein folding through the formation of ionic bonds. By refusing to become trapped in the hydrophobic interior of the protein, they function much more often to define the external surface of the enzyme and promote its interaction with water.

Hydrogen-bonding Amino Acids

Two amino acids in the model peptide have side chains that are capable of forming hydrogen bonds—serine and

Schematic representation
of backbone of chymotrypsin

The hydrogen bond between Serine-195 and Histidine-57.

asparagine. The X-ray crystallographic data indicate that, in the folded enzyme, the OH group of Serine-195 is in a position to form a hydrogen bond with a nitrogen atom on the side chain of Histidine-57. This potential hydrogen bond is of particular interest, as we will see in the next chapter when we discuss the mechanism of enzyme action. The other hydrogen-bonding subunit in the model peptide, Asparagine-204, is located at the surface of the protein and interacts with solvent water molecules.

In proteins generally, about half of the side chains capable of forming hydrogen bonds are located in the interior, where their potential to form hydrogen bonds is almost always achieved. The other half lie on the surface and, through hydrogen bonding with water molecules, promote the solubility of the protein in the aqueous environment of the cell.

Of the 200 hydrogen bonds in chymotrypsin, most are contributed not by amino acid side chains but by backbone N—H and C=O groups organized into α-helix and β-sheet structures. Indeed, 10 of the 12 amino acids in our model peptide participate in the formation of β-sheets. In our earlier theoretical discussion, we predicted that hydrogen bonds could also occur between amino acid side chains on the one hand and C=O and N—H groups of the protein backbone on the other. The model peptide provides one example of this type of hydrogen bond. Glycine-193 uses its backbone C=O to form a hydrogen bond with the side chain of Histidine-40.

In sum, from both side-chain and backbone interac-

tion, there is a preponderance of hydrogen bonds in chymotrypsin, which generate a web of interactions that contributes greatly to protein structure.

Hydrophobic Side Chains

The remaining six amino acids in our model peptide contain hydrophobic side chains: three glycines, a proline, a leucine, and a valine. The three glycine subunits have only a single hydrogen as their side chain, and they serve primarily as flexible backbone spacers where no larger amino acid would fit (for example, Glycine-193 and Glycine-196, each of which occurs in the middle of a β-turn). In contrast, the larger side chains of proline, leucine, and valine are active in protein folding. Oriented toward the interior of the enzyme, they form hydrophobic interactions with other similar amino acid side chains. Much of the interior of chymotrypsin (and other proteins) becomes filled with hydrophobic amino acid side chains; they occupy most of the space not taken up by the polypeptide backbone. Because hydrophobic interactions involve a large percentage of the protein's amino acids, and allow the release of a correspondingly large number of water molecules (which are then free to participate in entropically and energetically more favorable water lattice structures), it is believed that hydrophobic interactions provide the major driving force behind protein folding.

To sum up, the cumulative effect of all the side-chain and backbone interactions is the folding of chymotrypsin into its final, correct structure, yielding a biologically active enzyme. Because enzymes like chymotrypsin are such complicated molecules, they cannot possibly be understood through memorization. This is why we have focused on the general themes of construction (like the α-helix, the β-sheet, and the β-turn) and, in addition, on the folding behavior of a short representative segment of the protein—the model peptide. By studying the model peptide and referring to X-ray crystallographic data, we have been able to verify our general assumptions about protein folding, and we can now feel confident that we understand the theoretical basis of protein structure. This emphasis on protein architecture is well placed, for, as we will see, the function of an enzyme is intimately related to its structure.

The Effect of Temperature and pH on Enzyme Structure and Function

Because much of the folding of a protein depends on relatively weak interactions—ionic bonds, hydrogen bonds, and hydrophobic interactions—enzymes are somewhat fragile molecules. This is why many enzymes are inactivated when the temperature is raised only a few degrees above normal (page 64). The higher temperature imparts to various areas of the protein chain enough extra kinetic energy and random thermal motion to disrupt the delicate network of noncovalent bonds that holds the enzyme together; the protein loses both structure and function. This sensitivity to temperature is one reason why high fever poses a threat during illness: key enzymes involved in sustaining the metabolism of the organism begin to lose the structural integrity necessary for their function. Because the cell's enzymes are organized into relay teams, carrying out successive modifications that move substrate molecules along metabolic pathways, the loss of any one enzyme can bring the entire machinery to a halt.

A similar danger arises if the concentration of acid or base in an enzyme's environment changes significantly. When the pH changes, hydrogens are added to or removed from certain amino acid side chains. The electrical charges of the side chains are altered, preventing the formation of key ionic bonds. As in the case of temperature perturbation, the folding of the protein is disrupted, and, as the structure of the enzyme falters, its biological activity disappears. Changing pH can also destroy enzyme function by subtler means. In some cases, the overall structure of the enzyme is not changed but a particular amino acid is affected. If the amino acid is intimately involved in the catalytic reaction mechanism, a change in its electrical charge will prevent the enzyme from functioning. We will see examples of both of these effects of pH on chymotrypsin in the next chapter.

The Folding Process

It is apparent that the structure of an enzyme like chymotrypsin is extremely intricate and precise. Each of 241 amino acid subunits occupies its own specific location in the final folded structure, held in place by a three-dimensional network of backbone and side-chain interactions. Although we can now feel confident that we understand the bonding forces that hold the protein in its final form, the folding process itself poses a distinct and interesting problem.

Given the huge variety of possible disulfide bonds, ionic bonds, hydrogen bonds, and hydrophobic interactions, it might appear that thousands of different folding patterns would compete with each other as the correct structure was sought. Yet we know, from the very fact that crystals of the enzyme give clear X-ray diffraction pictures, that all chymotrypsin molecules fold up in precisely the same way. We must assume, therefore, that this particular conformation allows the chymotrypsin polypeptide chain to achieve the greatest possible number of bonds and the closest fit for van der Waals interactions. In this conformation, the enzyme attains the lowest possible energy state, and the surrounding solution the most favorable entropy state (due to the release of water molecules that were originally bound to the protein chain but now become free to interact with each other).

How the correct folding pattern is achieved is not yet known. If a protein chain were to arrive at its correct folded structure by testing each possible conformation, each amino acid would have to be tried in several orientations, and it would be necessary to allow some period for each test, say one-tenth of a trillionth of a second per test—the amount of time required for the rotation of a typical molecular group. For a protein chain with 250 subunits, testing only three orientations for each amino acid would require an imponderably long time:

3^{250} conformations $\times 10^{-13}$ seconds per conformation
$$= 10^{106} \text{ seconds}$$

This is equivalent to 3.2×10^{98} years. In comparison, the age of the universe is only 20 billion (2×10^{10}) years, and life on earth has existed for only 3.5 billion years. Thus, if trial and error were the method of protein folding, not a single protein could be expected yet to have found its proper three-dimensional structure.

In fact, after their synthesis in the cell, protein chains fold up within a matter of minutes. It therefore follows that folding cannot proceed simply by a random testing of all possible conformations. Although such a testing procedure may play a part in protein folding, there must be a way to enhance its efficiency.

The key to understanding protein folding is to realize that the process involves intermediate stages. For example, in the first stage hydrophobic amino acids might group together to form the core of the protein, while the charged amino acids remain on the outside because of their affinity for water and their reluctance to become trapped singly in the hydrophobic interior. These external amino acids define those segments of the polypeptide chain that will constitute the future surface of the enzyme. This first stage divides the nascent protein into two general areas and greatly reduces the number of subsequent folding possibilities. A second major stage in folding would then come as some smaller regions of the protein chain are successful in finding α-helix, β-sheet, or β-turn structures. Such events could occur in many different orders, analogous to the multiple ways in which a jigsaw puzzle may be solved. At first, some folding units may be incorrect, but they will be replaced eventually by more stable interactions. The correct interactions not only reinforce each other but, most important, are able to serve as an increasingly stable scaffold upon which the rest of the structure can be progressively built.

That the linear amino acid sequence of a protein does, in fact, contain all the information needed to direct the folding process is known from a classic experiment performed by Christian Anfinsen and his colleagues, Michael Sela and Fred White, at the National Institutes of Health (see top of page 150). They began with a digestive protein rather similar to chymotrypsin and completely disrupted its three-dimensional structure by immersing the protein in chemical reagents that (1) broke disulfide bonds and (2) disrupted backbone hydrogen bonding. As a consequence, the enzyme was reduced to a random coil. In this conformation, of course, the protein had no biological activity. In the key step of the experiment, the unfolded or denatured protein was incubated under a variety of empirically chosen near-physiological conditions of pH, salt concentration, and temperature. When the right conditions were found, the random-coil polypeptide chain refolded and regained full biological activity over a period of

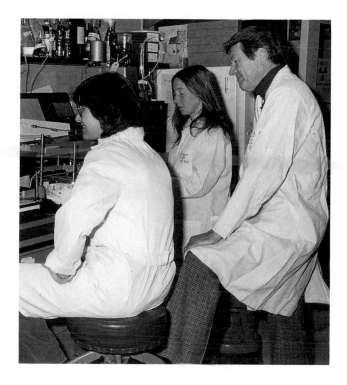

Christian Anfinsen (1916–) in his laboratory, with two students. For his demonstration that linear protein chains are inherently capable of folding into biologically active enzymes, and for his introduction of the techniques of protein denaturation, Anfinsen shared the 1972 Nobel prize in chemistry.

hours. The significance of this experiment was its definitive demonstration that *the bonding capacities of the amino acids as they occur in the linear polypeptide chain are sufficient to fold a newly synthesized amino acid chain into an architecturally complex and biologically active protein.* No outside agency or force in the cell is needed.

Perspective

The working out of the structure of enzymes like chymotrypsin was a milestone in the study of biological chemistry. The invisible agents of change—bringing about the

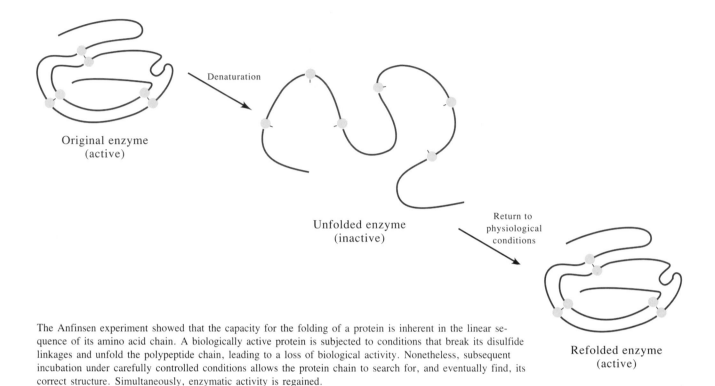

Original enzyme
(active)

Denaturation

Unfolded enzyme
(inactive)

Return to
physiological
conditions

Refolded enzyme
(active)

The Anfinsen experiment showed that the capacity for the folding of a protein is inherent in the linear sequence of its amino acid chain. A biologically active protein is subjected to conditions that break its disulfide
linkages and unfold the polypeptide chain, leading to a loss of biological activity. Nonetheless, subsequent
incubation under carefully controlled conditions allows the protein chain to search for, and eventually find, its
correct structure. Simultaneously, enzymatic activity is regained.

chemical transformations upon which all life depends—
had finally acquired a precise geometrical reality. Though
still invisible in a literal sense, enzymes could now be
precisely seen by those scientific instruments that are extensions of the human eye and mind.

When we think back to Fischer's lock-and-key hypothesis and the structure–function relationship for enzyme action that it implies, we realize that we now have a
detailed knowledge of the structure of both the lock and
the key components—in this case, the chymotrypsin enzyme and the target protein chain that it will cleave. This
knowledge in itself, however, does not allow us to discern
the enzyme's mechanism of catalysis. Even after a careful
analysis of the structure of chymotrypsin, it is not clear

why this molecule has the capacity to promote a specific
chemical reaction. Somewhere locked within its structure
is the secret of the enzyme's activity—but that secret is
not displayed in an obvious way. Nonetheless, with our
knowledge of enzyme structure, this is the problem we are
now prepared to solve.

In the next chapter we will explore the surface of
chymotrypsin, searching for the critical area of the enzyme that constitutes the active site where catalysis occurs. Once we have located this site, we will be able to
design a set of experiments to determine how the amino
acids in this region induce a target substrate molecule to
undergo reaction. We are now at the point of identifying
the real "vital force" of the enzyme.

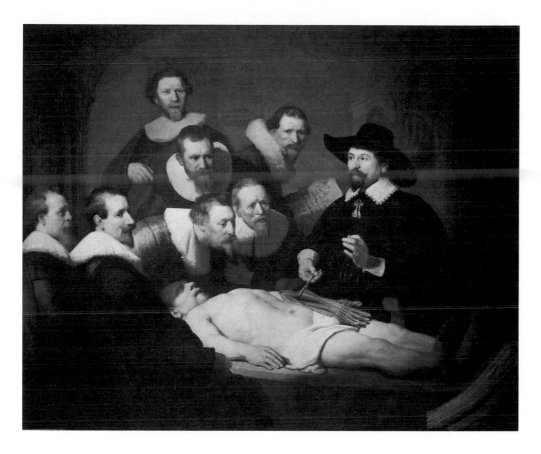

The relationship between structure and function is the key to understanding living systems. In this famous 1640 painting by Rembrandt, *The Anatomy Lesson of Dr. Tulp*, the members of the medical guild of Amsterdam are observing the early exploration of the structure of the human body. Traditionally, anatomy was studied at the level of organisms. But, as the molecular dissections of chymotrypsin in this chapter demonstrate, it is now possible to study anatomy at the molecular level, and all of the cell's important molecules, including its enzymes, are now being precisely characterized.

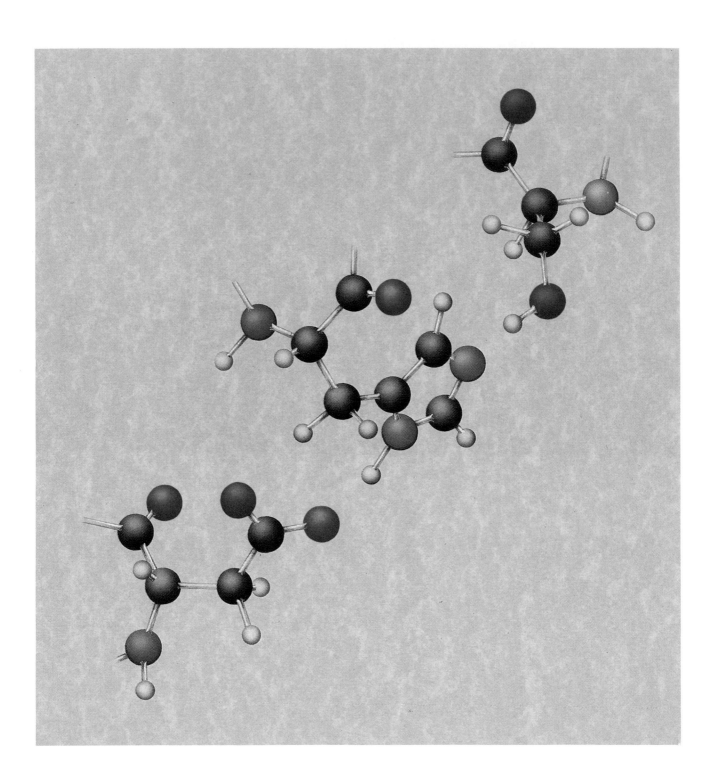

5

The Physiology
of an Enzyme

The catalytic triad of chymotrypsin—
Serine-195, Histidine-57, and Aspartic
acid-102—lies at the heart of the enzyme's active site.

S ince an enzyme emerges unchanged from the reaction it catalyzes, we do not directly see *how* it works. We see only *what*
the enzyme has done, by comparing the nature of the starting
material to the products. Of course, what we would really like
to know is the *precise mechanism* by which the enzyme brings about the
reaction.

Using chymotrypsin as our window onto the world of enzymes, we
took our first step toward understanding the relationship between enzyme
structure and function in the last chapter, when we determined the exact
architecture of chymotrypsin. As a result of this lesson in molecular anatomy, we could see the way in which the protein was held together in a
precise, three-dimensional structure by a network of bonds between the
amino acid subunits. However, we could not directly identify that special
part of the enzyme, known as the *active site*, at which the chemical
reaction actually takes place. Logic tells us that this site must be confined
to a small area on the surface of the enzyme: chymotrypsin is immense,
while the target peptide bond it will break lies in a thin thread of protein
only a few angstroms in diameter. It follows that only a few amino acids
of the enzyme can actually be in direct physical contact with the substrate. These amino acids, however, will play the crucial role. They will
hold the substrate in a lock-and-key configuration with the enzyme and,

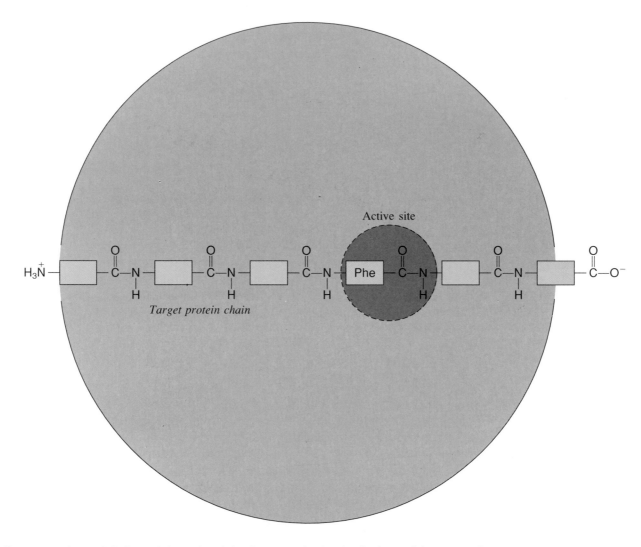

Enzymes are characteristically much larger than their substrates, and only a localized part of the enzyme, the active site, is directly involved in catalysis. In this schematic representation, an unwound (denatured) protein chain is about to be cleaved at one of its phenylalanine (or, alternatively, tyrosine or tryptophan) subunits by chymotrypsin.

ultimately, promote the chemical changes that occur in the substrate. To correctly identify the active-site amino acids and learn how they are responsible for catalysis is the final and most difficult peak scaled by the protein chemist studying the mechanism of enzyme action.

Approaching the Active Site

We will begin with an analysis of the chemical problem faced by chymotrypsin when it undertakes the cleavage of a peptide bond in a target protein chain. The task of the enzyme is to disrupt an ordinarily stable C—N bond and allow the carbon and nitrogen atoms that form it to interact instead with OH^- and H^+ groups, which are derived from water.

This reaction is a *hydrolysis*—literally, the breakage of a chemical bond by water. As can be seen above, the reactive components of water (H^+ and OH^-) are added to the N—H and C=O groups of the peptide bond, allowing the formation of free amino and carboxyl groups.

Since water contains hydrogen and hydroxide ions—although at a low concentration of one in 10 million—we might therefore suppose that the hydrolysis of peptide bonds would occur spontaneously. In fact, this is the case. Even without chymotrypsin present, hydrolysis can be initiated when an OH^- (derived from the dissociation of a water molecule: $H_2O \rightarrow H^+ + OH^-$) uses its free electrons to attack (reach out and establish a new covalent bond with) the C=O group of the peptide linkage.

The attack occurs because of an inherent property of the target C=O group. As discussed on page 131 in Chapter 4, when carbon and oxygen share electrons they do so unequally, with the electrons spending more of their time in the vicinity of the electronegative oxygen atom. This leaves the carbon with a slight deficit of electron density and therefore a partial positive charge (δ^+).

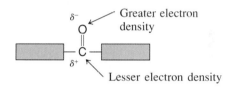

Its electron deficit makes the carbon atom in the peptide bond particularly susceptible to an attack by any chemical group with unshared electrons—that is, a *nucleophile* (from the Greek, a group seeking a positive center). In non-enzymatic hydrolysis, the *nucleophilic attack* comes from a hydroxide ion (OH^-), and as it proceeds, a nascent COOH group develops.

For the nucleophilic attack to succeed, one pair of electrons originally associated with the carbonyl carbon

The first modern insights into physiology—the analysis of how biological systems function—came about 300 years ago through studies carried out at the level of organisms. The painting shows the English physician William Harvey, noted for his pioneering studies on the circulation of the blood, describing his work to King Charles I, circa 1640. Today we are able to explain physiology in molecular terms, including the way in which important cellular molecules, such as enzymes, function. In this chapter, we explore the molecular physiology of chymotrypsin.

atom must retreat from that atom, allowing the carbon to retain its normal complement of four covalent bonds. Thus, in response to the incoming OH$^-$ group, one pair of electrons in the C=O double bond initially retreats to the oxygen atom, giving this atom a formal negative charge. This electron rearrangement also causes the initially planar peptide bond region to take a structurally very different form—a *tetrahedral configuration,* in which the four bonds radiating from the target carbon extend outward to the corners of a tetrahedron. The tetrahedral intermediate is shown below, although, for convenience, in a planar format.

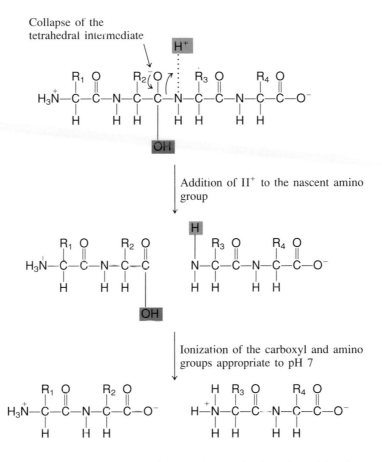

The final event in peptide-bond hydrolysis is the acquisition of H^+ by the electrons that formerly participated in the peptide C—N bond.

The tetrahedral intermediate—with its oxygen atom carrying a full negative charge—is unstable and short-lived. It has two possible fates: the exiled electrons must eventually return from the oxygen atom, and they can either repel the attacking nucleophile or force the withdrawal of the electrons that form the peptide (C—N) bond. Thus, in the final step of a successful hydrolysis reaction, the peptide bond breaks and the pair of electrons forming the C—N bond become completely associated with the nitrogen atom. As these electrons then acquire an H^+ (de-rived from the dissociation of water), the N—H group develops into a free amino group, completing the cleavage of the peptide bond.

The Need for an Enzyme

Although the non-enzymatic reaction we have just considered generates the same products as are produced when the peptide-bond cleavage is catalyzed by chymotrypsin, the two reactions differ in one important respect. The enzymatic reaction occurs about a billion times more rapidly, proceeding at a rate of about 100 peptide bonds cleaved per second per molecule of enzyme. This difference is important—in fact, it is what enzymes are all about. The problem is to appreciate the significance of a billion in an age in which such numbers are often used but seldom appreciated. Perhaps this will help: one billion hours ago, our ancestors lived in caves. If a cellular device could be found that would speed up progress a billionfold, it would be as remarkable as a time machine. In fact, this is exactly what enzymes do.

The acceleration of the rate of chemical transformations is not, of course, a problem faced by chymotrypsin alone. All of the physiological processes of life require that ordinarily stable molecules be induced to undergo whole series of chemical reactions at rates far higher than would occur spontaneously. In the absence of such rate acceleration, life would not merely be slow, it would be nonexistent.

The Energy Barrier

The first key to unlocking the secret of the enzyme is to appreciate why the non-enzymatic breakage of a peptide bond occurs so slowly. One reason, of course, is the low concentration of free hydrogen and hydroxide ions in water. At pH 7, the concentration of these ions is only 10^{-7} M; that is, only one water molecule in 10^7 is dissociated into H^+ and OH^- at any one time.

But an even more important reason for the inherent slowness of the non-enzymatic reaction turns on the issue of energy. That the uncatalyzed reaction can occur at all is only because the energy of the final state, a free COOH group and a free NH_2 group, is lower than the energy of the initial state, a peptide bond (see the figure on the facing page). But before the reacting peptide bond can undergo the necessary changes to form the products and achieve the more favorable energy state, it must first pass through one or more *transition states*. These are intermediate structures in which new chemical bonds are being formed and old chemical bonds are being broken. The

tetrahedral intermediate that occurs in the middle of peptide bond hydrolysis is a particularly well-defined transition state.

The important point is that transition states are high-energy configurations. In these structures, the electrons involved in bonds that are being formed or broken temporarily occur in more strained and improbable—and therefore energetically unfavorable—arrangements. At any one moment, only a few reactant molecules will, by virtue of random vibrational motion, have enough energy to exist in the transition state. Yet only these molecules have the capacity to undergo a chemical reaction leading to a lower final energy state. So, even though the products of the hydrolysis reaction ultimately exist at a lower energy level than the reactants, the reaction is prevented from proceeding rapidly by the existence of a *transition-state barrier*. The fact that in order to react virtually all molecules must pass through a high-energy transition state explains why the world is populated by essentially stable molecules.

Catalysis in the World of the Chemist

Given these problems, how do enyzmes function? Our first clue about how enzymes work comes from the ways that laboratory chemists have found to speed up chemical reactions. Two standard treatments lead to a moderate increase in reaction rate: the first of these is elevated temperature and the second is the addition of acid or base.

The use of heat is the most common method laboratory chemists use for accelerating ordinary chemical reactions. Typically, for each successive 10°C increase in temperature, there is a twofold increase in reaction rate. The effectiveness of elevated temperature lies in the extra energy it introduces into the system. This energy is harbored in the increased translational, vibrational, and rotational motions of molecules (see page 58). The average molecule has a higher kinetic energy, and this has two beneficial effects in promoting reactions. There is an increase in the frequency of collisions between molecules, which is important for reactions involving two or more species. And in addition, the higher energy content of the mole-

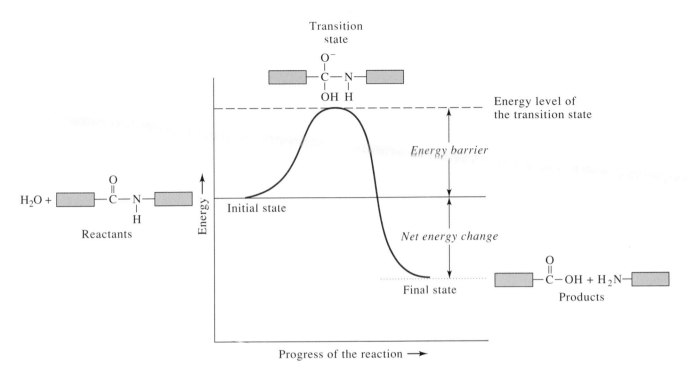

Molecules must pass through high-energy transition states during their reaction. Thus, although a favorable net energy change drives the reaction, the reaction is slowed by an energy barrier.

cules allows more of them to attain the transition-state structure required for reaction. Thus, in terms of a protein chain undergoing hydrolysis, as the various atoms vibrate more rapidly and extensively at higher temperature, there will be a greater percentage of peptide bonds in which the carbon, nitrogen, and oxygen atoms, and the electrons they share, occur in the tetrahedral transition state (as in the figure above). The result is that more molecules will be capable of successfully interacting with H^+ and OH^- ions from the surrounding water solution and undergoing reaction.

The second general approach taken by laboratory chemists to increase the rates of chemical reactions is to apply acid or base. Looking at the mechanism of peptide-bond hydrolysis, as shown on pages 155 to 157, it is apparent that an OH^- is essential for the first step in the reaction—the attack on the carbonyl carbon. One might

therefore expect that the reaction could be accelerated by adding OH^- ions to the system at a much higher level than they normally occur—that is, much higher than the 10^{-7} M concentration found in water. Increasing the concentration of OH^- is easily accomplished by adding a base, like NaOH, which immediately dissociates (NaOH $\rightarrow$ Na^+ + OH^-), and, indeed, this approach is successful—speeding up the hydrolysis reaction about 10- to 100-fold.

A similar increase in the rate of peptide-bond hydrolysis can be achieved with acid. The role of H^+ ions in the hydrolysis reaction is already familiar to us from page 157, and it is not surprising that an increased concentration of H^+ is able to help with a difficult part of the reaction—the acquisition of H^+ by the departing amino group.

Heat, or the use of acid or base, may be useful to the laboratory chemist as a catalyst, but they would be not be very useful to the cell. As the temperature is raised above the physiological range (for example, about 98.6°F or 37°C for humans), the increase in thermal motion begins to disrupt the complex network of noncovalent bonds that hold the cell's molecules, including its enzymes, in their proper three-dimensional conformations. As the enzyme molecules partially unfold, they begin to lose their specific geometry, and therefore their lock-and-key fit with their substrates. Indeed, if the temperature is raised to too great an extent, the enzymes will completely unfold, or denature, and lose their activity entirely. In the living cell, such denaturation would occur long before elevated temperature would have any significant effect in promoting chemical reactions (see graph on top of page 64).

Similarly, although a high concentration of acid or base would increase the rate of peptide-bond hydrolysis in a test-tube reaction, in the cell a variety of other problems would immediately follow. The altered pH would change the ionization state, and therefore the stability, of numerous cell molecules, including enzymes, and interfere with metabolism in general (see graph on bottom of page 64). Furthermore, the use of free acid or base as a catalyst would lack the specificity seen in an enzymatically catalyzed reaction—all peptide bonds would be subject to cleavage (not just those next to phenylalanine, tyrosine, and tryptophan), and similar bonds in other cell molecules would also be broken.

Catalysis in the World of the Cell

To sum up our discussion thus far, the conditions we have considered—elevated temperature and the use of acid or base—have given us only increases of 10-fold to 100-fold in reaction rate. However, when we compare the hydrolysis of a peptide bond in the presence of chymotrypsin with the rate of the corresponding non-enzymatic reaction, we find not merely a 100-fold, but a 1,000,000,000-fold acceleration in reaction rate. Moreover, *when we speak of enzymatic catalysis, as in peptide-bond cleavage by chymotrypsin, we mean not only that the reaction is enormously accelerated over the spontaneous rate, but also that it is restricted to well-defined areas in appropriate target molecules.* Our problem, then, is to determine how

the enzyme is able to both accelerate and bring specificity to a particular reaction without using elevated temperature or resorting to H^+ or OH^- concentrations that are too far from the neutral pH of the cell environment. If we can learn how the enzyme causes this enormous additional increase in reaction rate, and how it achieves substrate specificity, it will allow us to deepen our thinking about chemical reactions from the level of general chemistry to biochemistry. Moreover, we will be able to appreciate the power that enzymatic catalysis brought to nature as chemistry crossed over the threshold of life and biochemical systems evolved.

Inasmuch as elevated temperature and free acid or base would be injurious to the living cell, it is evident that enzymes must use their own devices to accelerate chemical reactions. And of course, the key to these devices must lie within the enzyme's amino acids, since there is no other physical source for the enzyme's catalytic vital force.

Might the enzyme be able to use any variation on the catalytic power of acids and bases to accelerate chemical reactions? This is an intriguing idea because of the properties of some of the amino acid side chains. As we have seen, although most amino acids have unreactive hydrocarbon side chains of various shapes and sizes, and are designed to serve structural or space-filling functions, the side chains of some amino acids in fact have the capacity to function as acids or bases. Among this group are arginine, aspartic acid, cysteine, glutamic acid, histidine, lysine, serine, and tyrosine. These amino acids have in common the ability of their side chains to gain or lose a hydrogen ion (see the figure on the facing page). And after all, the functional definition of an acid is any compound that can donate a hydrogen ion and, correspondingly, a base is any compound that can accept a hydrogen ion. Thus, our knowledge of the fundamental principles of chemistry may provide our first insight into how amino acids could indeed promote chemical reactions by some version of acid-base catalysis. But as we will see, this aspect of enzyme function will involve a surprising degree of sophistication because it requires the synergistic cooperation of a number of amino acids in the active site.

As we set out to explore the active site, we might also wonder whether the enzyme is able to use any variation on

Amino Acid	Structure	pH at which H⁺ dissociates

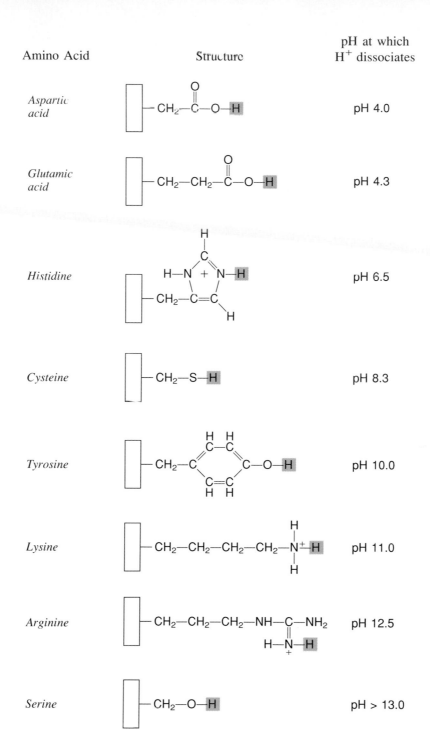

		pH at which H^+ dissociates
Aspartic acid		pH 4.0
Glutamic acid		pH 4.3
Histidine		pH 6.5
Cysteine		pH 8.3
Tyrosine		pH 10.0
Lysine		pH 11.0
Arginine		pH 12.5
Serine		pH > 13.0

Amino acid side chains as acids and bases. The structures show acidic (containing H^+) forms of aspartic acid, glutamic acid, histidine, cysteine, tyrosine, lysine, arginine, and serine. When the hydrogen in blue dissociates, the base form of the amino acid side chain results. The right-hand column shows the pH at which the change in ionization state occurs.

the rate-accelerating power of elevated temperature to promote chemical reactions. There is, in fact, no physical embodiment of elevated temperature in the side chains of amino acids. But, as we also will see, through its own remarkable devices, the enzyme is able to use amino acids to achieve the same catalytic benefits that would be derived from heat. And once again—as in the case of acid and base side chains—this catalysis is achieved without endangering the cell.

Our study of the reaction mechanism of chymotrypsin has ultimately come down to a search for those specific amino acids of the amino acid's *active site* that are directly involved in promoting the cleavage of a peptide bond. We wish to know what these amino acids are and how they function. As these amino acids are elucidated one by one, the enzyme's catalytic mechanism will come into focus.

By the end of this chapter, it will be clear how certain key amino acids of the enzyme, in the environment of the active site, work cooperatively to give the enzyme its special power—how they allow it to accelerate a specific transformation to a level that is a billionfold higher than would occur spontaneously, a level necessary to sustain life.

The analysis we will describe was, of course, carried out to define the mechanism of action of chymotrypsin, but the strategies involved are general ones and are used by enzymologists to study the active sites in numerous other enzymes.

The Involvement of Histidine-57

The histidine subunit at position 57 is one of the key amino acids in the chymotrypsin active site. This amino acid was identified as part of the active site by means of a time-honored technique that is often used in active-site analysis. Quite simply, the activity of the enzyme is measured at various pH values. It is expected that, as the pH is changed over a wide range of H^+ and OH^- concentrations, certain amino acid side chains will be altered, leading to a change in enzyme activity.

When such a *pH profile* was determined for chymotrypsin, it was found that the enzyme cleaved peptide

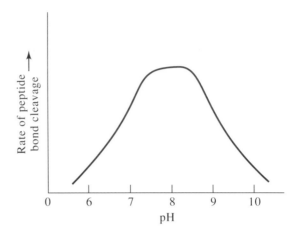

The rate of peptide-bond cleavage by a fixed amount of chymotrypsin, at various pH values. Each one-unit increase in pH corresponds to a 10-fold decrease in hydrogen ion concentration. And as the concentration of H^+ moves away from the optimal value (pH 7–9), certain amino acid side chains gain or lose a dissociable hydrogen, leading to a change in enzyme structure and function.

bonds most rapidly as the pH was raised from 6 to 7, and that its activity quickly fell off as the pH rose much above 9. This result, obtained almost half a century ago, opened the study of the enzyme's active site. It suggested that two amino acids that undergo a change in ionization state—one at about pH 7 and the other at about pH 9—each play an important role either in determining the overall structure of the enzyme or in carrying out the function of the active site itself.

Of the 20 amino acids in proteins, only one changes its ionization state around pH 7. This is histidine, which

pH 6   pH 7

changes its charge from positive to neutral when its side chain loses a hydrogen ion as the pH is raised from 6 to 7. The same reaction is shown below in an abbreviated notation that emphasizes only the two nitrogen atoms and the ring structure of the histidine side chain:

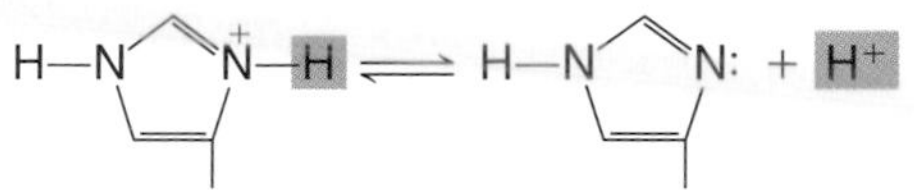

pH 6   pH 7

The pH profile can imply, but it cannot prove, the presence of a particular amino acid at the active site. That a histidine subunit is, in fact, present in the chymotrypsin active site was shown directly by Eliot Shaw in the 1960s. He developed an approach known as *affinity labeling*, or *active-site labeling*, which relied on a small laboratory-synthesized molecule, *tosyl-phenylalanine chloromethylketone*. This molecule was designed to possess two particularly useful properties. First, it was structurally related to the normal substrate of the enzyme—note how the high-

Tosyl group

Chemically active area

Phenylalanine-like component

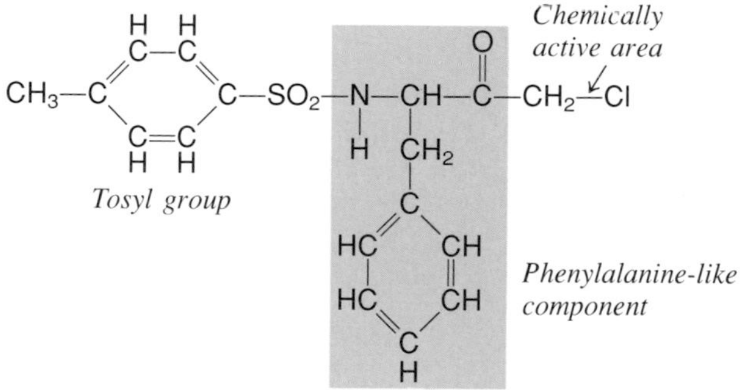

The structure of tosyl-phenylalanine chloromethylketone, a molecule for labeling the active site of chymotrypsin. This molecule is structurally analogous to the normal substrate of chymotrypsin (because of its phenylalanine component) and also incorporates a chemically reactive group (the CH_2—Cl component). The rather odd tosyl group was used because it made the laboratory synthesis easier. In the reaction of tosyl-phenylalanine chloromethylketone with the chymotrypsin active site, chlorine is eliminated, and the remainder of the affinity label becomes covalently linked to the nearby side chain of Histidine-57.

lighted area in the figure below resembles a phenylalanine subunit in a protein chain. Second, the synthetic molecule contained a chemically reactive group: CH_2—Cl. Thus the affinity-labeling reagent could be expected to (1) bind specifically to the enzyme's active site, and (2) upon binding, use its reactive group to form a covalent bond with a nearby amino acid side chain, thus "labeling" an amino acid in the enzyme's active site.

When tosyl-phenylalanine chloromethylketone was added to a solution of chymotrypsin, the desired coupling reaction quickly occurred, and subsequent analysis showed a covalent linkage between the added compound and one of the enzyme's two histidine subunits. The analysis of enzymatically produced peptides showed that it was the histidine at position 57, and not the histidine at position 40, that had become tagged.

It may have already occurred to the reader that it is particularly intriguing that a histidine subunit should be located in the enzyme's active site. As we saw in the figure on page 161, it is the only amino acid able to change its ionization state in the neutral pH environment of the cell. In the range of pH 6 to pH 7, its side chain is poised either to gain or to lose a hydrogen. The equilibrium for this change is such that at pH 6 the majority (about 90 percent) of the histidine subunits carry the extra hydrogen; just the opposite is true at pH 7 (90 percent of the histidines have given up their dissociable hydrogen). An equally valid and often more useful way to think about the ionization of histidine is that at pH 6 the molecule carries an extra hydrogen 90 percent of the time; at pH 7, however, the hydrogen is present only about 10 percent of the time. *The important consequence of this behavior is that histidine is potentially able to serve as an acid or base at the pH maintained by living systems.*

We discussed earlier how free acids and bases promote chemical reactions by accepting or donating hydrogen ions, and it is a simple matter to demonstrate that pure histidine has similar catalytic properties. If the side chain of histidine, called imidazole, is added to a solution containing a molecule with a peptide (or peptidelike) bond, it will increase the rate of hydrolysis about 10-fold. In the figure on the following page, the area shaded in red shows the side chain of histidine serving as a *general base catalyst*. An unshared pair of electrons on a nitrogen atom can

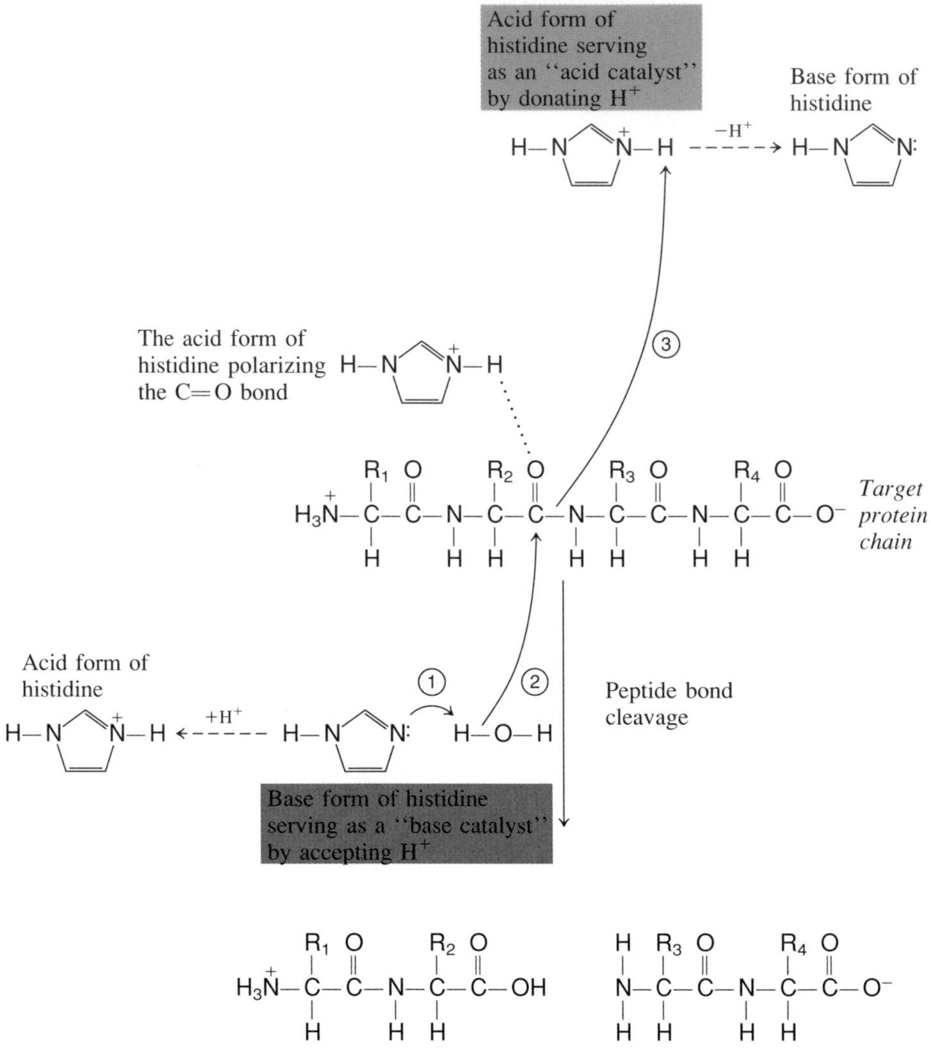

The use of amino acid side chains as physiological acid and base catalysts. The figure shows (in red) histidine promoting the hydrolysis of a peptide bond by serving as a general base catalyst: it promotes the ionization of water by accepting H^+ from H_2O. The generated OH^- attacks the carbon of the peptide bond. Histidine is also able to serve as a general acid catalyst (shown in blue) by donating a hydrogen to the departing amino group. In its acid form, histidine may also function as an electronic catalyst by hyperpolarizing the target $C=O$ group and making it more subject to nucleophilic attack. The important point is that histidine—unlike free acid or base—is a physiologically acceptable molecule that can function without changing the pH of the cell away from neutrality.

accept a hydrogen ion (H^+) from water, leaving a hydroxide ion (OH^-) behind in solution. The OH^- then attacks the carbonyl ($C=O$) carbon of the peptide bond.

It is at this point that the special properties of histidine as a potential catalyst at neutral pH can best be appreciated. At the same time that one histidine molecule is serving as a general base catalyst, another molecule can serve as a *general acid catalyst*. This molecule will be a histidine whose side chain carries an extra H^+. As shown in the blue shaded area at the top of the figure, the positively charged histidine is able to release an H^+ into solution. This hydrogen can be used in peptide-bond hydrolysis to donate H^+ to the developing amino group.

The important point is that there is a strong analogy between the acid and base catalysis of peptide bonds discussed in connection with the figures on pages 155–157 and the catalysis shown by histidine on page 164. In both cases, we see (1) the abstraction of H^+ from a water molecule, leaving an OH^- nucleophile behind in solution, and (2) the donation of H^+ to the leaving amino group. It is evident that there are at least two ways in which histidine could, in principle, play a catalytic role in the active site of chymotrypsin.

Indeed, both these possibilities were considered in the wake of the experiments showing that histidine was present in the active site. However, it is generally true in the study of enzymes that one cannot predict a complete reaction mechanism from first principles. The strategy of catalysis usually involves several amino acids working cooperatively in the active site to effect catalysis. Thus, not surprisingly, even while the pH and affinity-labeling experiments were showing the involvement of Histidine-57 in the active site of chymotrypsin, other experiments were leading to the conclusion that a second amino acid also played a crucial role.

Before considering this second amino acid, however, let us finish our discussion of the pH activity curve in the graph on page 162. Why does the activity of chymotrypsin fall off as the pH is raised above 9? This is a case where a change in the ionization state of an amino acid causes a major change in the structure of an enzyme. Physical studies have shown that in chymotrypsin an important change occurs at about pH 9 in Isoleucine-16, the first amino acid of the chymotrypsin B chain. The pure hydrocarbon side

chain of Isoleucine-16 is unaffected by pH. But because the subunit falls at the beginning of a chain, its free amino group does have a hydrogen ion that can dissociate. As the pH changes from 9 to 10 and the H^+ concentration falls by an order of magnitude, this amino group gives up a hydrogen ion to the surrounding solution and changes its charge from positive to neutral.

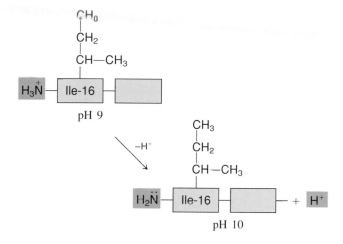

The loss of the single positive charge from this precise point in the protein has a dramatic effect on the structure of the enzyme. Normally, the positive charge on Isoleucine-16 interacts with a complementary negative charge carried on the side chain of Aspartic acid-194.

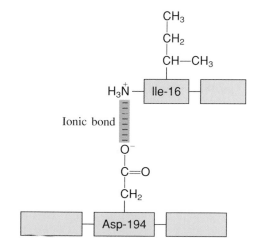

The resulting ionic bond holds the end of the B chain in position on the surface of the enzyme. Without the ionic bond, the end of the B chain swings free. Thus, with the loss of the ionic bond above pH 9, the enzyme loses an important aspect of its structure and becomes inactive. Indeed, the positive charge on Isoleucine-16 plays a key role in the activation of chymotrypsin, following its synthesis as an inert precursor protein by the cells of the pancreas, as will be explained in Chapter 6.

The Involvement of Serine-195

The involvement of a second amino acid, Serine-195, in the active site of chymotrypsin came into focus in a way totally different from the role of Histidine-57. The story begins in the laboratory of Dr. Willi Lang in Germany just after World War I. Chemists preparing a new group of phosphorus-containing organic compounds, such as diisopropyl phosphofluoridate, became disoriented and developed a constriction of the larynx and a painful inability to focus the eye.

There appeared to be a paralysis of striated muscle, among other effects. And indeed, years later, in several countries, these very compounds were developed into powerful nerve gases, as will be discussed in Chapter 7.

Our present interest in the organophosphates is less diabolical. When experiments with these synthetic compounds were begun in the 1940s to discover the basis of their toxic effect, it became clear that the compounds inhibited a number of enzymes, including chymotrypsin. For example, upon incubation in the presence of diisopropyl phosphofluoridate (or DPF), chymotrypsin was rapidly and irreversibly inactivated. Chemical analy-

sis of the inactivated enzyme by acid hydrolysis, which reduces the protein to its amino acid subunits, showed that the diisopropyl phosphate group had become covalently attached to one of the enzyme's serine residues. Furthermore, by subjecting the chymotrypsin treated with DPF to *limited* hydrolysis with proteolytic enzymes, the sequence of amino acids surrounding the modified serine could be established as Gly, Asp, *Ser*, Gly. This result localizes the reactive serine to the region of amino acids 193 to 196 in the linear sequence of chymotrypsin (page 114). Although there are 27 other serines in chymotrypsin, significantly, none of them reacts with DPF.

Now, many enzymes are destroyed by exposure to one or another reactive chemical. Usually their inactivation results from nothing more than a chance contact between the small molecule and a particular amino acid that happens to be exposed on the surface of the enzyme and is not informative in terms of the enzyme's mode of action. In the case of the reaction between DPF and chymotrypsin, however, evidence began to accumulate that the inactivation *was* related to the enzyme's catalytic mechanism.

1. DPF does not react with denatured (unwound) chymotrypsin. Thus, Serine-195 is not reactive as an individual amino acid, but only in the context of the properly folded protein chain. Since protein folding is intimately related to the formation of the active site and the development of catalytic activity, the most straightforward interpretation of this result is that

Diisopropyl phosphofluoridate reacts with a specific amino acid, Serine-195, in chymotrypsin. As the result, the enzyme is inactivated.

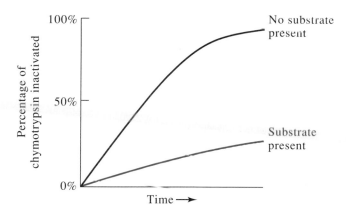

Chymotrypsin is protected from DPF by the presence of normal substrate.

Serine-195 is uniquely reactive because of its inclusion in the enzyme's active site.

2. A second experiment provides *direct* evidence that Serine-195 is in fact located in the enzyme's catalytic center. If one attempts to inactivate chymotrypsin

with DPF, *but in the presence of a normal substrate for the enzyme* (a protein chain), the inactivation is slowed. The most straightforward interpretation of this result is that the occupation of the active site by the normal substrate interferes with the access of DPF to chymotrypsin, implying that both the substrate and the inactivating chemical compete for the same small area on the surface of the enzyme.

What can the inactivation of chymotrypsin by DPF—a reaction that at first sight appears to be so far removed from the hydrolysis of a peptide bond—tell us about the mechanism of catalysis by chymotrypsin? The answer begins to become clear when we consider the basic chemistry of the DPF reaction, which can be inferred from the structure of the final product. As shown in the reaction at the bottom of the page, for DPF to become covalently coupled to Serine-195, electrons from the serine hydroxyl group must mount a nucleophilic attack on the phosphorus of the small molecule, forming a new covalent bond and displacing the fluoride ion. This attack occurs because the phosphorus atom carries a partial positive charge (δ^+)—the result of its unequal sharing of electrons with oxygen. The important point is that DPF has in common with the

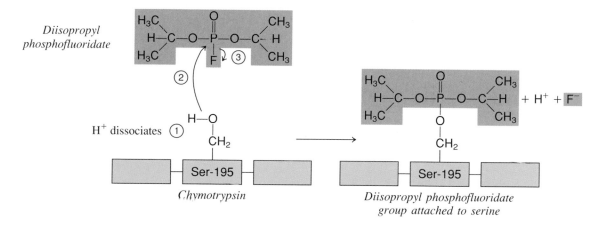

The reaction between chymotrypsin and DPF. The side chain of Serine-195, using a pair of electrons associated with its oxygen atom, initiates a nucleophilic attack on the phosphorus of the small molecule, displacing the fluoride ion. This results in the formation of a covalent bond between the enzyme and the diisopropyl phosphate component.

normal substrate of chymotrypsin (a peptide bond) the presence of an atom (the phosphorus) with a relative positive charge. Normally, chymotrypsin catalyzes a nucleophilic attack on the partially positively charged carbon atom of a C=O group, "displacing" an amino group with its attached polypeptide chain. But when DPF is the target, it is a phosphorus atom that is attacked and a fluoride ion that is displaced. *Thus, the organophosphate compound can be considered a pseudosubstrate—that is, a molecule that reacts with the enzyme in a manner mechanistically analogous to that of a true substrate.* The basic difference is that with DPF the reaction is never completed; although the target molecule is cleaved into two parts, one part remains bound to the enzyme, permanently inactivating it.

The Isolation of an Intermediate during the Chymotrypsin Cleavage Reaction

The reaction between chymotrypsin and DPF strongly suggests that a particular amino acid—Serine-195—is located in the enzyme's active site and is able to carry out a nucleophilic attack, potentially related to the attack that leads to the hydrolysis of a peptide bond. Still, a twinge of uncertainty surrounds this conclusion because the method by which Serine-195 is detected as an active-site amino acid is inherently somewhat abnormal—involving as it does a highly unusual laboratory-produced compound— and might not accurately reflect the actual mechanism of peptide-bond cleavage. Specifically troubling is the fact that the pseudosubstrate becomes *permanently* attached to Serine-195 and as a consequence inactivates the enzyme. This differentiates DPF from a true substrate and limits our confidence in any inferences we might draw from its behavior.

Given our basic understanding of how an enzyme works—that it catalyzes a reaction and returns to its original state—we would have much more confidence in the relevance of the DPF result if we could see the enzyme use Serine-195 to initiate a similar partial reaction on a model compound, but then continue, complete the reaction, and emerge unchanged.

This is exactly what two British scientists, Brian Hartley and B. Kilby, were able to achieve through the use of the small molecule *nitrophenyl acetate* as a pseudosubstrate. In this case the enzyme is able to work on a compound that is neither too far removed from the real substrate nor too close to it, and, as a result, the reaction occurs in a particularly revealing way.

Nitrophenyl acetate is a small organic molecule that has an *ester bond*, which is closely related in structure to a peptide bond.

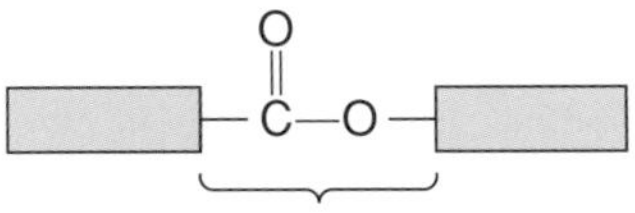

Ester bond

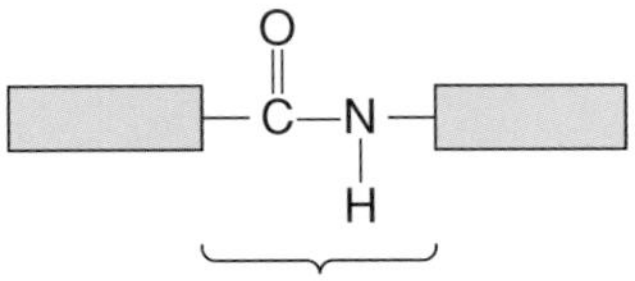

Peptide bond

When chymotrypsin is allowed to work on nitrophenyl acetate, the result is a slow but complete hydrolysis, yielding equal amounts of the two products, acetic acid and nitrophenol.

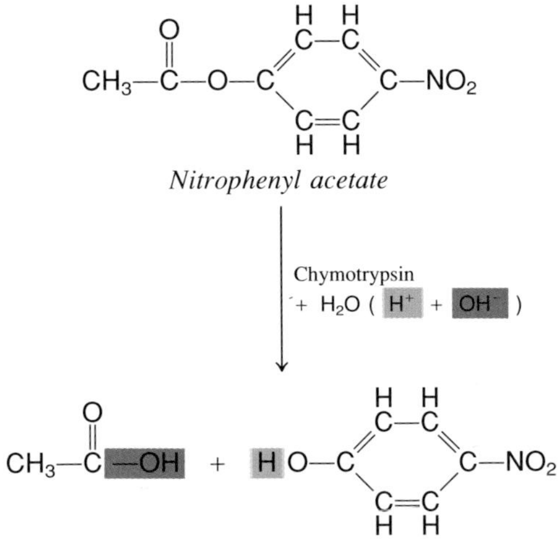

Nitrophenyl acetate

Chymotrypsin
+ H₂O (H⁺ + OH⁻)

Acetic acid   *Nitrophenol*

It is precisely because the reaction is so slow that Hartley and Kilby were able to make their important observation. They found that the two products of the reaction did not arise at the same time. After the enzyme was mixed with the substrate—and before any acetic acid could be detected—there was an immediate production of nitrophenol. During this *initial burst* stage of the reaction, one molecule of nitrophenol was produced for each molecule of enzyme. Thereafter nitrophenol was generated at a much slower, steady rate, along with an equal amount of acetic acid.

In explaining this result, Hartley and Kilby were led to the first general description of how the chymotrypsin active site works. They proposed that chymotrypsin carries out its reaction in two steps. In the first step, the substrate, nitrophenyl acetate, forms a complex with the enzyme and is immediately cleaved into two pieces. One of these pieces (nitrophenol) is released, while the other (*acyl*, or $CH_3CO—$, related to acetic acid) is temporarily retained in the active site of the enzyme.

STEP 1

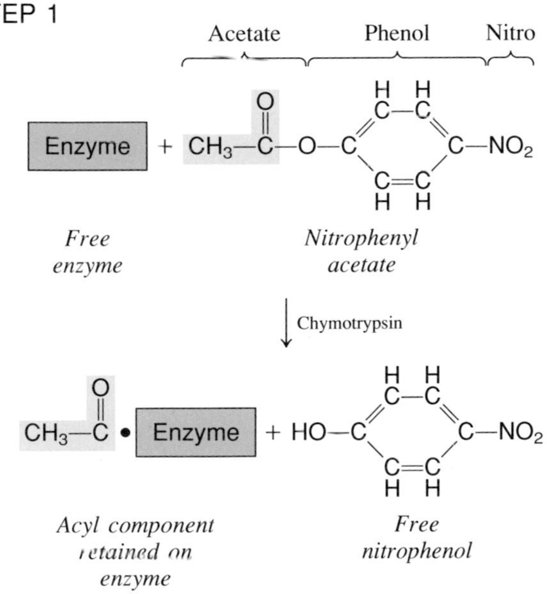

Acetate   Phenol   Nitro

$Enzyme$ + $CH_3—C$(=O)$—O—C$ ... $C—NO_2$

Free enzyme   *Nitrophenyl acetate*

↓ Chymotrypsin

$CH_3—C$(=O) • $Enzyme$ + $HO—C$ ... $C—NO_2$

Acyl component retained on enzyme   *Free nitrophenol*

A second, distinct step completes the reaction. In this step the acyl part of the substrate is released from the enzyme and emerges as free acetic acid (CH_3COOH). This part of

the reaction carries the substrate to its final state of lower energy and also restores the enzyme to its original form.

STEP 2

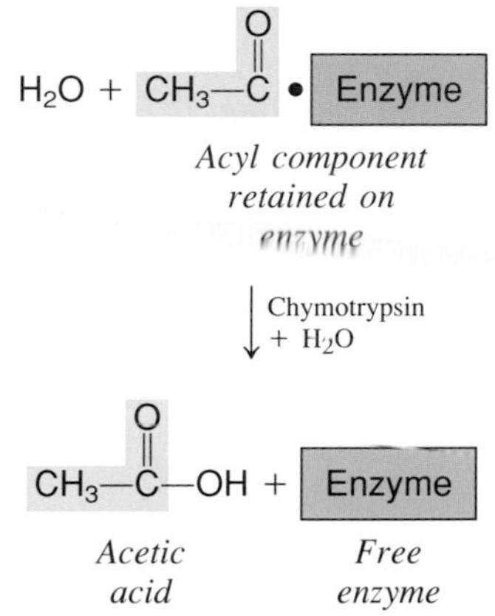

H_2O + $CH_3—C$(=O) • $Enzyme$

Acyl component retained on enzyme

↓ Chymotrypsin + H_2O

$CH_3—C$(=O)$—OH$ + $Enzyme$

Acetic acid   *Free enzyme*

Evidently, the reaction has a rate-limiting step. This is deduced from the shape of the curve in the figure below, which shows that, after an initial burst, nitrophenol pro-

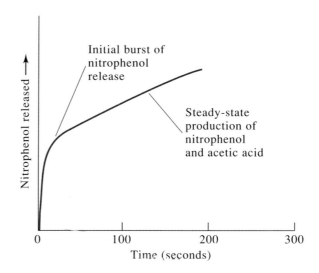

During the cleavage of nitrophenyl acetate by chymotrypsin, nitrophenol is produced quickly at first, then at a slow, constant rate.

duction must proceed at a slower rate. The reason is that the catalytic site becomes available for the binding of new substrate molecules only slowly, at a rate that reflects the removal of the acyl component of the substrate from the enzyme.

In terms of our interest in the mechanism of chymotrypsin action, the important question is the location of the "missing" part of the substrate—the acyl group—which is not initially released from the enzyme. *Hartley and Kilby proposed that in the early stage of the reaction the acyl group becomes attached to one of the amino acids of the enzyme, forming a covalent acyl–enzyme intermediate:*

$$CH_3-\overset{\overset{\displaystyle O}{\|}}{C}-\boxed{\text{Enzyme}}$$

Here the dash denotes the covalent bond between the acyl group and the side chain of an amino acid. Subsequent experiments in the 1960s and 1970s fully confirmed this prediction.

The simplest of these experiments to describe makes use of the slight difference in geometry between the ester bond and a peptide bond. Because of this difference, and the relatively slow removal of the acyl group from the enzyme, the cleavage of nitrophenyl acetate pauses at the intermediate stage of the acyl–enzyme complex. In fact, under certain conditions the reaction can be made to stop

altogether at this point. If one lowers the temperature to $-55°C$, hydrolysis comes to a halt after the initial release of nitrophenol. The acyl part of the substrate remains attached to the enzyme, and neither the release of acetic acid nor the further processing of new substrate molecules occurs. The blocking of the reaction makes it possible to take what is essentially a slow-motion moving picture of the cleavage reaction and put it on freeze-frame. As a result, the acylated form of the enzyme can be caught in midreaction and studied.

Analysis of the acyl–enzyme intermediate supports the conclusion that the CH_3CO- portion of the substrate is, in fact, bound to the hydroxyl group of Serine-195. This, of course, is the same amino acid that reacted with the much less successful substrate analogue diisopropyl phosphofluoridate (DPF). But by using nitrophenyl acetate as a substrate, it is possible to proceed further and obtain evidence that the Serine-195 attachment is in fact meaningful, not just the result of an aberrant side reaction unrelated to the mechanism of catalysis. In a final experiment the temperature is raised and the enzyme comes back to life. The acyl group is severed from the serine side chain and appears as free acetic acid. The enzyme becomes able once again to work on new substrate molecules. Thus, having seen chymotrypsin carry out all the steps of a full, if still a bit artificial, cleavage reaction, we can conclude with confidence that Serine-195 is a central amino acid involved in catalysis.

Clear as it is, this result actually came as a surprise to many investigators. For, on first principles, it appears to

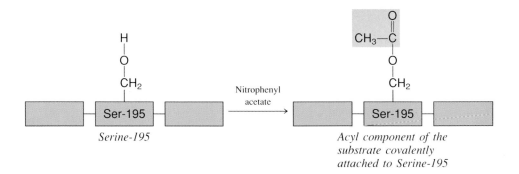

Serine-195   Nitrophenyl acetate   *Acyl component of the substrate covalently attached to Serine-195*

The proposed covalent intermediate formed between chymotrypsin and part of its substrate.

pose a chemical problem. Serine is a poor candidate for mounting the nucleophilic attack needed to initiate peptide-bond hydrolysis. The ionization of the serine side chain to lose H^+ and free up electrons for the required nucleophilic attack seems chemically improbable in the neutral pH environment of the cell. In fact, as we saw in the figure on page 161, a pH of at least 13 must normally exist before there are enough OH^- ions in solution to draw H^+ away from a serine hydroxyl group. Yet experimental evidence clearly indicates that the side chain of Serine-195 *does* mount a nucleophilic attack during catalysis.

We have here one of those classic and most intriguing of conflicts in science—a conflict between theory and fact. How shall we proceed? In the resolution of other problems in science, there is precedent for the solution going either way. The fact could be wrong, or it could be true but misleading. Or the theory could be in need of modification. As we will now see, the nucleophilicity of Serine-195 is genuine, and, indeed, it is the key to the reaction. Once we understand how this key works, we will be able to unlock the catalytic mechanism of the enzyme.

How Chymotrypsin Works: The Reaction Mechanism

A detailed model for how chymotrypsin works was proposed in 1970 by David Blow and his colleagues Brian Hartley, Jens Birktoft, Brian Matthews, and Paul Sigler of the Medical Research Council in Cambridge, England. They had available all of the experimental evidence showing the involvement of Serine-195 and Histidine-57 as active-site amino acids, and they had a knowledge of the crystallographic structure of chymotrypsin, which they had just solved. In addition, of course, they knew the theoretical problem of invoking Serine-195 to carry out the nucleophilic attack that initiates peptide-bond cleavage.

A Catalytic Triad in the Active Site

Looking at the three-dimensional structure of chymotrypsin, Blow and his colleagues first analyzed the positions of

In David Blow's laboratory, Jens Birktoft builds the first model of chymotrypsin based on the X-ray data of Brian Matthews and Paul Sigler.

the two amino acids known to be involved in catalysis. Serine-195 was found on the surface of the enzyme at the rim of a pocket formed by an infolding of the polypeptide chains. Importantly, although Histidine-57 was quite far away in the linear amino acid sequence of chymotrypsin, it was actually adjacent to the critical serine after protein folding. (See the figure on page 172.) And, significantly, one of the nitrogens of the histidine ring was only about 3 Å away from the oxygen atom of Serine-195, indicating that these two amino acids might be hydrogen bonded.

Why is the potential formation of a hydrogen bond between Histidine-57 and Serine-195 so important? Its significance lies not just in its contribution to the network of bonds holding the protein together, but in its ability to indirectly affect the chemical reactivity of Serine-195. The effect of the hydrogen bond is to begin to draw the hydrogen away from the OH group of the serine side chain, leaving the oxygen atom of Serine-195 with a relatively high electron density, which could be used to mount a

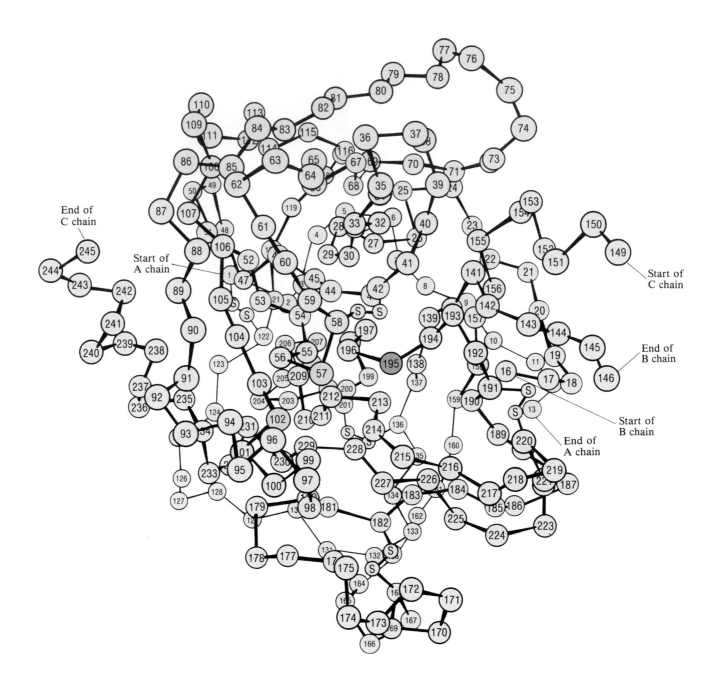

The proximity of Serine-195 and Histidine-57 in the folded structure of chymotrypsin. For reasons that will become clear in just a moment, the adjacency of a third amino acid, Aspartic acid-102, is also shown.

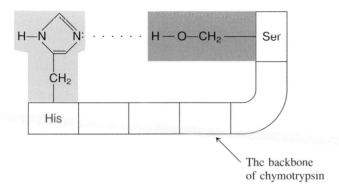

The proposed hydrogen bond between Serine-195 and Histidine-57 in the active site of chymotrypsin. Even if the hydrogen were only partly drawn away from the serine side chain, the nucleophilicity of the serine oxygen would still be significantly increased.

nucleophilic attack. *In other words, through its formation of a hydrogen bond, the histidine subunit would enhance the intrinsic nucleophilicity of Serine-195.* As noted by Blow, the idea that histidine might play such a role had been put forward some years earlier by Cunningham and by Wang and Parker. But in the absence of any precise data about the three-dimensional structure of the enzyme, this suggestion had been only that—a possibility. Now, the X-ray crystallographic data gave this idea form and reality.

One further result of the X-ray analysis strengthened the idea that the role of Histidine-57 was to increase the nucleophilicity of Serine-195. A third amino acid in the region of the active site, Aspartic acid-102, was in a position to aid histidine in the removal of the hydrogen from the serine hydroxyl group. The negatively charged Aspartic acid-102 is not directly exposed in the active site—it lies buried in the hydrophobic interior of the protein. Its side chain, however, is in contact with the more superficial Histidine-57, and is positioned so that its COO^- group forms a hydrogen bond with the N—H group on the histidine side chain. Thus, Aspartic acid-102 is able to stabilize the orientation of the planar Histidine-57 side chain and ensure that the ring projects its N: component in a direction that allows it to participate in a hydrogen bond with Serine-195.

The net effect of the hydrogen bonds between the three amino acids is to set up an *H^+ shuttle* or *charge relay system*, which exerts a continual pull on the hydrogen of the Serine-195 hydroxyl group. It is this pull that increases the nucleophilicity of Serine-195 and makes its involvement in the catalytic reaction mechanism chemically acceptable.

Thus we see theory and fact reconciled. The difficulty with the earlier analysis, which suggested that Serine-195 should not be able to mount a nucleophilic attack, was that the theory was viewed too narrowly. The ionization of Serine-195 must be considered within the more general context of its environment. As a free amino acid in a water solution, serine would indeed need a pH of at least 13 to lose H^+, yet in the microenvironment of the enzyme's active center, the proximity of a stationary ac-

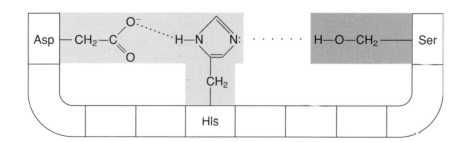

The contact between Histidine-57 and Aspartic acid-102 in the active site of chymotrypsin. The presence of a hydrogen bond serves to orient the Histidine-57 side chain.

ceptor for H^+ makes a high concentration of free OH^- unnecessary.

The Catalytic Triad in Action

As envisioned by Blow, the triad of Serine-195, Histidine-57, and Aspartic acid-102 constitutes the catalytic core of the enzyme's active site. The triad is able to govern the overall catalytic process because of its remarkably fluid electron arrangements. Before the reaction, Serine-195 lies at one end of the active site, its OH group not ionized and seemingly inert. But unlike serine side chains in other environments, this serine is close to ionization. As the reaction begins, Histidine-57 either shares or accepts the H^+ from Serine-195, and the negatively charged Aspartic acid-102 side chain stabilizes the overall rearrangement by sharing the hydrogen on the "other side" of the Histidine-57 ring. And as the serine side chain loses H^+, the electrons left behind become able to attack the C=O group of a susceptible peptide bond. This nucleophilic attack, of course, is the step that initiates peptide-bond cleavage, as diagramed in panel A of the figure on pages 176–177. As the attack proceeds, the tetrahedral transition state is formed (panel B) and, when it collapses, the C-terminal portion of the target protein chain is released. The N-terminal portion is retained as part of an acyl–enzyme intermediate (panels C and D).

The important point to note in this reaction mechanism is that the nucleophilic attack has been made possible because, within the enzyme, Aspartic acid-102 and Histidine-57 have acted together as a general base catalyst—abstracting a hydrogen from Serine-195. In fact, they constitute a powerful base, since when they effectively ionize the serine OH group they cause an event to happen at pH 7 that would otherwise not occur unless the pH were 14. Nevertheless, this process is entirely safe. The histidine–aspartic acid base, buried as it is in the interior of the chymotrypsin molecule, poses no general threat to the cell. It exerts its influence only on Serine-195, turning it into a potent nucleophile. And, in turn, Serine-195 is only able to act on a substrate molecule that is complementary in shape to the active site of the enzyme and thus can approach the reactive OH group.

The Release of the First Part of the Substrate

Abstracting H^+ from the serine is not the only function of the histidine–aspartic acid couple. As the cleavage reaction proceeds, it plays a second, equally important role. After the formation of the acyl–enzyme intermediate, the C-terminal part of the polypeptide chain must be free to leave the active site, and its release requires the availability of a hydrogen to allow the developing N—H group to become a free amino group (panel D). If this H^+ had to come from the surrounding water solution, the limited supply would greatly slow the reaction. But once again the amino acids of the enzyme's active site come into play to keep the reaction proceeding smoothly. The aspartic acid–histidine couple has been holding an extra H^+ ever since it took one from the Serine-195 side chain. Thus the necessary H^+ is already available—held within the active site of the enzyme precisely where it is needed. In terms of acid catalysis, we can view the H^+ as being present in "high concentration" in the area of the reaction, although the presence of the ion in no way contributes to (or depends on) a high concentration of acid in the cell in general.

To transfer the required H^+ to the leaving N—H group, the hydrogen shuttle simply works in reverse. The Histidine-57 side chain releases its stored H^+ (functions as a general acid catalyst) and changes back from being positively charged to neutral. With the acquisition of the hydrogen by the N—H group, the peptide bond is cleaved, and the carboxyl end of the substrate polypeptide chain now leaves the surface of the enzyme (panel D).

The important point is that chymotrypsin draws its power from the ability to function sequentially as both an acid and a base catalyst. This ability is built into the enzyme's geometry. No sooner does the histidine–aspartic acid pair serve as a general *base catalyst* (abstracting H^+ from Serine-195) than it reverses itself and continues the reaction by functioning as a general *acid catalyst;* the H^+ stored on the histidine ring is passed on to the leaving amino group, facilitating the breakage of the peptide bond.

Their ability to function as both acid and base is fundamental to understanding how enzymes promote electron rearrangements—and therefore chemical reactions. As we saw in our discussion at the beginning of the chapter, while peptide-bond cleavage can be promoted by high concentrations of either free acid or base, the effect is only moderate, speeding up the hydrolysis reaction by a factor of 10 to 100. It may have occurred to the reader that, if it were possible to use both acid and base at the same time, the reaction might have been accelerated to much higher levels. But in standard chemistry this of course is impossible since the addition of both H^+ and OH^- to a reaction mixture would simply result in their neutralization to produce water: $H^+ + OH^- \rightarrow H_2O$. The final concentration of H^+ and OH^- would be determined by the intrinsic ionization properties of water and would remain low at one molecule in ten million. In contrast, the enzyme *does* have the capacity to serve as both an acid and a base catalyst. *Through the chemical properties of amino acid side chains such as histidine, which can function as an acid and then a base in rapid, alternating succession, the enzyme achieves a multiplicative and synergistic (that is, much greater than the basic 10- to 100-fold) effect on catalysis.*

To sum up, through its amino acid side chains, the enzyme makes full use of the principles of acid and base catalysis—but does so in a remarkable way that is at once more subtle, more powerful, and more controlled than the effect achieved when acids or bases are used as free ions in solution.

Lowering the Energy of the Transition State

The catalytic triad of Serine-195, Histidine-57, and Aspartic acid-102 is remarkable, but it is not the only way in which chymotrypsin promotes the peptide-bond cleavage. X-ray analysis has revealed another geometrical feature of the active site that plays a role in catalysis.

Previously, we have discussed the fact that all reactions must proceed through a high-energy transition-state intermediate. In the case of peptide-bond cleavage, this is the *tetrahedral intermediate* shown in green in panels B and C of the figure on pages 176–177. This structure is formed when the nucleophilic attack on the target carbon atom causes a pair of electrons to retreat entirely onto the

$C=O$ oxygen. The four covalent bonds radiating out from the target carbon are released from the constraint imposed by the $C=O$ double bond and can now rotate into a more open tetrahedral configuration.

Blow's colleague Richard Henderson, in examining the three-dimensional structure of chymotrypsin, realized that, in the enzyme–substrate complex, the target peptide bond is positioned so that its $C=O$ oxygen might be able to form hydrogen bonds with two N—H groups from the backbone of the enzyme—the N—H groups of the nearby amino acids Glycine-193 and Serine-195. These hydrogen bonds appeared to be weak (in a strained orientation) when the substrate peptide bond was in its original planar configuration. But Henderson noted that the bonds would be stronger when the substrate assumed the intermediate tetrahedral configuration. They therefore would be able to stabilize the transition state and favor its formation (see the figure on page 178).

This observation—the ability of the enzyme to stabilize the transition state—provides another insight into how enzymes work. Essentially, *it is the device that enzymes use to accomplish the same accelerating effect as elevated temperature*. We recall that elevated temperature raises the average energy level in the system so that a higher percentage of the molecules have the energy necessary to exist in the transition state and undergo reaction. The enzyme, by having an active site that is complementary in geometry to the transition state, allows additional bonds to form beyond those that form when the substrate initially binds. By stabilizing the transition state, these additional bonds allow it to exist at a lower energy level than would otherwise be possible. In effect, the energy needed to surmount the transition-state barrier is lowered, and more substrate molecules are able to react. It is as if a tunnel had been created through the mountain raised by the energy barrier (see the graph on page 179).

Thus, in our now enlarged view of the enzyme's active site, we see that at the very time that Aspartic acid-102 and Histidine-57 are serving as a general base, and then acid, catalyst (to abstract and then pass on H^+ from the side chain of Serine-195), Glycine-193 and Serine-195 are functioning to help stabilize the transition state (to make the peptide bond more subject to nucleophilic attack).

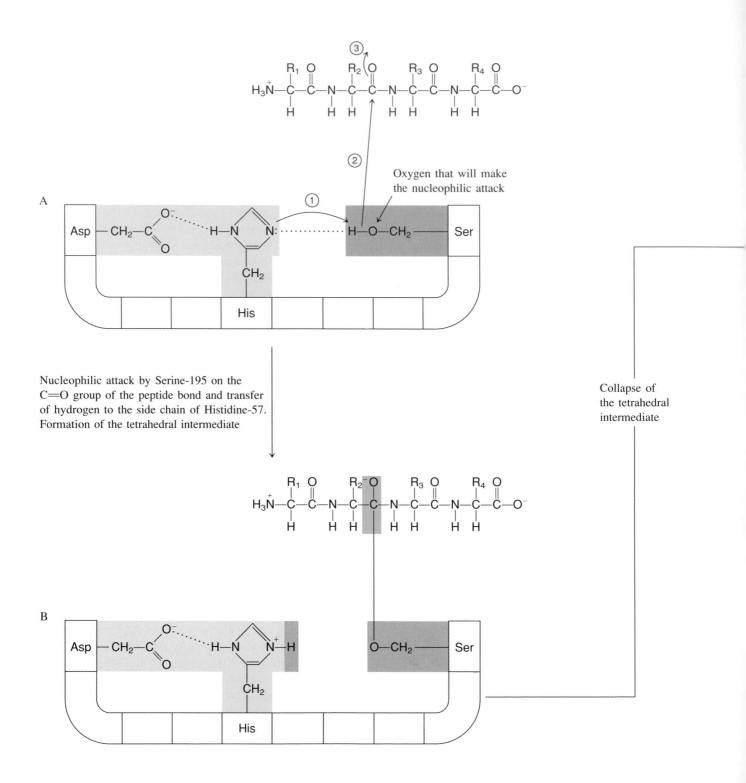

Nucleophilic attack by Serine-195 on the C=O group of the peptide bond and transfer of hydrogen to the side chain of Histidine-57. Formation of the tetrahedral intermediate

Oxygen that will make the nucleophilic attack

Collapse of the tetrahedral intermediate

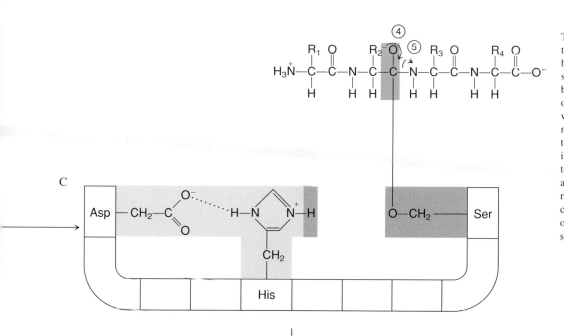

The role of the catalytic triad in the hydrolysis of a peptide bond by chymotrypsin. The diagram shows the involvement of acid-base catalysis in the first stage of the cleavage reaction, in which the acyl–enzyme intermediate is formed. The sequential numbering of the arrows is intended to make it easier to trace the logic of the reaction, although, in reality, the electron rearrangements may occur in a concerted way so that the various steps in the reaction occur simultaneously.

Formation of the acyl–enzyme intermediate and transfer of H^+ from Histidine 57 to the departing amino group. Restoration of the histidine side chain to its original base (N:) state

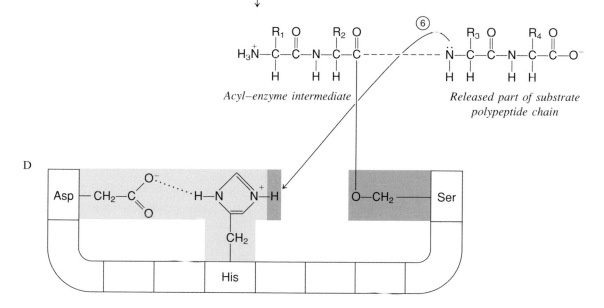

Acyl–enzyme intermediate

Released part of substrate polypeptide chain

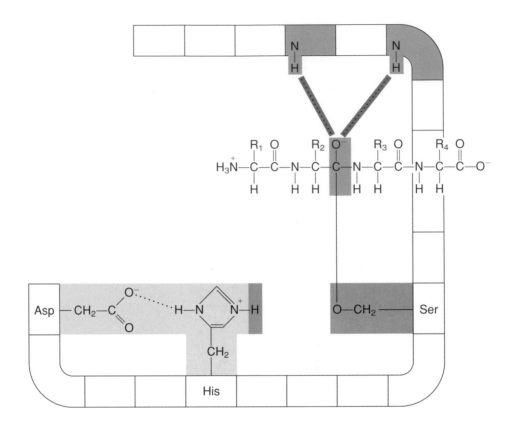

The enzyme stabilizes the substrate as it undergoes reaction by forming two hydrogen bonds with the intermediate tetrahedral transition state (top). The stabilizing hydrogen bonds cannot properly form in the initial enzyme–substrate complex, where the oxygen atom is constrained in a planar structure and is not advantageously
oriented toward the backbone N—H groups. (Note that, for the sake of clarity, these hydrogen-bond connections are drawn at the top of the diagram, whereas in reality one of the participating amino acids is Serine-
195, seen at the right.)

The Breakdown of the Acyl–Enzyme Intermediate

In the second stage of peptide-bond hydrolysis (see the figure on pages 180–181), the acyl-containing part of the substrate is removed from the enzyme. In essence, the chemically reactive components of water (H^+ and OH^-) must be added to dissolve the covalent linkage between the substrate C=O group and the serine side chain. This

reaction would be exceedingly slow if the enzyme were not a direct participant.

After the carboxyl-terminal component of the substrate diffuses away, a water molecule takes its place in the active site. The catalytic triad with its hydrogen shuttle now works a second time. The aspartic acid–histidine couple draws an H^+ away from the water molecule (whereas before it removed a hydrogen from the serine side chain). This results in the formation of a free OH^-

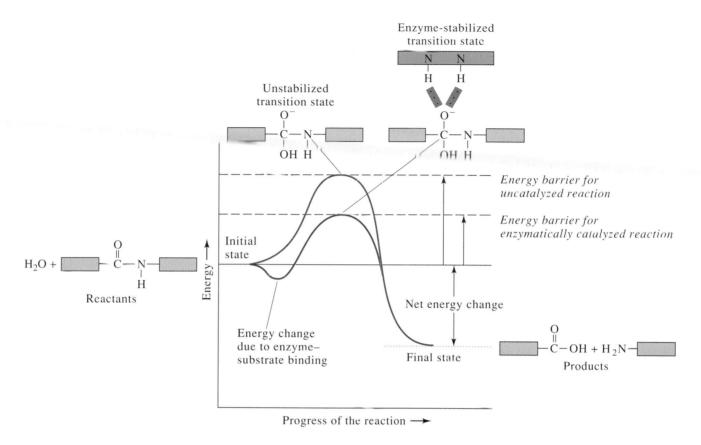

During the course of a chemical reaction, the starting molecule must pass over a significant energy barrier. The reaction is slow because very few molecules have the energy necessary to form the high-energy transition state. By forming additional chemical bonds with the substrate when it enters the transition state, the enzyme lowers the energy required to establish this intermediate structure, allowing more molecules to surmount the energy barrier, and thus accelerates the chemical reaction.

within the active site of the enzyme. The OH^- now attacks the $C=O$ group of the acyl–enzyme intermediate. A tetrahedral transition state and then a carboxyl group are created, and the remaining portion of the original substrate protein is released from the Serine-195 side chain. As the H^+ stored on the aspartic acid–histidine couple (recently taken from the water molecule) is transferred to the serine side chain, the enzyme is returned to its prereaction state, ready for another round of catalysis (panel D, page 181).

In the second stage of the chymotrypsin reaction, as in the first, the enzyme has functioned as both an acid and a base catalyst.

A Challenge from the Devil's Advocate

We can take considerable satisfaction in the state of our knowledge about chymotrypsin—and in what it tells us

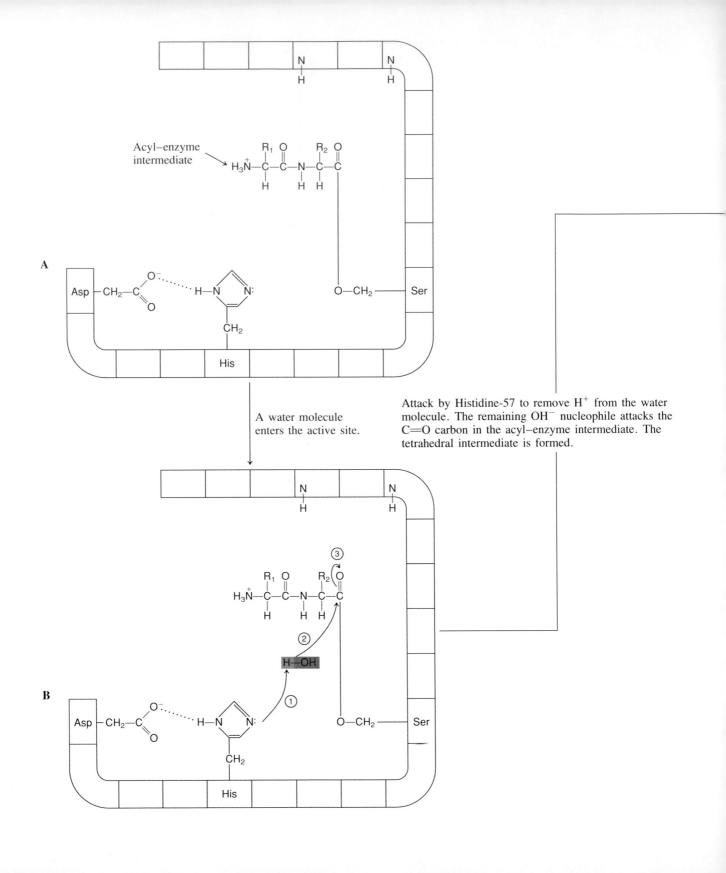

Acyl–enzyme intermediate

A

A water molecule enters the active site.

Attack by Histidine-57 to remove H⁺ from the water molecule. The remaining OH⁻ nucleophile attacks the C=O carbon in the acyl–enzyme intermediate. The tetrahedral intermediate is formed.

B

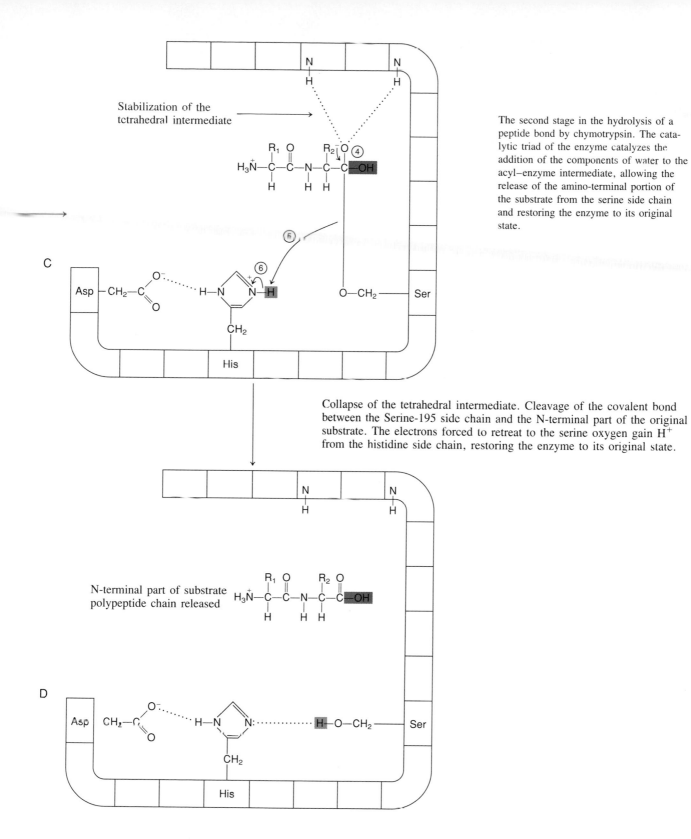

Stabilization of the
tetrahedral intermediate

The second stage in the hydrolysis of a peptide bond by chymotrypsin. The catalytic triad of the enzyme catalyzes the addition of the components of water to the acyl–enzyme intermediate, allowing the release of the amino-terminal portion of the substrate from the serine side chain and restoring the enzyme to its original state.

C

Asp

His

Collapse of the tetrahedral intermediate. Cleavage of the covalent bond between the Serine-195 side chain and the N-terminal part of the original substrate. The electrons forced to retreat to the serine oxygen gain H^+ from the histidine side chain, restoring the enzyme to its original state.

N-terminal part of substrate
polypeptide chain released

D

Asp

His

Ser

about enzymes more generally. We have progressed from a state where we had no idea how an enzyme worked to the state where we have a highly detailed reaction mechanism that, at the very least, provides a sensible molecular way for thinking about how enzymes function as catalysts in promoting specific chemical transformations. But is the mechanism God's truth, or only our truth?

The Devil's Advocate (the critical observer whose job it is to search for flaws in an argument) would tell us that our model for how chymotrypsin works is just that— only a model, rooted as much in our imagination and logic as in objective fact. Much of our proposed reaction mechanism for chymotrypsin is based on indirect evidence— what might be called circumstantial evidence in a court of law. Only one experimental result provides direct support for the reaction mechanism: the trapping of an acyl or similar chemical group on the side chain of Serine-195. All the other aspects of our intricate reaction mechanism

- The use of Histidine-57 and Aspartic acid-102 as a general base catalyst to increase the nucleophilicity of Serine-195

- The shuttling of H^+ ions from Serine-195 to the developing amino group, and, later, from water back to Serine-195

- The use of Glycine-193 and Serine-195 to stabilize the transition state with its tetrahedral configuration

have been inferred from the chemical properties of the amino acids and from the specific geometry of the active-site region.

We might reply, in answer to the Devil's Advocate, that these uncertainties do not mean that our work is unsubstantial. Scientific progress occurs by a series of successive approximations—models that become closer and closer to the truth. A model is built around certain key facts; other features are inferred. It is not always possible, at a given moment in time, to provide direct experimental support for each feature of a model, and this is why scientific investigation is a continuing endeavor. We could continue—somewhat defensively—that it is extremely un-

likely that our model for catalysis by chymotrypsin is totally inaccurate—it must contain some elements of truth. Too many independent observations point in the same direction, and the chemical logic of our model corresponds too well with the basic chemical properties of amino acid side chains and the theoretical requirements for catalysis.

Were we to take this approach, then with a limited appeal to reason and an extensive rhetorical flourish, we would have dispensed with the Devil's Advocate as if he were a fine fellow who one should always take seriously, but who is not actually likely to show up and cause any great difficulty. Nothing could be further from the truth. It took thirty years from the time of the pH profile studies of chymotrypsin and the pseudosubstrate experiments to the X-ray crystallographic analysis of the enzyme and the proposal for its catalytic mechanism. But following the proposal, the Devil's Advocate marked just three years before making his appearance. He came, of course, not as a black-cloaked personal demon but as a metaphorical representation of the inward skepticism of good scientists.

What every Devil's Advocate knows is that, when an elaborate model has been proposed that explains a great amount of data, the model also becomes a captive of those data. So much about the organization of a particular part of the universe is specified that any new discovery must be consistent with the model, fitting neatly in its expected place. If a new and important observation does not fit, it is the model, and not the fact, that must give way. In our case, everything turns on the position of a single hydrogen atom among the 4000 atoms of chymotrypsin. To a great extent, the chymotrypsin reaction mechanism depends on the unusually high nucleophilicity of Serine-195 and the way in which the structure of the enzyme is organized to achieve this enhanced nucleophilicity. In the model of Blow and his colleagues, the increased reactivity of Serine-195 is brought about by Histidine-57 forming a hydrogen bond with the Serine OH group, thereby beginning to draw the hydrogen ion away from Serine-195. The Devil's Advocate simply asks: Does the crucial hydrogen bond exist?

The problem is that the exact position of hydrogen atoms cannot be determined through the X-ray crystallographic analysis of proteins. While carbon, nitrogen, sulfur, and oxygen atoms are sufficiently electron-rich to bend and scatter the electromagnetic waves of X rays, this is not true for the hydrogen atom (which contains only a single electron). The position of hydrogens in the structure is therefore inferred from the ways in which they are known to be attached to other atoms. Specifically, to determine whether a hydrogen bond is present, two aspects of molecular geometry must be considered. First, the distance between the two atoms that share the hydrogen must be suitably short. Second, the hydrogen must project from the donor atom (in our case the serine oxygen) toward the unshared pair of electrons on the recipient atom (in this case the N: on histidine). Examples of hydrogen bonds of the type we are concerned with here, namely

$$\boxed{}\!-\!O\!-\!H\;\blacksquare\blacksquare\;N\!-\!\boxed{}$$

can be found in the crystallographic analysis of many simple model compounds and also complex proteins. The generalization has emerged that the distance between the oxygen and nitrogen atoms is about 2.8 Å, and in no case more than 3.1 Å. Thus, if two atoms such as oxygen and nitrogen are close together in the protein, it is *possible* that a hydrogen bond exists between them. But very high resolution data are required in order to determine this with

certainty. What is eminently reasonable on first principles cannot be taken for granted when the point is crucial—as in the case of the proposed hydrogen bond between Histidine-57 and Serine-195.

It was with these thoughts in mind that Joseph Kraut and his co-workers at the University of California at San Diego carried out a high-resolution X-ray crystallographic study of the chymotrypsin-like enzyme produced in bacteria, *subtilisin*. Focusing on the structure of the active site, they came to the conclusion that the crucial hydrogen bond between serine and histidine did *not* exist, or at least that any such hydrogen bond is severely distorted and weak. The distance between the oxygen of the serine side chain and the nitrogen of the histidine side chain was 3.7 Å, far too long for a strong hydrogen bond. Moreover, the angle at which the hydrogen projected from the serine oxygen did not properly orient it toward the N: of the histidine ring. A reexamination of the most reliable crystallographic data available for chymotrypsin confirmed the finding—the distance between the oxygen and nitrogen atoms was again too long, 3.8 Å.

The diagram of the catalytic triad on this page takes into account this new finding. The key hydrogen bond that would allow Histidine-57 to begin to draw the hydrogen away from Serine-195 and increase its nucleophilicity is missing. In terms of chemical reactivity, the serine side chain becomes nothing more than a stationary water molecule.

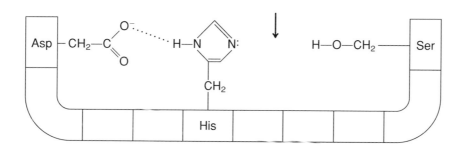

The absence of a hydrogen bond between histidine and serine in the catalytic triad of subtilisin, as determined by very high resolution X-ray crystallographic studies. Note that a strong hydrogen bond, with normal geometry, still exists between Aspartic acid-102 and Histidine-57.

What remains of the catalytic mechanism if the serine–histidine–aspartic acid triad is no longer functional? If one pulls the pin of a hand grenade, soon nothing remains. However, in our case, the whole theory does not explode.

While challenging an essential part of the original model, Kraut was able to confirm a second contribution of the enzyme to catalysis, which had been somewhat briefly considered before by Henderson and Blow. Recall the proposal that two hydrogen bonds might be formed between the backbone NH groups of Glycine-193 and Serine-195 and the C=O group in the target peptide bond (as shown below and in the figure on page 178).

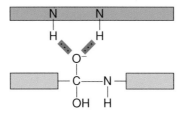

Kraut and his colleague David Matthews determined that these hydrogen bonds do, in fact, exist. Although distorted and weak in the initial enzyme–substrate complex, they form effectively when the substrate is passing through the tetrahedral transition state. Kraut elaborated a model for catalysis that emphasized this structural feature of the enzyme.

The centerpiece of the model is the theory of *transition-state stabilization,* the fundamental thesis of which is that *an enzyme functions by furnishing a template complementary to the substrate in its transition-state configuration.* In the case of the chymotrypsin reaction mechanism, this means that when the substrate binds to the enzyme (and even before the nucleophilic attack occurs), the planar area of the peptide bond becomes distorted so that the atoms attached to the target carbon are coaxed by their interactions with the enzyme into a tetrahedral conformation. This pyramidalization of the reactive carbonyl group induces electronic strain in the substrate, carrying it up the

energy curve toward the transition-state level (see the figure on page 179). But, because of the extra, stabilizing bonds that are formed between the enzyme and the substrate, the final energy level of the tetrahedral intermediate is lower. Even so, in its strained configuration the substrate is still fully susceptible to attack by a nucleophile.

In this view of the catalytic mechanism, Serine-195 is not intrinsically nucleophilic, but instead reacts simply because it is in the best position to attack the tetrahedrally distorted C=O group in the substrate. In fact, the serine can be considered more or less equivalent to a water molecule permanently held in position in the active site of the enzyme and ready, if somewhat reluctantly, to give up H$^+$ to become a nucleophile.

Overall, in this view of the chymotrypsin active site there is a lessened role for acid–base catalysis. Because the crucial hydrogen bond is missing, the histidine–aspartic acid couple cannot serve as a general base catalyst to increase the nucleophilicity of Serine-195. Instead it facilitates the transfer of a hydrogen, either from the attacking serine oxygen to the leaving amino group in the first step of the reaction, or from an attacking water molecule to the serine oxygen in the second step of the reaction. In other words, the histidine–aspartic acid remnant of the original catalytic triad can be regarded as a hydrogen relay station, serving as a site for binding the H$^+$ that Serine-195, and later a water molecule, must give up to hydrolyze a peptide bond.

To summarize the position of the Devil's Advocate, (1) the critical hydrogen bond between Serine-195 and Histidine-57 does not exist, consequently (2) Serine-195 is not especially nucleophilic, and (3) the enzyme's main function is therefore to bind to the transition state more strongly than the substrate itself, thereby promoting its formation and drawing the reaction forward. In this way—even without the catalytic triad—the enzyme can still provide the basis for catalytic rate enhancement.

Restoration of the Catalytic Triad

Now, even Devil's Advocates have Devil's Advocates. Or, as Jonathan Swift put it in 1733,

So, naturalists observe, a flea
Hath smaller fleas that on him prey;
And these have smaller fleas to bite 'em,
And so proceed *ad infinitum*.

In our case, the new Devil's Advocate actually seeks to defend the original formulation of the chymotrypsin active site based on the increased nucleophilicity of Serine-195. He agrees that in the absence of the serine–histidine hydrogen bond, and the catalytic triad, it is eminently reasonable to suppose that catalysis is driven by stabilization of the substrate in its tetrahedral transition state. And, indeed, no less an authority than Linus Pauling had considered the primary role of enzymes to be their stabilization of transition-state structures. The basis of catalysis, he said, lies to a great extent, in

an active region on the surface of the enzyme, which is closely complementary in structure not to the substrate molecule itself in its normal configuration, but rather to the substrate molecule in a strained configuration, corresponding to the transition state [in our case, the tetrahedral intermediate].

Nonetheless, while accepting this aspect of enzyme function, it can be argued in defense of the original model that the Devil's Advocate's attack on the catalytic triad has a hidden assumption. The careful reader will no doubt have noticed it. It is the assumption that the critical hydrogen bond between Serine-195 and Histidine-57 must be found in the resting enzyme. But, to be rigorous, this hydrogen bond need not be present until the substrate has actually bound to the enzyme and the cleavage reaction is ready to begin. Thus, the new Devil's Advocate points out that if we looked at the "proper situation"—that is, chymotrypsin bound to a substrate—then we might well find that the catalytic triad is intact. After all, it would not really be too surprising if an enzyme were able to undergo a slight conformational change upon binding its substrate, so as to realign nearby groups. The issue then becomes to examine the crystallographic structure of chymotrypsin (or subtilisin) bound to a target protein chain or an appropriate substrate analogue and to determine the status of the catalytic triad in that structure.

Several contemporary studies with chymotrypsin-like enzymes, complexed with target molecules, have indicated that the catalytic triad, although absent in the resting enzyme, does form upon substrate binding. In one such study, the nitrogen-to-oxygen contact in the resting enzyme is found to be 3.3 Å, and the histidine ring is poorly oriented for forming a hydrogen bond. When the enzyme binds to a peptidelike substrate, however, the histidine ring shifts slightly so that it now makes a strong hydrogen bond (2.9 Å) with serine. In another study, the hydrogen bond between serine and histidine, which in the free enzyme has a length of about 2.9 Å but a strained orientation, becomes perfectly linear and of optimal length, 2.6 Å, when the enzyme–substrate complex is formed.

The fortunes of war can ebb and flow, of course, but at present it looks very much as if the original formulation of the catalytic triad remains valid. A hydrogen bond with Histidine-57 does indeed serve as a general base catalyst to increase the nucleophilicity of Serine-195.

A Strengthened Model

In Shakespeare's *King Lear,* the Duke of Albany advises Lear's eldest daughter not to supervise too closely the old king's activities. "Striving to better, oft we mar what's well." Or, in the words of the American aphorism of the twentieth century, "If it ain't broke, don't fix it." Science never works this way. As we have seen, it is from the constant interplay between theories and experiments, between models and hard facts, that the truth slowly evolves. In our case, the original model for the active site of chymotrypsin was challenged by the Devil's Advocate, who found that, at least in the resting enzyme, the hydrogen bond necessary to increase the nucleophilicity of Serine-195 was missing. In the end the model survived this challenge, because it was found that the crucial hydrogen bond *was* formed when the enzyme bound the substrate. Moreover, during the course of this challenge, additional evidence was obtained establishing the ability of the enzyme to promote the reaction through stabilization of the transition state. In sum, through the process of thesis, antithesis, and synthesis, the entire conceptual framework of the science was strengthened.

Molecular Surgery

Our model for the active site of chymotrypsin involves several amino acids that work cooperatively to promote catalysis. The final studies we will consider confirm all of the conclusions we have reached so far concerning the chymotrypsin reaction mechanism and also allow us to assess the relative importance of the roles played by acid–base catalysis and transition-state stabilization.

Imagine the opportunities for experiments that would result if only it were possible to perform some sort of microsurgery on chymotrypsin that would allow us to remove one or another active-site amino acid and determine its individual contribution to catalysis. Of course such surgery on a mature protein is impossible in the literal sense but, in fact, with very modern technology, an equivalent result can be achieved. The key is *molecular genetics,* and the general idea is to change the structure of the gene that directs the synthesis of the enzyme so that one or more of its 241 amino acids are replaced in the final protein.

The past thirty years of research have shown that the information for the construction of enzymes resides in the cell's DNA. In a Morse code–like fashion, each sequence of three nucleotides forms a *codon* specifying a particular amino acid. Thus, the order of A, G, T, and C nucleotides in a particular segment of DNA can be read out three at a time, from the beginning of a gene to the end, to align in sequence the amino acids for a specific protein. As shown in the table on this page, the nucleotide triplets TCA and TCG are among those that code for the amino acid serine. Similarly, CAC and CAT code for histidine, and GAC and GAT code for aspartic acid. This relationship between nucleic acids and proteins constitutes the central dogma of molecular biology: in the flow of genetic information in the living cell, the sequence of nucleotides in DNA is ultimately translated into the sequence of amino acids for a protein. The details of this process are not of primary concern to us here. What is important is that the modern techniques of recombinant DNA technology now allow us to identify individual genes and study how they code for their respective proteins. In fact, the gene that codes for the chymotrypsin-like enzyme produced by bacteria, subtilisin, has been isolated. The order of its nucleotide subunits has been determined, and they can be seen to

The Genetic Code

Amino acid	Triplet codon in DNA	Amino acid	Triplet codon in DNA
Alanine	GCT	Leucine	CTC
Alanine	GCC	Leucine	CTA
Alanine	GCA	Leucine	CTG
Alanine	GCG	Leucine	TTA
		Leucine	TTG
Arginine	AGA		
Arginine	AGG	Lysine	AAA
Arginine	CGT	Lysine	AAG
Arginine	CGC		
Arginine	CGA	Methionine	ATG
Arginine	CGG		
		Phenylalanine	TTT
Asparagine	AAT	Phenylalanine	TTC
Asparagine	AAC		
		Proline	CCT
Aspartic acid	GAT	Proline	CCC
Aspartic acid	GAC	Proline	CCA
		Proline	CCG
Cysteine	TGT		
Cysteine	TGC	Serine	AGT
		Serine	AGC
Glutamic acid	GAA	Serine	TCT
Glutamic acid	GAG	Serine	TCC
		Serine	TCA
Glutamine	CAA	Serine	TCG
Glutamine	CAG		
		Threonine	ACT
Glycine	GGT	Threonine	ACC
Glycine	GGC	Threonine	ACA
Glycine	GGA	Threonine	ACG
Glycine	GGG		
		Tryptophan	TTG
Histidine	CAT		
Histidine	CAC	Tyrosine	TAT
		Tyrosine	TAC
Isoleucine	ATT		
Isoleucine	ATC	Valine	GTT
Isoleucine	ATA	Valine	GTC
		Valine	GTA
Leucine	CCT	Valine	GTG

code, triplet by triplet, for the 275 amino acids of the subtilisin protein. The figure on the following page shows the key regions of the subtilisin gene, emphasizing those that code for the amino acids of the catalytic triad in the active site.

Once a gene has been isolated in the laboratory, it becomes possible to perform a certain type of surgery. The DNA molecule can be broken into pieces and reconstructed so that a particular segment of three nucleotides is altered to code for a different amino acid. For example, if we choose the triplet that codes for the active-site serine in subtilisin and change the nucleotides that constitute this triplet from TCA to GCA, the resulting enzyme will contain alanine in the active site in place of serine. The same general process of *site-directed mutagenesis* can be applied to change the codons of the active-site histidine or aspartic acid. In such experiments, it is particularly advantageous to use alanine as the replacement amino acid in the active site. The reason is simple: its side chain is a relatively small, nonreactive hydrocarbon, CH_3. Thus, alanine does not have the ability to function in the manner of serine and carry out a nucleophilic attack after losing a hydrogen ion. Nor can it behave as an acid or base catalyst or form new ionic or hydrogen bonds that might distort the structure of the enzyme.

When the surgically altered subtilisin gene is reintroduced into a living cell, it directs the synthesis of a variant form of the enzyme lacking, for instance, the serine, histidine, or aspartic acid of the active site. It thus becomes possible to assess the consequences on enzymatic activity of replacing one or another active-site amino acid.

What catalytic properties do such altered enzymes have? If the dominant enzymatic role is played by the catalytic triad with its acid/base properties, the activity of the enzyme should be dramatically reduced, or eliminated, when serine, histidine, or aspartic acid is replaced by alanine. On the other hand, if the catalytic triad is *not* of primary importance, and the basis of the enzyme's activity lies in the stabilization of the tetrahedral transition state, then the altered enzyme should still be active. Finally, if both factors constitute the molecular machinery responsible for catalysis, then the enzyme's activity will be reduced in steps as the active-site amino acids are eliminated.

This is exactly the effect observed in the 1988 experiments of James Wells and Paul Carter of Genentech, one of the world's foremost companies dedicated to the commercial development of biotechnology. The table below shows the results of their study comparing normal subtilisin with several genetically engineered variants whose active-site serine, histidine, or aspartic acid (or a combination of these subunits) had been replaced by alanine. It can be seen that making no change in the enzyme other than to replace the active-site serine with alanine

Catalysis by Genetically Engineered Mutant Subtilisins

Change in enzyme	Active-site configuration			Relative catalytic activity
	Ser	His	Asp	
No change	+	+	+	10,000,000,000
Ser → Ala	−	+	+	5,000
His and Asp → Ala	+	−	−	37,000
Asp → Ala	+	+	−	330,000
Ser and His and Asp → Ala	−	−	−	4,000
Asn → Leu*	+	+	+	10,000,000
No enzyme				1

*In subtilisin, the stabilization of the transition state is carried out, in part, by the asparagine subunit at position 155. This subunit uses an N—H group on its side chain to form a hydrogen bond with the oxygen atom in the C=O group of the target peptide bond. The hydrocarbon side chain of leucine in the mutant enzyme cannot participate in such an interaction.

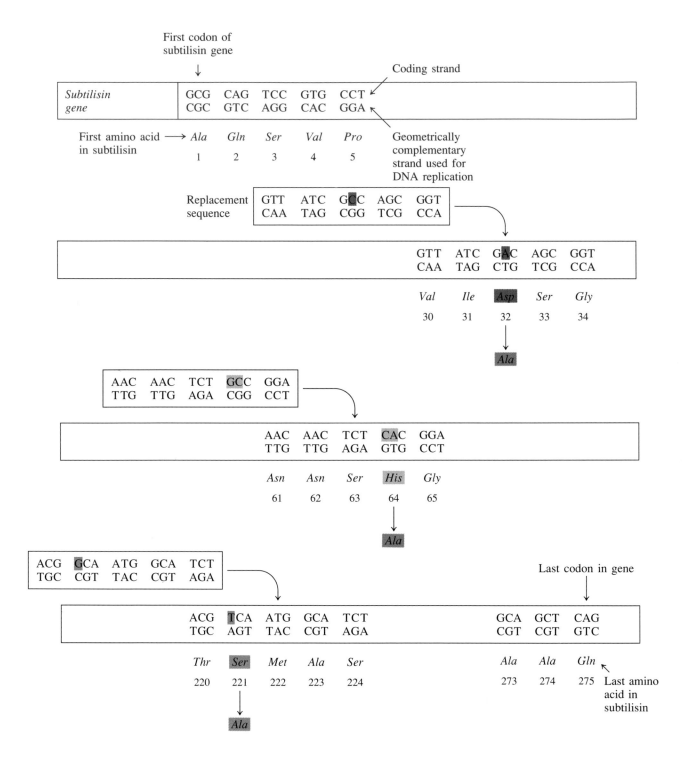

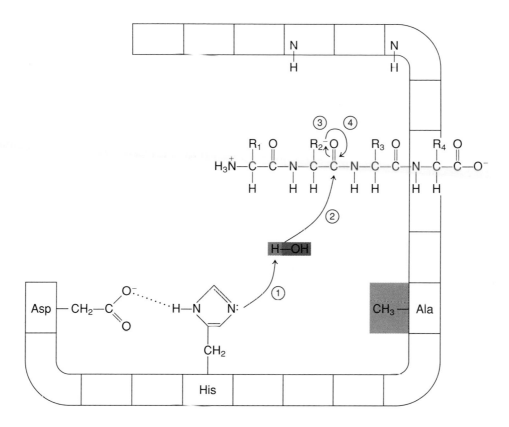

A variant of subtilisin produced by genetic engineering. Because serine has been replaced by alanine in the active site, the reaction cannot proceed by the usual serine acyl–enzyme intermediate. Instead, aided by the histidine–aspartic acid couple, a direct attack by an ionized water molecule (H^+ and OH^-) on the peptide bond must occur to produce a tetrahedral intermediate that, in a single step, collapses to give the cleaved protein chain. The reaction is slow because, unlike serine, the water molecule is not held in place in the active site so that histidine and aspartic acid can efficiently remove H^+ from it.

An abbreviated version of the subtilisin gene, showing the beginning and end of the gene and three internal regions that code for the amino acids of the catalytic triad in the active site. Beneath each DNA triplet codon is the amino acid it encodes, for instance GCG for alanine at position one. The replacement sequences represent altered DNA segments that, when substituted in the gene through recombinant DNA techniques, change the amino acids coded for.

reduces enzymatic activity by a factor of 2,000,000 (see line 2 of the table). The loss of the histidine–aspartic acid couple has a similar but 10-fold less severe effect: the activity of the enzyme drops by a factor of 300,000. And when only aspartic acid is replaced by alanine, the enzyme's activity is reduced by a factor of 30,000. These results are fully consistent with the view we have developed of the active site and of the relative importance of the amino acids in the catalytic triad. Serine carries out the primary nucleophilic attack on the substrate. Histidine serves as a general base catalyst to increase the reactivity of serine, and the role assigned to aspartic acid is primarily that of a helper, stabilizing the orientation of histidine to increase its effectiveness.

Interestingly, even after all three members of the catalytic triad have been removed by site-directed mutagenesis, the enzyme still retains a significant level of activity. Although about a millionfold less active than the normal enzyme, it still accelerates peptide-bond cleavage 4000-fold over the uncatalyzed reaction. Thus, even without the catalytic triad, the remaining amino acids of the enzyme active site are able to promote the reaction, presumably by stabilizing the tetrahédral transition state. Indeed, an experiment of Poulos and Bryan at the Genex Corporation has shown that, when Asparagine-155 (an amino acid known by X-ray crystallographic studies to use its side chain to stabilize the tetrahedral intermediate) is replaced by leucine, there is a 10- to 100-fold drop in catalytic activity, most likely because one of the side-chain hydrogen bonds involved in binding the transition state is lost.

To sum up, the amino acids in the catalytic triad have a strongly synergistic effect and contribute a factor of about 2,000,000 to the total catalytic rate enhancement of nearly 10,000,000,000. The residual activity, after complete replacement of the catalytic triad, results from stabilization of the transition state by contacts outside the catalytic triad.

An Insight into Enzyme Evolution

A century ago, Charles Darwin proposed that organisms evolve through a two-stage process involving *mutation*

and *natural selection*. In each generation, mutations occur randomly in the genetic material (the DNA), leading to the production of variant proteins and generating diversity within the species. In addition, the individual members of the population are always subjected to a process of natural selection in which those organisms best fitted to survive the pressures of the environment prosper and leave behind more offspring. The net effect over time is the "evolution" of new and modified species.

If we look again at the table on page 187, we can see that the same potential exists for the evolution of enzymes. All that is required is that variants of an enzyme like chymotrypsin arise that can be selected by environmental pressure. The evolution of subtilisin is particularly easy to envision. The enzyme is produced by certain strains of bacteria and secreted into the immediately surrounding environment. There the enzyme scavenges amino acids for the cell by initiating the degradation of any extracellular proteins that it might encounter from decaying organic matter. The resulting short peptides are then further degraded by other proteases, and the final products enter the cell and aid in its growth. Clearly a cell that produces a better subtilisin may divide more rapidly and leave behind more descendants, eventually becoming numerically dominant in the population. All other things being equal, virtually every individual in the species will, in time, contain the improved subtilisin.

Now suppose that the earliest, most primitive subtilisin was simply a protein that could bind to a target polypeptide chain—and, in particular, bind especially tightly to the tetrahedral transition state that occurs during peptide-bond cleavage. This early enzyme could have accelerated peptide-bond hydrolysis some 4000-fold over the velocity of the non-enzymatic reaction, even though it had none of the members of the catalytic triad we now associate with the active site. *Inherent in this experimental result is an important general proposal about the origin of enzymes: they may have arisen initially as substrate-binding proteins with no specific catalytic amino acids.*

The cell that produces the primitive subtilisin will prosper, relative to other cells, because of its enhanced food supply. And, importantly, it can proceed forward on the stepwise course of evolution if further mutations lead to the introduction of the members of the catalytic triad. How can this occur?

From an evolutionary point of view, it is extremely unlikely—and unnecessary—that the catalytic triad arose in one step. Rather, it doubtless evolved through a series of intermediate stages in which each successive form of the enzyme had an increased catalytic effectiveness. Assuming that the present-day enzyme is a reasonable model of its ancestor, we might imagine that the next step was the introduction of serine into the appropriate position in the active site. As the data of the table show, this would boost catalysis an order of magnitude over the 4000-fold acceleration already provided by transition-state stabiliza-tion—creating an enzyme that hydrolyzes peptide bonds some 37,000 times more rapidly than the uncatalyzed reaction (compare lines 3 and 5 in the table). Installing histidine and then aspartic acid gives further increases in the catalytic strength of the enzyme. The important point is that mutational studies show that inserting the catalytic triad amino acids in a stepwise fashion can produce intermediate enzymes with successively higher levels of activity, progressively benefiting the organism in its struggle for existence.

In sum, enzymes, on the molecular scale, like organisms, on a more visible level, have the ability to evolve through the process of mutation and natural selection.

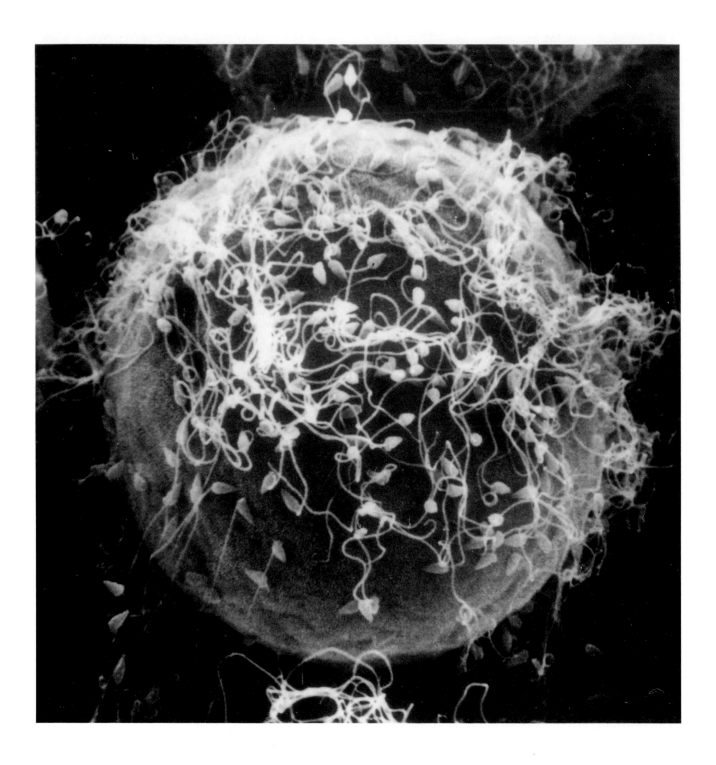

6

The Evolution of an Enzyme Family

Chymotrypsin represents an entire family of enzymes, whose members are involved in a wide variety of physiological processes, including the union of egg and sperm that is the key step in fertilization.

Why is it useful to know so much about a particular enzyme such as chymotrypsin? One reason is that the basic principles of the enzymatically catalyzed reaction are universal. Chymotrypsin shares with other enzymes (1) the nature of the forces that fold an enzyme into its biologically active form, (2) the complementarity in shape that allows an enzyme to bind to its substrate in a lock-and-key fashion and distort it toward a reactive transition state, and (3) the use of amino acid side chains as acid and base catalysts to promote the electron rearrangements involved in reaction. In learning about chymotrypsin, we come to understand enzymes in general. Even the few new RNA catalysts that are now known follow these very same principles—using nucleotides instead of amino acids to achieve the requisite structure-function relationships.

But there is another, and equally important, reason for having a thorough working knowledge of one enzyme. The English poet William Blake put it best:

> To see a World in a Grain of Sand . . .
> And Eternity in an hour.

The first insights into evolution were obtained only a little over a century ago and set forth in the brilliant
writings of the English naturalist Charles Darwin, whose study is shown above. Careful observation of fundamental similarities in anatomical structure and physiological function between different organisms impelled the
idea that living creatures were not the products of unique and separate creations, but rather had evolved from
common ancestors through a process of variation and natural selection. As we will see in this chapter, evolution can now be studied at the molecular level, and it is possible to see how enzymes, through their own process of variation and natural selection, have diversified and evolved ever greater capabilities so as to generate
the range and scope of life forms that fill the earth.

After nature has invented a pattern, it is often used over and over again as a leitmotif—developed and changed, as in music, to achieve an amazing variety of effects. In biology the end result is that most impressive of all adventures—the continuing evolution of living cells and their molecules.

Chymotrypsin is, in fact, representative of an entire family of enzymes that evolved from a common ancestor and now occur throughout nature. All of these enzymes carry out a basically similar type of cleavage reaction but work on different substrates. Each enzyme is a variation on the same theme, and their evolution has allowed the cell to venture into chemical transformations involved in processes ranging far afield from digestion—into areas as diverse as blood clotting and neurobiology. The important point to note in the discussion that follows is that an enzyme like chymotrypsin first evolved and then was modified as the cell and the organism enlarged the scope of its activities. In fact, the very processes we considered in the Prologue, though seemingly so different, are all carried out, in whole or in part, by members of the chymotrypsin family.

Chymotrypsin: The Substrate Binding Site and Enzyme Specificity

We begin our analysis of enzyme evolution by walking through a new part of the now-familiar territory of the structure of chymotrypsin. Up to this point we have been focusing our attention on the catalytic active site, with its group of amino acids that act synergistically to promote peptide-bond cleavage. Judged from its reaction mechanism alone, this catalytic active site should be able to cleave any peptide bond in any protein—and indeed a variety of closely related chemical bonds in other molecules. Yet chymotrypsin, like all enzymes, is remarkably specific with respect to the compounds it works on. Along with its tremendous catalytic power, this specificity is the second hallmark of enzyme action.

The key to an enzyme's specificity lies in an area of the protein adjacent to the catalytic active site, an area that we will call the *substrate recognition site*. This is the part of the enzyme that is responsible for the lock-and-key fit with the substrate. In chymotrypsin, for example, it ensures that only peptide bonds adjacent to phenylalanine, tyrosine, or tryptophan will be cleaved.

Like the catalytic active site, the substrate recognition site cannot be identified by direct inspection of the convoluted surface of the enzyme. Rather, it is located by applying X-ray analysis to enzyme crystals that have been exposed to a compound resembling a substrate. The small molecule diffuses into the crystal, where it eventually finds its binding site. When the three-dimensional structure of the complex is then determined, one sees where the miniaturized substrate has bound to the enzyme. Tom Steitz, Richard Henderson, and David Blow carried out

just such an analysis for chymotrypsin, using the substratelike compound formyl-tryptophan. The addition of the formyl group (CH=O) to tryptophan converts the otherwise free amino acid into a structure with a peptide

Formyl-tryptophan

bond. The modified amino acid, formyl-tryptophan, thus contains all of the structural properties needed to be a substrate of chymotrypsin—a peptide bond adjacent to a flat, hydrophobic side chain.

The figure on page 197 shows the binding of formyl-tryptophan to chymotrypsin. The major contact occurs near the catalytic triad through the placement of the flat, hydrophobic side chain of the compound in a well-defined pocket in the surface of the enzyme. This pocket is formed primarily by amino acids 214 to 220, together with 189 to 192, and it is lined entirely with hydrophobic side chains. It is about 10 to 12 Å deep and about 4 Å by 5 Å in cross section, which gives a very snug fit, since the side chain of tryptophan is about 6 Å long and 3.5 Å thick. The binding pocket thus provides a geometrically and chemically suitable environment for interacting with the flat, hydrophobic side chain of tryptophan and the similarly shaped side chains of phenylalanine and tyrosine. The hydrophobic pocket offers no basis for interaction with small amino acid side chains and actively excludes branched or charged side chains.

Studies with compounds related to formyl-tryptophan have led to a complete description of the substrate recognition site in chymotrypsin. In addition to providing a hy-

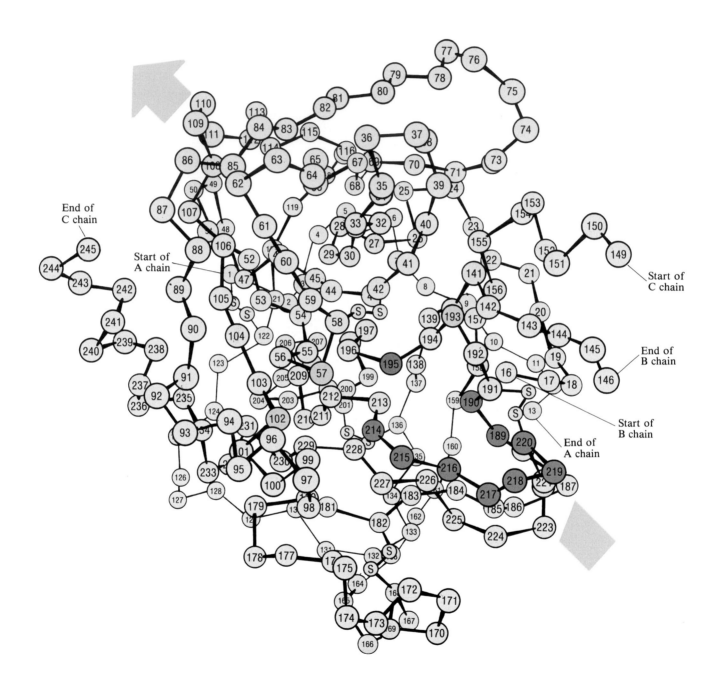

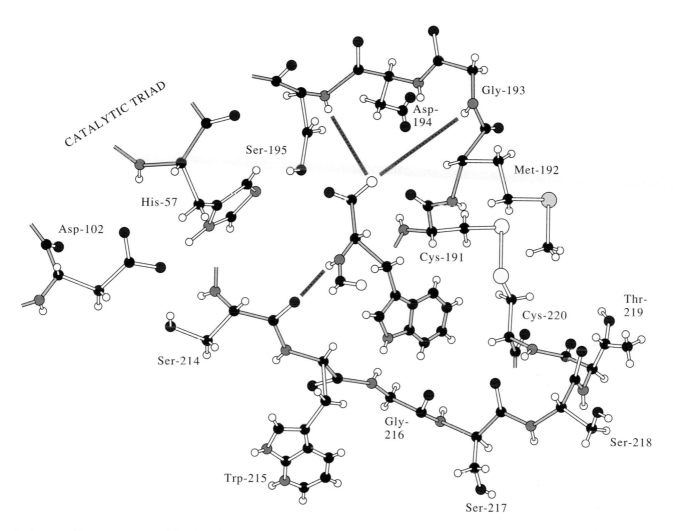

CATALYTIC TRIAD

Gly-193

Asp-194

Ser-195

Met-192

His-57

Asp-102

Cys-191

Thr-219

Cys-220

Ser-214

Gly-216

Ser-218

Trp-215

Ser-217

Facing page: The substrate recognition site of chymotrypsin (as shown in green) is responsible for establishing the enzyme's lock-and-key fit with the substrate. The target protein chain lies across the surface of the enzyme, running from the lower right to the upper left, with its target phenylalanine, tyrosine, or tryptophan inserted downward into the complementary-shaped binding pocket. The amino acids in the binding pocket form interactions with the substrate that align it in such a way that its target peptide bond is exposed to the enzyme's catalytic active site. This is made up of the key catalytic amino acid Serine-195 (shown in blue), flanked by Histidine-57 and Aspartic acid-102 on the left, which increase the serine's nucleophilicity, and Glycine-193 on the right, whose backbone NH group stabilizes the transition state during the course of the reaction. *Above:* A ball-and-stick representation of the binding site of chymotrypsin, occupied by the substratelike compound formyl-tryptophan. The figure shows the approximate placement of the hydrophobic side chain of tryptophan in the binding pocket, and the stabilization of the "substrate" through a hydrogen bond formed between its backbone NH group and the C=O group of Serine-214 in the enzyme. Also shown are the two hydrogen bonds between the target C=O group and the backbone N—H groups of Glycine-193 and Serine-195 that will form when the substrate passes through the high-energy transition state during the course of the reaction.

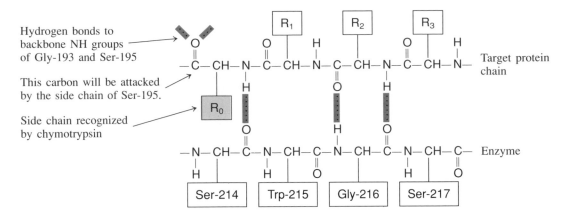

Hydrogen bonds to backbone NH groups of Gly-193 and Ser-195

This carbon will be attacked by the side chain of Ser-195.

Side chain recognized by chymotrypsin

Target protein chain

Enzyme

The formation of hydrogen bonds that stabilize the enzyme–substrate complex formed between chymotrypsin and a target polypeptide chain.

drophobic pocket for the target amino acid, the enzyme also forms several hydrogen bonds with the substrate. These serve both to increase the strength of binding and to tie down the substrate in position for reaction. The orientation of the substrate is stabilized by an important hydrogen bond that forms between the backbone N—H group of the target amino acid and the nearby backbone C=O group of Serine-214. This hydrogen bond initiates a short region of antiparallel β-sheet between Serine-214, Tryptophan-215, and Glycine-216 of the enzyme and the three amino acids of the substrate protein chain leading up to the target peptide bond. In sum, when these hydrogen bonds are formed and phenylalanine, tyrosine, or tryptophan is properly seated in the hydrophobic pocket, the C=O group of the target amino acid is positioned where it can be attacked by Serine-195.

Thus, as it settles in, the target protein chain becomes tightly bonded to the enzyme and loses much of its translational and rotational freedom. Without these ties, the substrate would rattle around in the active site, and the efficiency of catalysis would be low. The figure on the opposite page illustrates this point. It shows how the catalytic efficiency rises for a set of related substrates that make an increasing number of contacts and become ever

more precisely oriented on the surface of the enzyme. For the larger compounds, which make the most favorable contacts, the rate of catalysis is enhanced a millionfold.

In sum, our analysis of the substrate recognition site has shown that the enzyme and the substrate are complementary in shape, proving Fischer's original lock-and-key concept. *It is this aspect of the design of the enzyme that not only promotes the reaction—by positioning the substrate near several amino acid side chains that function as acid and base catalysts—but also geometrically restricts the enzyme's powerful catalytic groups so that they work only on the appropriate substrate.* The cell thus maintains a high degree of control over the chemical transformations that occur within it.

Trypsin

We are now ready to see our first example of the evolution of a new enzyme—it is both dramatic and subtle. Let us consider the enzyme trypsin, which, like chymotrypsin, is a digestive protein produced by the pancreas. It differs from chymotrypsin with respect to the amino acids it at-

	SUBSTRATE	RELATIVE CATALYTIC EFFICIENCY

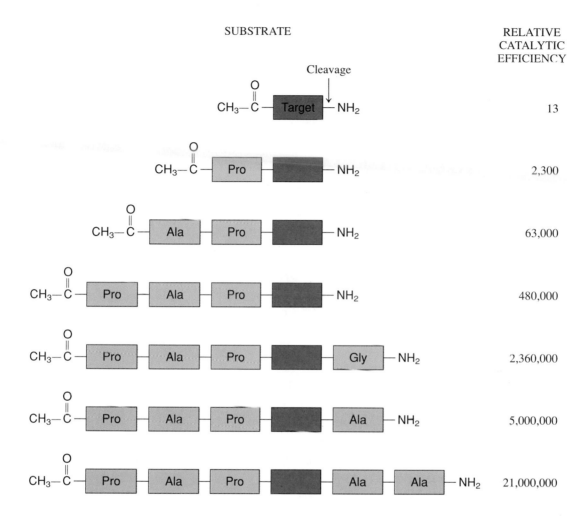

Relative rates of cleavage by a chymotrypsin-like enzyme acting on a series of related substrates. When the substrate is very small (top line), it is not firmly anchored and rattles around in the enzyme's active site. Most of the time its target C=O group is not properly oriented with respect to Serine-195, and catalysis is slow. Adding a second and third amino acid allows the formation of hydrogen bonds between the substrate and amino acids 214 to 216 in the enzyme. The target protein chain is more firmly anchored, and the relative catalytic efficiency is increased as much as a millionfold.

In general, the relative catalytic efficiency is a function of two factors: (1) how perfectly the substrate is held in the binding site, exposing it to the active-site catalytic amino acids, and (2) the chemical properties of the substrate itself—for example, whether the "departing group" is more or less easily displaced. As the bottom of the figure shows, a simple —NH$_2$ is harder for a chymotrypsin-like enzyme to displace, slowing cleavage, than is —Ala—NH$_2$ or —Ala—Ala—NH$_2$, both of which are better able to stabilize the extra electron pair from the peptide bond that must be carried off during cleavage.

1st amino acid of chymotrypsin →

9th amino acid of trypsin →

Cys	Gly	Val	Pro	Ala	Ile	Gln	Pro	Val	Leu	Ser	Gly	Leu	Ser	Arg	15
Phe	Val	Ala	Ala	Ala	Leu	Ala	Ala	Pro	Phe	Asp	Asp	Asp	Asp	Lys	
Ile	Val	Asn	Gly	Glu	Glu	Ala	Val	Pro	Gly	Ser	Trp	Pro	Trp	Gln	30
Ile	Val	Gly	Gly	Tyr	Asn	Cys	Glu	Glu	Asn	Ser	Val	Pro	Tyr	Gln	
Val	Ser	Leu	Gln	Asp	Lys	Thr	Gly	Phe	His	Phe	Cys	Gly	Gly	Ser	45
Val	Ser	Leu	—	—	Asn	Ser	Gly	Tyr	His	Phe	Cys	Gly	Gly	Ser	
Leu	Ile	Asn	Glu	Asn	Trp	Val	Val	Thr	Ala	Ala	His	Cys	Gly	Val	60
Leu	Ile	Asn	Glu	Gln	Trp	Val	Val	Ser	Ala	Gly	His	Cys	Tyr	Lys	
Thr	Thr	Ser	Asp	Val	Val	Val	Ala	Gly	Glu	Phe	Asp	Gln	Gly	Ser	75
Ser	Arg	Ile	Gln	Val	Arg	Leu	—	Gly	Glu	His	Asn	Ile	Glu	Val	
Ser	Ser	Glu	Lys	Ile	Gln	Lys	Leu	Lys	Ile	Ala	Lys	Val	Phe	Lys	90
Leu	Glu	Gly	Asn	Glu	Gln	Phe	Ile	Asn	Ala	Ala	Lys	Ile	Ile	Arg	
Asn	Ser	Lys	Tyr	Asn	Ser	Leu	Thr	Ile	Asn	Asn	Asp	Ile	Thr	Leu	105
His	Pro	Gln	Tyr	Asp	Arg	Lys	Tyr	Leu	Asn	Asn	Asp	Ile	Met	Leu	
Leu	Lys	Leu	Ser	Thr	Ala	Ala	Ser	Phe	Ser	Gln	Thr	Val	Ser	Ala	120
Ile	Lys	Leu	Ser	Ser	Arg	Ala	Val	Ile	Asn	Ala	Arg	Val	Ser	Thr	
Val	Cys	Leu	Pro	Ser	Ala	Ser	Asp	Asp	Phe	Ala	Ala	Gly	Thr	Thr	135
Ile	Ser	Leu	Pro	Thr	Ala	Pro	Pro	—	—	Ala	Thr	Gly	Thr	Lys	
Cys	Val	Thr	Thr	Gly	Trp	Gly	Leu	Thr	Arg	Tyr	Thr	Arg	Ala	Arg	150
Cys	Leu	Ile	Ser	Gly	Trp	Gly	Asn	Thr	Ala	Ser	Ser	Gly	Ala	Asp	
Thr	Pro	Asp	Arg	Leu	Gln	Gln	Ala	Ser	Leu	Pro	Leu	Leu	Ser	Asn	165
Tyr	Pro	Asp	Glu	Leu	Gln	Cys	Leu	Asp	Ala	Pro	Val	Leu	Ser	Gln	
Thr	Asn	Cys	Lys	Lys	Tyr	Trp	Lys	Thr	Lys	Ile	Lys	Asp	Ala	Met	180
Asp	Lys	Cys	Glu	Ala	Ser	Tyr	Pro	Lys	Lys	Ile	Thr	Ser	Asn	Met	
Ile	Cys	Ala	Gly	—	—	Ala	Ser	Lys	Val	Ser	Ser	Cys	Met	Gly	193
Phe	Cys	Val	Gly	Phe	Leu	Glu	Gly	Gly	Lys	Asp	Ser	Cys	Gln	Gly	
Asp	Ser	Gly	Gly	Pro	Leu	Val	Cys	Lys	Lys	Asn	Gly	Ala	Trp	Thr	208
Asp	Ser	Gly	Gly	Pro	Val	Val	Cys	Asn	Gly	Gln	—	—	—	—	
Leu	Val	Gly	Ile	Val	Ser	Trp	Gly	Ser	Ser	Thr	Cys	Ser	Thr	Ser	223
Leu	Gln	Gly	Val	Val	Ser	Trp	Gly	Asp	Gly	Cys	Ala	Gln	Lys	Asn	
Thr	Pro	Gly	Val	Tyr	Ala	Arg	Val	Thr	Ala	Leu	Val	Asn	Trp	Val	238
Lys	Pro	Gly	Val	Tyr	Thr	Lys	Val	Tyr	Asn	Tyr	Val	Lys	Trp	Ile	
Gln	Gln	Thr	Leu	Ala	Ala	Asn									
Lys	Asn	Thr	Ile	Ala	Ala	Asn	Ser								

← Last amino acids of chymotrypsin and trypsin

tacks in a target polypeptide chain, cleaving after positively charged amino acids, such as lysine and arginine, rather than after the large, planar amino acids phenylalanine, tyrosine, and tryptophan. Trypsin complements chymotrypsin in the degradation of proteins in the small intestine. When ingested proteins are acted upon by both trypsin and chymotrypsin, they are broken into smaller fragments than would be produced by either enzyme alone. The increased number of free ends generated gives the final protein-digesting enzymes in the system greater access to the fragments, from which the enzymes remove subunits one at a time to yield an assortment of free amino acids. It goes without saying that the faster an organism can extract metabolic building blocks from food, the faster it can grow, and thus there is clear evolutionary pressure for the development of an enzyme like trypsin.

How trypsin arose is apparent from an examination of the structure of the protein. The amino acid sequence of trypsin is strikingly similar to that of chymotrypsin. Once again the enzyme is a protein of about 250 amino acids, and if the sequences of the two enzymes are laid out next to one another, an almost perfect alignment can be achieved. At each position, there is an approximately 50 percent chance that the amino acids are the same, as indicated by the orange shading in the diagram on the facing page. This degree of homology is even more impressive than it appears, for when the three-dimensional structures of trypsin and chymotrypsin are examined they are found to be virtually identical. In fact, it would be difficult to distinguish the two enzymes simply by looking at the path traveled by the polypeptide backbone. Most of the subunits that differ between trypsin and chymotrypsin occur

on the surface of the three-dimensional structure—where the projecting amino acid side chains interact only with the surrounding solvent and are thus "free" to change. The amino acids in the interior are the ones that are conserved—evidently because their specific interactions are responsible for achieving the final folded structure of the enzyme.

The similarity in structure between trypsin and chymotrypsin foreshadows a similarity in mechanism of action. In both enzymes, the key catalytic amino acid, serine, is located at approximately position 195 in the linear amino acid sequence and, in the folded enzyme, is brought together with the other amino acids of the catalytic triad, histidine and aspartic acid. Moreover, both enzymes are inactivated by diisopropyl phosphofluoridate (DPF), which becomes covalently attached to the active site serine. In fact, the inhibitory effect of DPF provides a quick method of identification for the members of the chymotrypsin family. Collectively, these enzymes are referred to as *serine proteases,* denoting their reliance on a key serine that lies at the heart of the catalytic active site.

The difference in substrate specificity between trypsin and chymotrypsin results from a modification in the binding pocket adjacent to the catalytic active site. In chymotrypsin, the pocket is lined entirely with hydrophobic amino acids—appropriate for interaction with phenylalanine, tyrosine, and tryptophan. In trypsin, one of the amino acids at the bottom of the pocket, Serine-189 (page 202), has changed to aspartic acid. Because this amino acid is negatively charged, it promotes the binding of a positively charged side chain. Thus, trypsin binds to and cleaves polypeptide chains after lysine and arginine subunits.

The variation in protein structure shown by trypsin and chymotrypsin is basically conservative—the architecture of both proteins is very similar. Nevertheless, we have seen that subtle differences are sufficient to allow the two enzymes to react with different substrates while continuing to use the same catalytic mechanism.

The way in which new enzymes evolve is explained by the science of genetics. In essence, the information for properly aligning the amino acids of a protein is permanently stored in the sequence of nucleotides in the chromosomal DNA. During evolution, random mutations occasionally alter the nucleotides in the DNA of a gene—for example, through mistakes made during copying. This

When aligned, the amino acid sequences of chymotrypsin and trypsin show a high degree of homology. Position by position, nearly half of the amino acids are identical (orange shading). At many other positions, changes occur but are conservative (gold shading), such as the change of arginine to lysine at position 15. Both of these amino acids are positively charged, thus the structure of the protein will not be significantly changed. The dashes represent "missing" amino acids in one protein or the other; they generally occur at shortened "loopouts" where the polypeptide chain turns and changes direction at the surface of the protein.

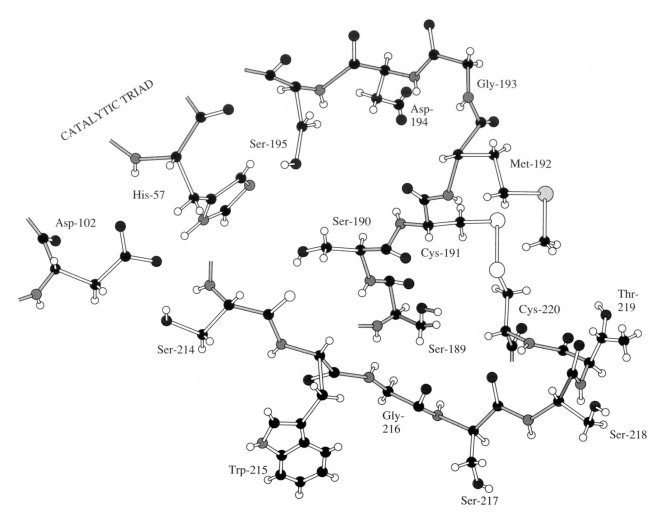

CATALYTIC TRIAD

Gly-193

Asp-194

Ser-195

Met-192

His-57

Ser-190

Asp-102

Cys-191

Thr-219

Cys-220

Ser-214

Ser-189

Trp-215

Gly-216

Ser-218

Ser-217

The binding pocket of chymotrypsin (shown above) contains a serine subunit at the bottom. In the enzyme trypsin, the serine has been replaced by a negatively charged aspartic acid, which by interacting with positively charged lysines and arginines alters the specificity of the enzyme.

leads to changes in one or another amino acid in the final protein. Some of these changes are harmful, disrupting the overall structure of the enzyme or removing an important catalytic amino acid. Many other changes are inconsequential—such as the change of a leucine for a very similar valine subunit at a particular position. As shown on page 202, the aligned sequences of chymotrypsin and trypsin show numerous examples of such conservative amino acid substitutions.

But occasionally, a single change in the DNA of a gene is of immense importance in terms of evolution, altering a critical amino acid and allowing the evolution of a new protein. Assuming that chymotrypsin was the primordial protease, a single mutation could have changed

Serine-189 to aspartic acid and created the enzyme trypsin. This change, however, would mean that the organism, while gaining trypsin, would have lost chymotrypsin and be no better off. There would seem to be no way to generate the diversity of proteases one finds today.

The resolution of this apparent paradox involves a process known as *gene duplication.* As shown in the diagram on the following page, genetic recombination (the exchange of DNA between homologous chromosomes) occurs normally each time an organism produces eggs or sperm. The precise alignment of the chromosomes at the time of this exchange of material ensures that the product chromosomes each have a full set of genes. Misalignment of the chromosomes at the time of their recombination, however (shown on the right side of the diagram), leads to unequal products—one short, genetically deficient chromosome and one long chromosome that contains a duplicated stretch of genes. *The important point is that both copies of a duplicated gene are free to mutate independently as the chromosome passes through successive generations.* Thus one copy is able to retain its original ability to code for chymotrypsin, while the other mutates randomly to produce a variant that eventually becomes trypsin. The stabilization of new mutations in the population occurs through the process of natural selection. If the new enzyme is useful, the organism possessing it will prosper and leave behind more progeny—in essence increasing the frequency of the new gene.

We assumed, for the sake of argument, that chymotrypsin came before trypsin in evolution. Actually, the logic of our argument does not tell us which enzyme arose first. And in fact, the appropriately correct statement is that the modern forms of both chymotrypsin and trypsin have evolved from a common ancestral protein.

Elastase

Whereas chymotrypsin and trypsin cleave after particular target amino acids in any protein, our next serine protease, *elastase,* has evolved a higher degree of specificity. It is not a generalized protease but, rather, is directed against a specific and rather remarkable protein, *elastin.*

Elastin is an important protein in a number of body tissues, serving as a major structural element in ligaments and in the walls of blood vessels. In marked contrast to all the other proteins we have considered, elastin does *not* have a well-defined three-dimensional structure. Rather it appears to have been designed to be easily deformed. Its flexibility is achieved through the highly unusual organization of the protein. The basic elastin unit is a 700–amino acid polypeptide chain with an extraordinarily high content of hydrophobic amino acids—for example, alanine, glycine, proline, and valine. Charged residues are few—mostly lysines—and serve a particular purpose. Their side chains are enzymatically joined together by unusual lysine–lysine cross-links so that thousands of individual elastin chains become parts of a larger covalently connected structure. The overall effect is that of a mesh net. The regions of elastin between the lysine cross-links contain the protein's hydrophobic amino acids, with the sequence Val-Pro-Gly-Val-Gly-Val occurring repeatedly. Within this sequence, the Val-Pro-Gly-Val segment is especially prone to forming β-turns (Chapter 4), and it is thought that the structure of the elastin protein between cross-links consists of a series of β-turns, compacting the protein chain into a spiral. These β-turns are probably responsible for the remarkable flexibility of elastin fibers, which can stretch to several times their original length and then rapidly return to their original size and shape when the tension is released. A disruption of the hydrogen bonds in the β-turns may be the mechanism that allows individual polypeptide chains to stretch out when stressed. At all times, covalent lysine cross-links maintain the overall strength and structural integrity of the intact elastin fiber.

The enzyme system that cross-links elastin fibers continues to work throughout life. As a consequence, the number of cross-links in elastin increases as we age. The resulting reduction in the elasticity of arterial walls may be a contributory factor to cardiovascular disease.

Naturally the food eaten by carnivorous animals contains blood vessels and ligaments whose elastins must be digested. In addition, the repair of damaged or diseased tissue requires the degradation of elastin fibers. These are the tasks of elastase—an enzyme made both by the pancreas for digestive functions and by white blood cells for tissue remodeling. This enzyme is another variation on the

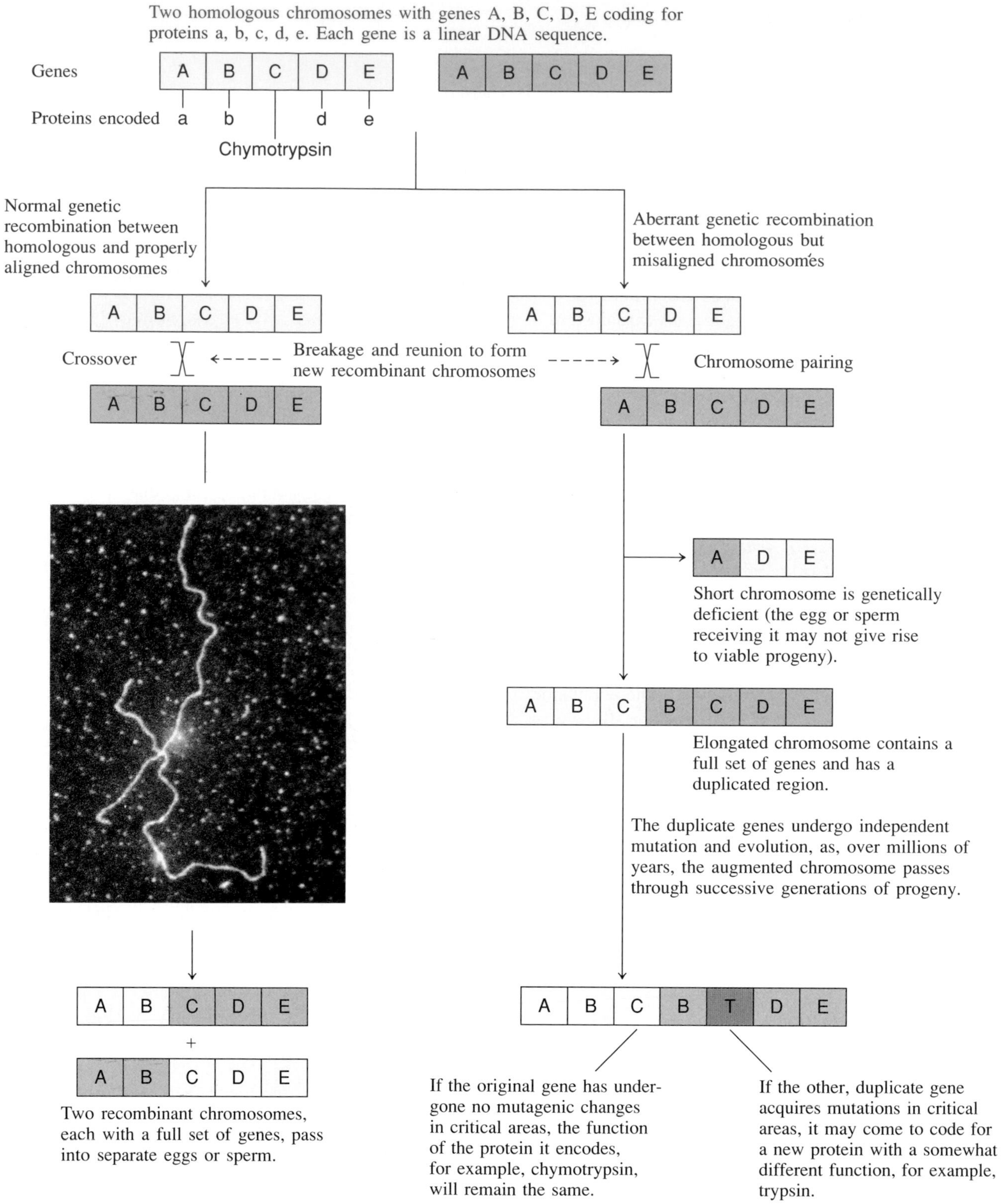

Two homologous chromosomes with genes A, B, C, D, E coding for
proteins a, b, c, d, e. Each gene is a linear DNA sequence.

Genes | A | B | C | D | E | | A | B | C | D | E |

Proteins encoded   a   b   d   e

Chymotrypsin

Normal genetic
recombination between
homologous and properly
aligned chromosomes

Aberrant genetic recombination
between homologous but
misaligned chromosomes

| A | B | C | D | E | | A | B | C | D | E |

Crossover   )(← - - - - Breakage and reunion to form   - - - - →   )(Chromosome pairing
                             new recombinant chromosomes

| A | B | C | D | E | | A | B | C | D | E |

| A | D | E |

Short chromosome is genetically
deficient (the egg or sperm
receiving it may not give rise
to viable progeny).

| A | B | C | B | C | D | E |

Elongated chromosome contains a
full set of genes and has a
duplicated region.

The duplicate genes undergo independent
mutation and evolution, as, over millions of
years, the augmented chromosome passes
through successive generations of progeny.

| A | B | C | D | E |
+
| A | B | C | D | E |

Two recombinant chromosomes,
each with a full set of genes, pass
into separate eggs or sperm.

| A | B | C | B | T | D | E |

If the original gene has under-
gone no mutagenic changes
in critical areas, the function
of the protein it encodes,
for example, chymotrypsin,
will remain the same.

If the other, duplicate gene
acquires mutations in critical
areas, it may come to code for
a new protein with a somewhat
different function, for example,
trypsin.

Mechanism for gene duplication and the evolution of new enzymes. Aberrant recombination between homologous chromosomes gives rise to a chromosome containing two identical gene segments. Subsequently, independent mutations accumulate in each of the duplicated genes and allow the development of proteins with altered function—that is, enzyme evolution. The electron micrograph shows two chromosomes in the process of genetic recombination.

evolutionary theme of chymotrypsin and trypsin. The 250 amino acids of elastase align impressively with chymotrypsin and trypsin, and its three-dimensional structure is superimposable with that of these proteases. Like chymotrypsin and trypsin, elastase is a serine-based protease, with a catalytic triad—histidine, serine, and aspartic acid—that is blocked by DPF. Elastase differs from chymotrypsin and trypsin in its substrate recognition site, which has been modified to accommodate the small amino acids glycine and alanine as targets. Within the binding pocket of elastase, two small glycine subunits that occur in both chymotrypsin and trypsin are replaced by much larger amino acids. Specifically, Glycine-216 has changed to

Homology of Serine Proteases in the Region of the Catalytic Triad

Enzyme	Active site serine
Chymotrypsin	Gly—Asp—Ser —Gly—Gly—Pro —Leu
Trypsin	Gly—Asp—Ser —Gly—Gly—Pro —Val
Elastase	Gly—Asp—Ser —Gly—Gly—Pro —Leu

Enzyme	Active site histidine
Chymotrypsin	Val —Thr —Ala —Ala —His —Cys—Gly
Trypsin	Val —Ser —Ala —Gly—His —Cys—Tyr
Elastase	Leu—Thr —Ala —Ala —His —Cys—Ile

Enzyme	Active site aspartic acid
Chymotrypsin	Thr —Ile —Asn—Asn—Asp—Ile —Thr
Trypsin	Tyr —Leu—Asn—Asn—Asp—Ile —Met
Elastase	Ser —Lys—Gly—Asn—Asp—Ile —Ala

valine, and Glycine-226 has become threonine. These more bulky side chains block off the entrance to the enzyme's substrate-binding pocket so that phenylalanine, tyrosine, tryptophan, lysine, and arginine cannot be accommodated. Small, hydrophobic amino acids can fit, however, and elastin with its high content of glycine (33 percent) and alanine (23 percent) becomes a primary target. In elastase, we begin to see the targeting of an enzyme with general catalytic capability so that it functions against a specific biomolecule.

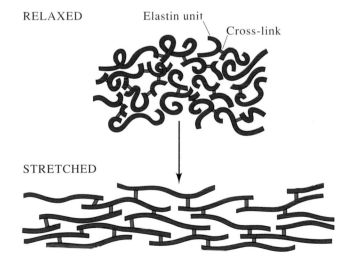

RELAXED

Elastin unit

Cross-link

STRETCHED

Thousands of covalently cross-linked elastin chains form a net that can stretch in various directions and then return to its original size and shape.

Perspective

In this chapter we see unfolding an overall strategy for extending enzyme function. The catalytic amino acids in a chymotrypsin-like enzyme are held constant while the adjacent surface of the protein changes its lock-and-key specificity to enable the enzyme to interact with different substrates. As we will now see, the extensive use of this strategy allows the enzyme to range far afield from its

THE CONTROL OF ENZYME ACTIVITY

Proteolytic enzymes are powerful agents and must be carefully controlled within the body. Two major forms of control for these enzymes have evolved. The first involves the synthesis and storage of the enzyme in an inactive form. The second control mechanism is based on the synthesis of *antiproteases*—non-enzymatic blocking proteins that specifically interact with their respective proteases in a lock-and-key manner, inactivating them.

Proenzymes

Most enzymes acquire their biological activity immediately after synthesis, as they spontaneously fold into their characteristic three-dimensional forms. For other enzymes, however, especially those that carry out potentially dangerous cleavage reactions, such a spontaneous activation process could lead to cell death. An example is provided by the serine proteases, which must be made by the cells of the pancreas without subjecting the cells to internal digestion. To solve this problem, such enzymes are synthesized as inactive precursors, known as *proenzymes* (or zymogens). Their activation occurs by the removal of a few specific amino acid subunits from the primary folded protein, which allows the remaining protein chain to fold up in a slightly different way, leading to activity.

Chymotrypsin, trypsin, and elastase provide a clear example of the proenzyme mechanism for the control of enzyme activity. For instance, when chymotrypsin is synthesized by the cells of the pancreas, it is initially a single chain of 245 amino acids, called chymotrypsinogen. This proenzyme is virtually devoid of enzymatic activity because its initial folded structure is slightly, but significantly, different from that of the mature active enzyme. Activation occurs after the proenzyme has been secreted from the pancreas into the intestine. The single chymotrypsinogen polypeptide chain is cleaved between Arginine-15 and Isoleucine-16 by an already active protease in the intestine with the appropriate specificity. After the initial cut, Serine-14 and Arginine-15 are trimmed back from the newly exposed carboxyl (COO$^-$) end. In a similar fashion, a second cut is made after Asparagine-148, and Threonine-147 and Asparagine-148 are excised. The resulting protein, still held together by disulfide linkages and a variety of noncovalent bonds, remains tightly folded

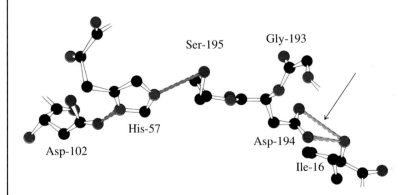

Ser-195

Gly-193

His-57

Asp-102

Asp-194

Ile-16

The arrow points to the linchpin ionic bond between Isoleucine-16 and Aspartic acid-194 in the active chymotrypsin enzyme. Prior to cleavage, the amino group of Isoleucine-16 is part of a peptide bond and not free to participate in this interaction.

and does not fall apart. Rather, it displays the three chains we associate with mature chymotrypsin: A, B, and C. However, the folded structure of the enzyme now has new degrees of freedom that allow it to undergo additional structural changes, at last arriving at the active form.

The conformational changes that lead to the active form of chymotrypsin have been elucidated by Joseph Kraut of the University of California, San Diego. He determined the three-dimensional structure of the chymotrypsinogen proenzyme and compared it to the structure of mature, active chymotrypsin. In both cases, all three members of the catalytic triad were in their proper places. The most important difference in structure involved the newly created beginning of the B chain, at Isoleucine-16. It could be seen that, in active chymotrypsin, the amino-terminal group of Isoleucine-16 (NH_3^+) had turned inward and established an interaction with the negatively charged Aspartic acid-194 subunit near the active site of the enzyme. This new electrostatic interaction triggers two conformational changes that result in a local reorganization of the protein and the final activation of the enzyme. Specifically, the boundary of the substrate recognition site formed by subunits 189 to 192 now becomes properly positioned. Moreover, Glycine-193, one of the two amino acids whose backbone NH group will help stabilize the tetrahedral transition state intermediate, is shifted into its exact final position.

The major point is that, in many instances, enzyme activity is controlled through the cleavage of a few specific peptide bonds in an inactive proenzyme, allowing the structural changes necessary for enzyme activation.

Antiproteases

The activation of a proenzyme is irreversible and hence can occur just once in the life cycle of a protein. Thus, another mechanism is required for later switching off the activity of such an enzyme. In the case of the serine proteases, this is accomplished by the synthesis of specific inhibitory proteins (called serpins, *serine protease inhibitors*) that bind very tightly to the protease active sites to block their activity.

A particularly interesting example of an antiprotease

is the one directed against elastase. It is produced in the liver and secreted into the blood, where it regulates the activity of the elastase released from white blood cells during the various repair and tissue remodeling activities that follow injury and infection. The antiprotease directed against elastase thus protects tissues from digestion after the tissue has healed and the elastase molecules have no further function. Some individuals carry a heritable mutation in their antielastase gene that reduces the secretion of the inhibitor from liver cells. There is a low (about 15 percent of normal) concentration of the inhibitor in the blood and correspondingly in the surrounding tissue fluids. As a consequence, an uninhibited overactive elastase destroys exposed elastin networks, particularly in the walls of the alveoli, the small, delicate air sacs in the lungs. Gradually, this condition develops into the disease emphysema, and an affected individual has a much reduced efficiency in breathing.

Cigarette smoking markedly increases the likelihood that this form of inherited emphysema will develop. Components in the smoke chemically react with a methionine subunit that projects from the surface of the inhibitor—a subunit essential for its lock-and-key interaction with elastase. The resulting methionine "sulfoxide" side chain is aberrant. Even though it results from just the introduction of one extra oxygen atom into methionine, the altered inhibitor is distorted and rendered nonfunctional. This is by

ANTIPROTEASE $\quad$ ▮— CH_2—CH_2—S—CH_3
Methionine

↓

$$\text{INACTIVE ANTIPROTEASE} \quad ▮— CH_2—CH_2—\overset{\overset{\text{O}}{\|}}{S}—CH_3$$
Methionine sulfoxide

no means the only harmful effect of smoking, but it does provide a clear molecular example of how agents from the environment can interact with important body components, disrupting enzyme systems and, ultimately, body physiology.

original role in digestion as it participates in increasingly complex physiological processes.

Our next examples of the chymotrypsin family of enzymes introduce an entirely different aspect of biochemical physiology now under the control of the serine proteases—blood clotting.

Enzymes and the Clotting of Blood

Like other organs of the body, blood is composed primarily of cells—red blood cells that carry oxygen and white blood cells that fight off infection. Unlike the cells of other organs, the cells of the blood are not organized into a solid tissue. Whereas normally cells are compacted together and separated only by a thin channel of interstitial fluid reinforced with structural proteins and carbohydrates that form a connective-tissue matrix, in blood the interstitial fluid is much more prominent and the matrix material is essentially absent. As a consequence, the cells of the

blood float freely in their surrounding fluid, called plasma. The plasma itself is a complex collection of substances—including, for example, dissolved, small metabolites absorbed from the intestine, large carrier proteins that transport oxygen, lipids, and other insoluble molecules, numerous enzymes that function in the extracellular spaces of the body, hormones secreted by one set of cells and destined to send signals to distant tissues, and a variety of antibody proteins that form lock-and-key interactions with invading microbes, inactivating them.

After this living river leaves the heart in a single large artery, the aorta, it is distributed into a ramifying network of ever-smaller tubes that finally become capillaries—channels so minute that the blood cells must pass through them in a single file. All along their length, the walls of the capillaries are formed by single cells, flattened and rounded into a tube. It is during the blood's passage through the capillaries, as they extend through the organs of the body, that the blood exchanges its many life-sustaining components with the surrounding cells. Oxygen and nutrients move through the thin capillary walls, perfuse into the surrounding tissue fluid, and are absorbed by the adjacent cells. At the same time, the cells release sub-

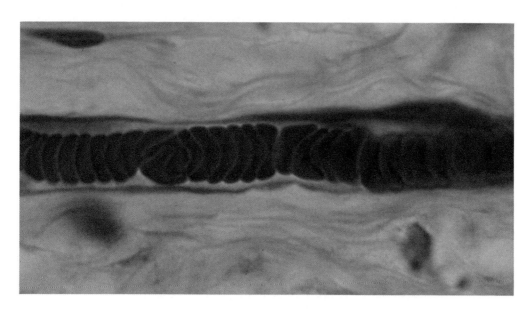

Red blood cells passing single file through the smallest of blood vessels, a capillary.

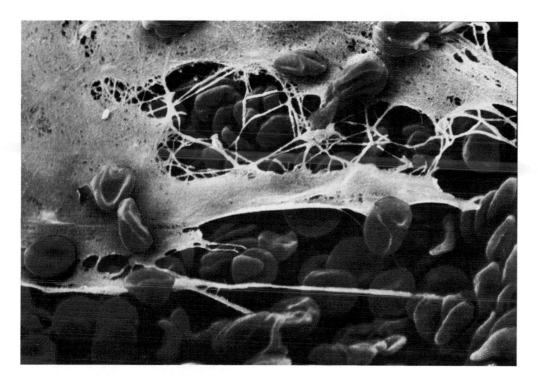

A blood clot is formed by fibrin filaments entrapping platelets and blood cells.

stances ranging from waste compounds to hormones, and this material travels the reverse journey through the tissue fluid into the capillaries, which then converge into ever-larger veins. In all, there are more than 50,000 miles of large and small vessels and capillaries through which the blood flows—an internal transport system kept in constant motion by a highly specialized muscle, the heart.

Whereas damage to most organs—even such important organs as the heart or the brain—can remain localized in a small area, damage to the circulatory system is inherently systemic, affecting the entire body, and can be immediately life threatening. A single break and the fluid, which is maintained under high pressure, begins to drain rapidly from the system.

Remarkably, however, the fluid mosaic structure of the blood is like a self-healing tapestry. Should it develop a tear, it is sewn back together by invisible needles—which are none other than serine proteases whose chymotrypsin-like activity has been specially evolved for this purpose.

Almost daily, breaks occur in the circulatory system as a result of ordinary wear and tear. Falls, cuts, splintering bones, and other accidents cause more serious damage. To cope with these ruptures, a highly sophisticated set of biochemical reactions has been evolved to bring about *hemostasis*, the cessation of bleeding. Basically, hemostasis can be divided into three steps. The first is *constriction* of the injured artery or vein, caused by the contraction of the layer of muscle tissue surrounding the vessel. Constriction of the vessel immediately reduces the flow of blood through the break. The second event is the clumping of *platelets*—small oval cells, derived by

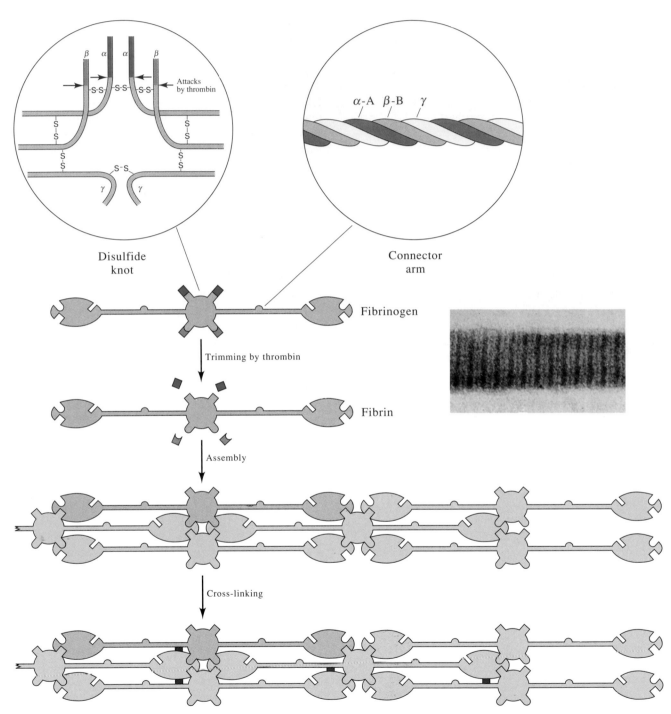

β α α β

Attacks
by thrombin

γ γ

Disulfide
knot

α-A β-B γ

Connector
arm

Fibrinogen

Trimming by thrombin

Fibrin

Assembly

Cross-linking

Fibrin filament

fragmentation of huge bone marrow cells called megakaryocytes, and present in the blood at about 5 percent the level of red blood cells. Platelets have the useful property of adhering tightly to the protein fibers that constitute much of the connective matrix between cells. They therefore aggregate in an area where blood is draining from a vessel into the surrounding tissue. Not only do platelets provide a temporary plug, they also mobilize the third and decisive event in hemostasis—the conversion of a usually soluble blood protein, *fibrinogen,* to an insoluble form, *fibrin*. The final result is the development of a three-dimensional network of protein filaments that forms a strong blood clot.

Although the sequence of events leading to the formation of a blood clot is highly complex, the last stage—the conversion of fibrinogen to fibrin—is well understood. It involves a controlled change in the structure of the fibrinogen molecule.

Fibrinogen

Before clotting occurs, fibrinogen is simply a large inert blood protein. Although X-ray crystallographic data are not yet available to define its structure precisely, electron microscopy and traditional methods of protein chemistry

have provided a general description of the molecule. Fibrinogen is constructed from six polypeptide chains of about 500 amino acids apiece. The molecule contains two copies each of chains designated types α-A, β-B, or γ. The amino terminals of the six chains are covalently linked to one another by a network of disulfide (S—S) bonds, forming a thick, central core. Projecting outward from the disulfide core, the α-A, β-B, and γ chains, on each side of the molecule, adopt an α-helical conformation and then wrap around each other to form two elongated arms that resemble triple-stranded ropes. After about 100 amino acids, these helical areas terminate, and the ends of the three chains ramify to form closely aligned but individually folded units that create thickened terminal domains. Thus, the overall fibrinogen protein is not spherical but extraordinarily elongated—a cylinder about 450 Å long and 100 Å in diameter. In essence, it is composed of three nodules joined by two connector arms.

As a free protein, fibrinogen is soluble as it floats through the blood. Both the disulfide cores and the terminal domains are negatively charged; the resulting mutual repulsion of fibrinogen molecules ensures that they remain as separate units. However, the properties of fibrinogen are dramatically changed if about 3 percent of the amino acids are removed from the central core—the area where the six chains are held together by disulfide bonds. The removal of about 20 amino acids from the N-terminals of both the α-A and β-B chains releases material with a high content of negatively charged aspartic and glutamic acid subunits. As a result, the core area of fibrinogen changes in net charge from -8 to $+5$, and its surface properties are altered significantly. The modified molecules, now called *fibrin monomers,* are no longer subject to electrostatic repulsion. On the contrary, the $+5$ charges on the central domains are now complementary to the -4 charges that exist on the terminal domains. Moreover, the proteolytic removal of the A and B peptides from the α and β chains exposes new amino terminals, which become the contact sites for lock-and-key interactions with pockets in the terminal domains of other fibrin monomers. Individual monomers now interact with one another in a side-by-side and end-to-end fashion—a "polymerization" event that leads to the formation of increasingly long and thick protein fibers.

In its inert form, the blood-clotting protein fibrinogen is composed of six polypeptide chains (two each of α-A, β-B, and γ) joined together at their amino-terminal ends by a network of covalent disulfide bonds (top). The enlargement in the left circle depicts the α-A $\rightarrow$ β-B $\rightarrow$ γ interchain connection in a schematic form. The enlargement in the right circle shows the structure of the connector arms, in which the α, β, and γ chains interwrap to form a triple helix. The conversion of fibrinogen to fibrin monomers results from the cleavage of four specific arginine–glycine bonds that occur near the amino terminals of both the α-A and β-B chains. After this cleavage, and the release of the negatively charged fibrinopeptides, the fibrin monomers are able to interact with each other and assemble into long protein filaments forming the backbone of the blood clot. The electron micrograph shows alternating light and dense regions every 230 Å, as would be expected for trinodular fibrin monomers that interacted in the staggered way shown in the diagram.

Fibrin monomers are held together initially only by weak noncovalent interactions, such as hydrogen and hydrophobic bonds. Gradually, however, the individual monomers become cross-linked to one another through the action of a blood-clotting enzyme that covalently connects the side chains of specific glutamine and lysine subunits.

The result of the polymerization of fibrin monomers and their subsequent cross-linking is the formation of long, stable fibrin cables. These form a mesh in the localized region of the circulatory system where fibrinogen has been converted into fibrin. As this three-dimensional lattice traps large numbers of platelets, red blood cells, and white blood cells, a strong, gelatinous blood clot forms. The effectiveness of the fibrin network in promoting clot formation is indicated by the fact that the final clot can contain as little as one-twentieth of 1 percent of fibrin protein, with the rest of the clot coming from the cells that have become trapped in the network of fibers.

Thrombin

The proteolytic cleavage that converts fibrinogen into fibrin monomers is the major "switch" thrown in the blood-clotting reaction, and it is carried out by an enzyme called *thrombin*. Thrombin is normally present in the blood in small amounts and in an inert form. The enzyme becomes activated at the end of a complex sequence of events, which is initiated by damage to the vascular system and leads to the aggregation of platelets at the site of the injury. Once activated, thrombin functions as a protease; it cleaves fibrinogen at four specific arginine-glycine sites that occur near the N-terminals of the α-A and β-B chains. The result, as we have seen, is the production of active fibrin and the formation of a blood clot.

Thrombin is a serine protease—a member of the chymotrypsin family of enzymes. In its structure and mechanism of action, it is closely related to the other proteases we have considered. Thus, the 250 amino acids at the carboxyl terminal of thrombin can be aligned with chymotrypsin, trypsin, and elastase. Moreover, there is a catalytic triad containing histidine, serine, and aspartic acid in their usual positions, and a surface to stabilize the transi-

Homology between Thrombin and Plasmin and Related Serine Proteases in the Region of the Catalytic Triad

Enzyme	Active site serine
Chymotrypsin	Gly —Asp—Ser —Gly —Gly —Pro —Leu
Trypsin	Gly —Asp—Ser —Gly —Gly —Pro —Val
Elastase	Gly —Asp—Ser —Gly —Gly —Pro —Leu
Thrombin	Gly —Asp—Ser —Gly —Gly —Pro —Phe
Plasmin	Gly —Asp—Ser —Gly —Gly —Pro —Leu

	Active site histidine
Chymotrypsin	Val —Thr —Ala —Ala —His —Cys—Gly
Trypsin	Val —Ser —Ala —Gly —His —Cys—Tyr
Elastase	Leu—Thr —Ala —Ala —His —Cys—Ile
Thrombin	Leu—Thr —Ala —Ala —His —Cys—Leu
Plasmin	Leu—Thr —Ala —Ala —His —Cys—Leu

	Active site aspartic acid
Chymotrypsin	Thr —Ile —Asn—Asn—Asp—Ile —Thr
Trypsin	Tyr —Leu—Asn—Asn—Asp—Ile —Met
Elastase	Ser —Lys—Gly—Asn—Asp—Ile —Ala
Thrombin	Asn—Leu—Asp—Arg—Asp—Ile —Ala
Plasmin	Phe—Thr —Arg—Lys—Asp—Ile —Ala

tion state. The specificity of thrombin (it cleaves only at arginine-glycine linkages) suggests that it most closely resembles trypsin, as in fact it does. A negatively charged aspartic acid subunit at the bottom of its binding pocket forms an ionic bond with the positively charged arginine of the target sequence. However, thrombin must be and is much more specific than trypsin. It cleaves only at arginine-glycine linkages, whereas trypsin will cleave protein chains following any lysine or arginine subunit. How is this increased specificity achieved? Apparently, the binding pocket of thrombin is so tightly arranged that it prevents the approach of any arginine not adjacent to glycine, the smallest amino acid, which has only a hydrogen as its side chain. Thus thrombin is a protease with a single substrate (fibrinogen with its exposed arginine-glycine sequences) and does not function as an enzyme of generalized protein degradation. *In thrombin, we see the full* evolution of a general digestive protease, such as chymo-trypsin, trypsin, or elastase, into a complex regulatory protease with a specific target. The use of a serine protease to convert fibrinogen to fibrin is a clear example of nature reaching out with a basic enzyme design to solve a new physiological problem.

A Special Domain for the Activation of Thrombin

Thrombin has another fascinating secret to disclose about the evolution of the chymotrypsin family of enzymes. This is most easily seen if we examine the structure of the pro-enzyme—the inactive form of thrombin that floats freely in the blood, waiting for the signal that will mobilize it into action against fibrinogen. The thrombin proenzyme is much larger than an ordinary serine protease. We recall that chymotrypsin is secreted from the cells of the pancreas as a polypeptide consisting of 245 amino acids, which is activated by the excision of four specific amino acids. Thrombin is secreted into the blood from liver cells, the site of its production, as a polypeptide chain of 600 amino acids, of which only the last 250 are relevant to its serine-protease activity. The N-terminal half of the protein chain folds into several independent domains, each with a different function. One of these domains is used to target the thrombin proenzyme to its site of action. For unlike the digestive serine proteases, which function at random locations throughout the intestine, the thrombin molecule must be able to find an unpredictable site of injury. It is the domain consisting of the first 35 amino acids that accomplishes this by recognizing and binding to the appropriate site.

Among the first 35 amino acids of prothrombin are 10 negatively charged glutamic acid subunits, an unusually high concentration. After the protein has been synthesized, these are converted to dicarboxylic acids by the addition of a second COO^- group to the ends of their side chains. This carboxylation reaction is carried out by an enzyme system that uses the small molecule vitamin K as a helper factor (see page 216). As this enzyme system builds up a core of doubly negatively charged amino acids

at the beginning of prothrombin, it confers a very important property upon the proenzyme. The doubly negatively charged side chains each bind a calcium ion (Ca^{2+}), which, in turn, enables prothrombin to bind to receptor molecules on the phospholipid membranes of aggregated platelets. Thus, the special modification of the initial portion of the thrombin proenzyme, to generate a calcium-binding domain, causes the whole protein chain to anchor at a specific site where blood platelets are responding to an injury. This binding brings prothrombin into close proximity with the blood-clotting enzymes on the surface of platelets that are responsible for its activation.

First amino acid
of prothrombin

Of the first 35 amino acids in prothrombin, 10 are glutamic acid subunits that are converted to doubly negatively charged carboxy-glutamic acid (Gla). This modification allows them to bind calcium, Ca^{2+}, and changes the structure of prothrombin so that it becomes anchored to platelet-derived membrane material. Several of the other enzymes involved in blood clotting contain the same type of domain and are similarly modified. All of these proteins then become localized in the same specific place where they are to function.

The activation of thrombin occurs when the 600-amino acid proenzyme is cleaved at two positions. The proenzyme is first cleaved at the arginine-isoleucine bond between subunits 323 and 324. This generates active thrombin in much the same way that chymotrypsin is activated (page 206). The cleavage allows the formation of an ionic bond similar to the one between the positively charged amino group of Isoleucine-16 and the negatively charged side chain of Aspartic acid-194. This new electrostatic interaction induces localized changes in the folded protein, forming the binding pocket and reorienting certain parts of the catalytic active site.

Nearly simultaneously, the second cleavage in the thrombin proenzyme occurs at the arginine-threonine bond between subunits 274 and 275, severing the terminal 300 amino acids from the rest of the protein chain. Freed of its anchor, the activated thrombin floats away from the site of injury and into the immediately surrounding blood. Now ready to carry out its job, it begins to trim fibrinogen molecules, converting them to fibrin monomers that are capable of spontaneous assembly into long filaments.

The additional domain found in thrombin illustrates a general principle. Enzymes can evolve not only to recognize new and different substrates, but also to acquire additional functions. The additional function acquired by thrombin, and seen in its prothrombin precursor form, is not catalytic; rather, the function of the extra N-terminal domain is to direct the protein to a specific site where it can carry out its physiological role more effectively.

Cascades and Signal Amplification

One might suppose that tissue damage would release a substance or enzyme that activates thrombin directly. This is not the case. On the contrary, a long sequence of events leads from vascular injury to thrombin activation, involving the sequential activation of about 10 "blood-clotting factors." Virtually all of these "factors" are members of the chymotrypsin family of enzymes—serine proteases. Like thrombin, they are all produced by the cells of the liver and released into the bloodstream as inert proenzymes. In response to tissue damage and the aggregation of platelets at the site of injury, the first blood-clotting

factor becomes active. Functioning as a serine protease, it carries out a cleavage reaction that activates the second factor, which, in turn, transmits the signal forward by activating the proenzyme of the third component in the system, and so forth—a chain reaction that ultimately culminates in the proteolytic activation of the key blood-clotting enzyme, thrombin.

Why are so many intermediate steps necessary for the activation of the thrombin proenzyme? Why is the initial response to injury not the direct activation of thrombin, so that it can immediately begin converting fibrinogen into fibrin?

If Paul Revere and William Dawes had stopped to fight the British themselves, Americans would still be paying the Stamp Tax. Instead, Revere and Dawes rode out across the New England countryside and recruited help. The final, if delayed, result was a mobilized response at Lexington and Concord. The blood-clotting mechanism represents a similar logic.

The original tissue damage could, itself, activate only a limited number of prothrombin molecules (say 10) in a given period of time. The 10 enzyme molecules would produce a relatively weak response, resulting in the conversion of a modest amount of fibrinogen to fibrin and allowing the formation of only an occasional thin fibrin filament, which would be swept away by the flowing blood. Instead, the tissue damage activates an equivalent number of molecules of an intermediate protein. Each of the 10 activated copies of this intermediate can then continue to expand the signaling process, yielding, say, 100 active copies of the next component in the system. This *enzyme cascade* thus allows *amplification* of the original stimulatory signal. With several steps in the cascade, the final amplification can be enormous. A virtual army of activated enzymes is mobilized, and increasingly large amounts of the last intermediate in the chain—thrombin—become available to work on the conversion of fibrinogen to fibrin. In sum, the exponentially increasing force that is brought to bear on the problem through the cascade ensures that the final step will be carried out rapidly and forcefully, which is precisely what is desired in forming a blood clot.

Like any other mechanical system, the blood-clotting cascade is subject to failure. This is exactly what happens in bleeder's disease, or *hemophilia*. In hemophiliacs, a

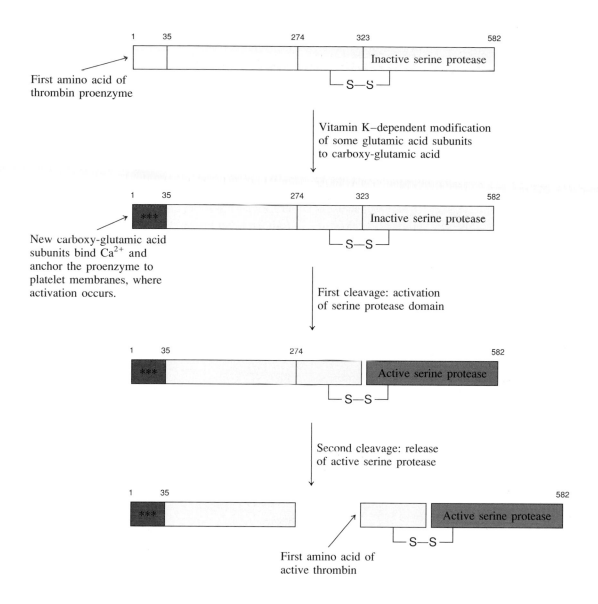

Thrombin is activated from an inert precursor form by two cleavages that occur between amino acids 323 to 324 and 274 to 275. These not only activate the serine protease domain that occurs at the carboxyl end of the protein, but also sever this domain from the calcium-dependent binding domain at the amino terminus that anchors the proenzyme to the site of injury. Two other, less well-defined domains occur in prothrombin between the calcium-binding domain and the serine protease domain. Although not well understood, it is generally believed that these domains also function to direct the precursor enzyme to specific tissue locations.

VITAMIN K AND BLOOD CLOTTING

Vitamins in general are helper factors for specific enzymes. The small molecule commonly known as vitamin K is used in an unknown way by the enzyme that recognizes and converts certain glutamic acid subunits to carboxy-glutamic acid in prothrombin. The thrombin proenzyme synthesized in the absence of vitamin K lacks Ca^{2+}-binding carboxy-glutamic acid subunits. As a consequence, the abnormal proenzyme cannot be properly anchored and activated on the surface of platelet membrane material and the blood-clotting process is inhibited.

A number of compounds occur in nature or can be synthesized in the laboratory that partially resemble vitamin K. Dicoumarol and warfarin are two such molecules. They recognize and bind to the enzyme system that modifies certain glutamic acid subunits, but, being unable to provide the assistance actually given by vitamin K, they thus inhibit the formation of normal prothrombin. This property is useful in both medicine and public health. Dicoumarol and warfarin are used to prevent blood clotting in patients prone to clot formation. In much higher doses, these vitamin K antagonists serve as effective rat poisons. Affected animals hemorrhage to death from the multiple small breaks that routinely occur in their capillaries due to wear and tear.

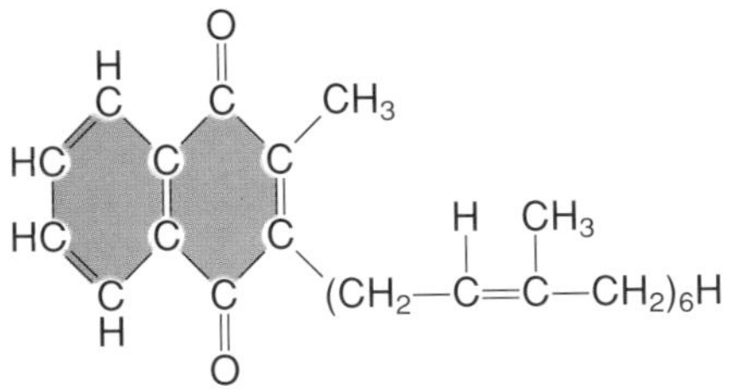

Vitamin K

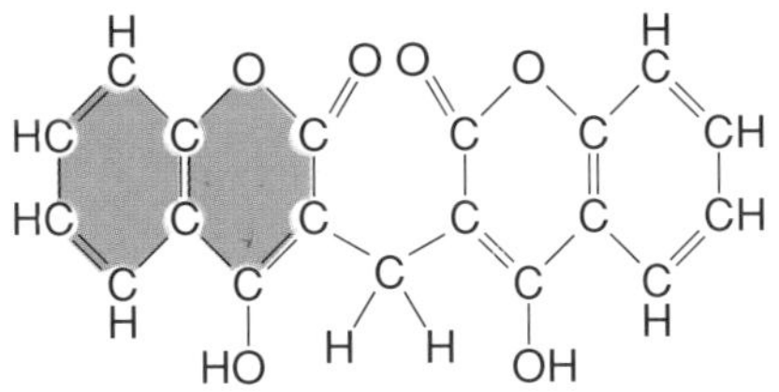

Dicoumarol

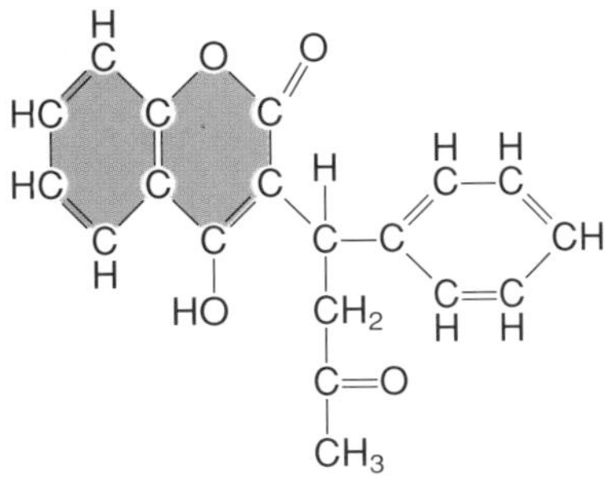

Warfarin

single component of the cascade mechanism, termed Factor VIII, is missing. As a result, thrombin is not activated and fibrinogen is not converted into fibrin clots. The affected individual experiences repeated episodes of internal and external bleeding; a serious accident can prove fatal. Because the enzymes and related proteins involved in blood clotting are so well understood, it is possible to purify Factor VIII from normal blood, administer it to hemophiliacs, and correct their blood clotting deficiency. Indeed, this therapy has proved so successful that whereas previously fewer than a quarter of hemophiliacs reached the age of 16, now most can lead a relatively normal life.

Unlike other diseases—for example, smallpox, diphtheria, or tuberculosis, which occur randomly—hemophilia occurs repeatedly in certain families, indicating that the disease results from the inheritance of a defective gene. In one well-studied case, a C → T mutation occurs in the 2207th triplet codon of the Factor VIII gene. As a result, what is normally a CGA triplet coding for arginine becomes a TGA triplet that fails to code for any amino acid. In the absence of an intact coding sequence, a complete Factor VIII protein cannot be assembled.

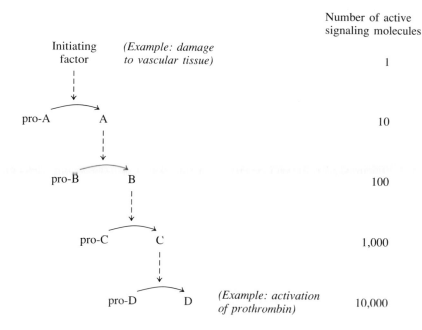

A small initiating event can set in motion an enzyme cascade, culminating in a powerful physiological response. Here a single initiating factor is assumed to convert 10 molecules of proenzyme A to their active state. The cascade then sets in as each of these 10 molecules of A converts 10 molecules of proenzyme B to their active state—for example, via a proteolytic cleavage reaction. The activation process continues exponentially as the 100 active molecules of B each work to activate proenzyme C, and so forth. Even in this simple four-stage sequence, 10,000 molecules of enzyme D have been mobilized by the end to carry out a terminal step—such as the cleavage of fibrinogen to allow the formation of a fibrin clot.

The important point, illustrated by hemophilia and of wide general applicability, is that all the various enzymes and proteins in the body are produced by genes, and when an individual has a mutant gene, the encoded protein may be missing or defective and some aspect of normal physiology will falter.

Controlling the Clotting Reaction

It is important that the rapid and powerful cascade leading to the activation of thrombin not be allowed to get out of hand. The conversion of fibrinogen to fibrin must be confined to the site of injury and not be permitted to extend into the surrounding circulation, where the result could be a massive clot. Several strategies are employed to keep the clotting process within bounds. The most important is based on the fact that the plasma at all times contains a substantial amount of an antiprotease known as *antithrombin*. Even while active thrombin molecules are being produced at the site of injury by the cascade and begin converting fibrinogen into fibrin, they are also being swept along with the blood and becoming progressively diluted. Molecule by molecule, the enzymes encounter antithrombin and are inactivated. Only at the original site of injury is the number of active thrombin molecules high enough to overcome the reservoir of antithrombin inhibitors so that fibrinogen is converted into fibrin. Fulfilling the same goal of limiting clot formation, antithrombin also blocks four earlier serine proteases in the blood-clotting system, effectively shutting down the cascade as the distance from the site of injury increases.

Blood Clotting, Heart Attacks, and Stroke

If blood clotting is the solution to one problem, it is the cause of another. The very same clot that can save a person's life also poses a life-threatening danger.

A small piece of a normal clot—an *embolus*—can break free and travel through the circulatory system, becoming lodged in a smaller blood vessel. The consequent blockage of the flow of blood in this vessel can be lethal. For example, an embolism in a coronary artery blocks the flow of blood carrying oxygen and nutrients to the heart, leading to heart attack and cell death (myocardial infarc-tion). Stroke, which results from an embolus that becomes trapped in a vessel in the brain, leads to the death of nerve cells and paralysis. And an embolism in the lung will re-strict the essential reoxygenation of blood.

Although such blockages in the circulation can arise during normal clotting after injury or surgery, more often they result from an abnormal clot that has formed sponta-neously at the wrong time and place. Most common and dangerous are clots that develop as a consequence of the cardiovascular disease process known as *atherosclerosis*. In this condition, the tissue surrounding the endothelial (innermost) lining of a blood vessel becomes thickened with lesions known as *plaque*. The plaque arises from lipid-protein aggregates containing cholesterol that is being transported by the blood to the liver (where choles-terol is metabolized) or to certain other organs (where cho-lesterol is used to construct steroid hormones). When this material fails to be removed from the circulation, it pro-gressively builds up in the walls of blood vessels. In a complex, long-term response, the normal structure of the

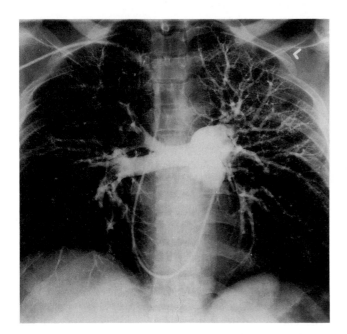 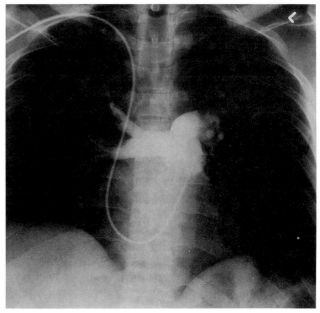

Left: The normal delivery of blood to the lungs. *Right:* The blockage of that circulation by a pulmonary embo-lism.

Causes of Mortality in the United States in 1984

Cause	Number of deaths	Percentage of total
Vascular diseases		
Heart disease	777,704	35
Cerebral vascular disease	154,327	8
Other	101,140	5
Total	987,171	48
All forms of cancer	459,928	23
Accidents and violence	145,012	7
Respiratory diseases	149,686	7
Infant mortality	39,580	2
Congenital abnormalities	13,039	1
Infectious diseases	24,500	1
Total	1,809,916	89
All causes	2,039,369	100

tissue becomes disrupted by abnormal cell proliferation and by the incursion of scavenger white blood cells, which attempt to internalize the cholesterol deposits. The result is a considerable amount of cell death (necrosis) and an accumulation of cell debris, extracellular protein fibers, and ever-larger cholesterol-containing deposits (diagrammed on p. 223). This collection of pathological material, or atheromatous plaque, with its semi-solid core and calcified fibrous crust, both constricts the internal diameter of the vessel and reduces its elasticity. Under the turbulence of high-pressure blood flow, abrasions may develop in the thin endothelial lining of the vessel as it is pressed against the underlying plaque. When this penetration occurs, the abnormal surface of the atheromatous plaque material can become a focus for platelet aggregation and the formation of a spontaneous blood clot (or *thrombus*). A similar type of clot-provoking damage to the lining of a vessel can be caused by the insertion of a catheter used to deliver tracer compounds during medical diagnosis. The important point is that, just as in the case of an embolism, a thrombus can block blood flow to critical organs, causing an interruption of the pulmonary, cardiac, or cerebral

circulation. As we age, the accumulation of atheromatous plaque material increases the likelihood of abnormal clot formation. Indeed, coronary thrombosis and stroke are the leading causes of death in the Western world.

To minimize the danger of embolism and to reduce the likelihood of thrombosis, as well as to clear away therapeutic blood clots formed during wound healing, a special enzyme is brought into play—and once again it is a serine protease. The enzyme responsible for dissolving blood clots is *plasmin*. This is yet another member of the chymotrypsin family that has become specialized as a complex regulatory protease. Its one target is the collection of polymerized fibrin units that make up a blood clot.

Plasmin and the Dissolving of Blood Clots

In contrast to other protein assemblies in living tissue, intended to be permanent, the fibrin filaments of blood clots appear to have been designed to allow their subsequent dismantlement. As we have seen, fibrin monomers are connected to one another through lock-and-key interactions between their nodular central and terminal domains. In this structure, the other portion of the fibrin units—the triple-helical arms—are less densely packed. As a result, the filaments of a blood clot consist of thick areas of interacting domains alternating with much less dense regions of connector rods. This design allows plasmin to diffuse into the interior of a blood clot and to destroy the fibrin filaments by attacking their accessible connector rods.

Like trypsin, plasmin cleaves protein chains randomly after lysine and arginine subunits. Each target fibrin unit contains a total of 360 such positively charged amino acids. Their exact positions are not known (because of the absence of X-ray crystallographic data), but it is likely that they project from surfaces of the protein to interact with the surrounding solvent and therefore are potentially accessible to cleavage. In fact, test-tube experiments with purified plasmin enzyme and fibrin monomers show that cleavage at five particular lysine and arginine

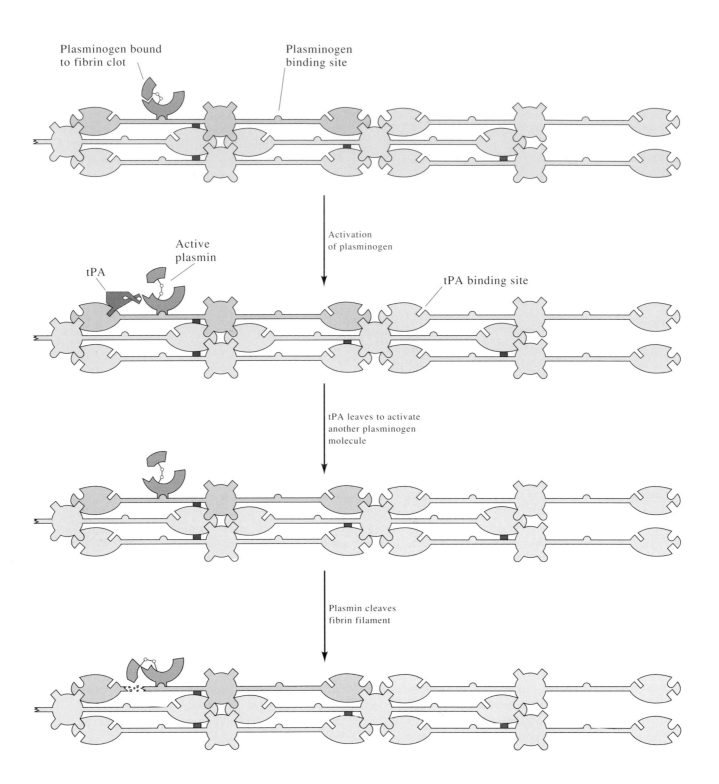

Plasmin, the enzyme that digests blood clots, is activated from an inert form by plasminogen activator. Both enzymes have serine protease domains; the one in the plasminogen activator is targeted specifically against the proenzyme form of plasmin, and the one in plasmin is targeted primarily against the thin connector arms of the fibrin units that compose a blood clot. Both enzymes also have nonenzymatic domains that anchor them to the fibrin filaments, bringing them together at the physiological site where they will function.

Like plasminogen, plasminogen activator, or tPA, is organized into several functional domains. One of these is a 250–amino acid serine protease domain, which carries out the cleavage reaction on plasminogen. Another, near the N-terminus, is responsible for anchoring the plasminogen activator to a fibrin clot, the physiological site where it is to carry out its function. Its placement on the clot brings tPA into close proximity with the plasminogen proenzyme it is to activate. The figure on the facing page shows the conversion of plasminogen to active plasmin by tPA, and the subsequent digestion of the blood clot by plasmin.

subunits occurs very rapidly. These preferred cleavage sites are all located in the thin connector arms and are believed to occur in particular areas where the triple-helical structure is disrupted by the presence of bulky amino acid subunits. It is the cleavage of the exposed lysine and arginine targets in a number of connector arms that is primarily responsible for reducing the fibrin filament to a set of nonfunctional fragments.

The diagram on the facing page shows how plasmin attacks fibrin blood clots at their most vulnerable points. But what restricts the enzyme's rather general proteolytic activity so that it only works on fibrin? The answer lies in the structure and activation of the plasmin proenzyme.

Like other serine proteases involved in blood clotting, plasmin is produced in the liver in an inert precursor form, *plasminogen*. Its activation involves the now-familiar proteolytic cleavage by another serine protease, of which there are a number in the body that can perform this function. One such enzyme, *plasminogen activator*, is perhaps the member of the chymotrypsin family best known to the general public, for a reason that will soon become apparent. It is known by the abbreviation tPA, where "t" stands for damaged *tissue*, the source of the enzyme. The job of tPA is to mobilize the 800–amino acid plasminogen proenzyme to action by introducing a single proteolytic cleavage at the arginine-valine linkage between subunits 560 and 561. This cuts the original plasminogen protein into two parts, which remain linked to each other through a disulfide bond. The smaller part contains an active 250–amino acid serine protease domain. It is the remaining larger part, with a domain that specifically binds to a surface on fibrin, that serves to anchor the protein to its substrate, the fibrin clot.

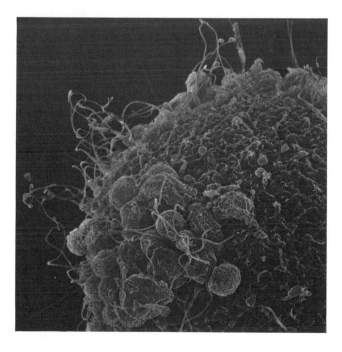

At the beginning of life, a serine protease plays an important role in the process of fertilization. The tip of the sperm consists of a specialized structure called the acrosome, which contains large amounts of a protease. During fertilization, this serine protease, *acrosin*, digests a hole through the protective, gelatinous layer surrounding the egg, allowing the sperm to make contact with the egg's outer membrane.

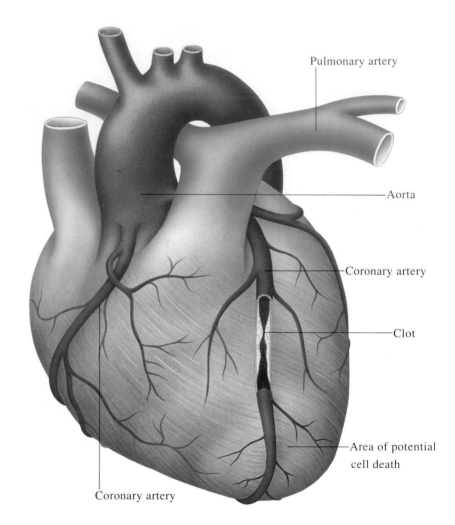

The heart under attack. Vascular disease has caused the accumulation of atheromatous plaque material, greatly narrowing a coronary artery. This narrowing, by itself, can gradually reduce the level of blood flow below that needed to nourish the heart tissue. But, in addition, when the plaque material penetrates the cavity of the vessel, it can become a focus for platelet aggregation and the formation of a lethal blood clot.

Enzyme Therapy

The dissolving of blood clots by plasmin has direct relevance to the treatment of heart attacks. When a clot is forming in an artery that supplies blood to the heart (coronary thrombosis), there is generally a period of several hours during which the open channel is being progressively closed off by the growing clot. This is the time when an oncoming heart attack often causes chest pain. Obviously, the body needs to bring into play its reservoir of plasmin proenzyme. Evidently, however, serious or fatal heart attacks result from a failure to achieve this goal. The likelihood of this scenario suggested to cardiologists a clot-dissolving (thrombolytic) therapy that would mobilize the requisite amount of plasmin by supplementing the relatively low amount of plasminogen activator in the blood.

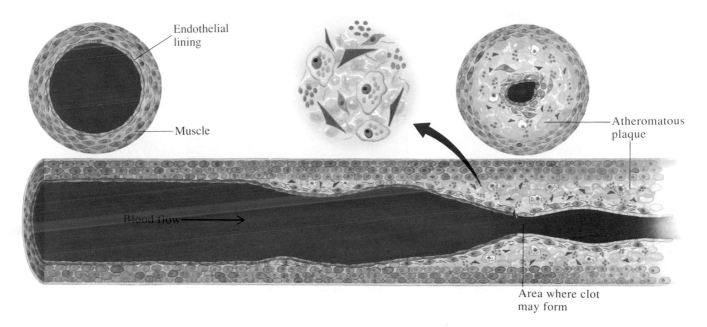

An artery partially damaged by vascular disease. *Left:* A normal part of the artery, with a wide channel allowing adequate blood flow. *Right:* In the diseased portion of the artery, thickening deposits of atheromatous plaque (initiated by lipid-protein aggregates containing cholesterol, shown in yellow; see pages 218 and 219) have begun to restrict the flow of blood.

In just the last few years, significant strides have been made in implementing this strategy, in large measure based on the new techniques of genetic engineering (page 186). The gene for plasminogen activator has been isolated by scientists at two leading biotechnology companies, Genentech and Genetics Institute. By means of gene-splicing techniques, the gene has been moved to a chromosomal environment that allows its expression at greatly enhanced levels. Large amounts of tPA are thus now available, produced in cells grown under laboratory conditions. For the first time it has become possible to administer this important enzyme to persons suffering from a heart attack. The most recent clinical trials of recombinant DNA–produced plasminogen activator, completed in March 1990, show that if tPA is administered intravenously within hours of the formation of a blood clot in a coronary artery, mortality is significantly reduced. Whereas the statistical fatality rate for heart attack is approximately 35 percent, with half of the individuals dying

before they reach the hospital, treatment with tPA or an equivalent thrombolytic agent such as streptokinase reduces mortality to just under 9 percent. The striking X-ray photographs on page 224 show the restoration of the coronary circulation following the administration of plasminogen activator during a heart attack.

Only a generation or two ago, heart attacks were thought of somewhat vaguely as failures of the cardiac muscle, which had become weakened after years of pumping blood through progressively constricted and inelastic arteries. But it is now realized that most heart attacks arise when an aged, lesion-ridden vascular system comes into conflict with the power of the blood-clotting system. Our understanding of the molecular biology of these events, and particularly of the serine proteases involved in blood clotting, has become so precise that it is now possible to intervene effectively in the process, through the production of a key enzyme and its administration at the moment of maximum danger.

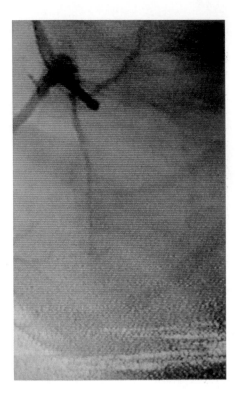 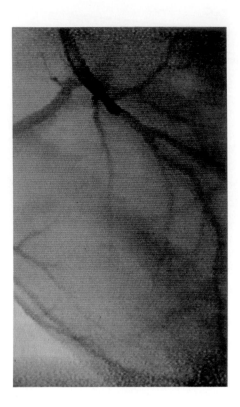

Left: The blood supply to a portion of the heart entering from the top has been cut off by a clot (thrombus). *Right:* The administration of plasminogen activator has stimulated the dissolution of the blood clot, restoring circulation and the delivery of oxygen and nutrients to the heart tissue.

Perspective

We began this chapter with a description of a simple digestive enzyme, chymotrypsin, searching for the mechanism by which it locates its appropriate substrate. We found the answer in a region of the enzyme adjacent to the catalytic active site, the substrate recongition site. Here, a group of amino acids formed a specific binding surface that interacted in a lock-and-key manner with the substrate.

The complementarity in shape between the enzyme's binding pocket and the substrate proved to be not only the explanation for enzyme specificity, but also an introduc-

tion to enzyme evolution. We saw how a single change in the amino acid composition of the binding pocket of chymotrypsin introduced a negative charge that allowed the modified enzyme, trypsin, to interact with a different target—the positively charged amino acids of a substrate polypeptide chain.

After finding (1) chymotrypsin and (2) the sibling enzyme trypsin, we saw, in a logical and orderly progression, (3) the first cousin elastase, (4) the second cousin thrombin, (5) the collateral enzymes plasmin and tPA, and (6) the distant relatives cocoonase and acrosin. The members of the chymotrypsin family that carry out the process

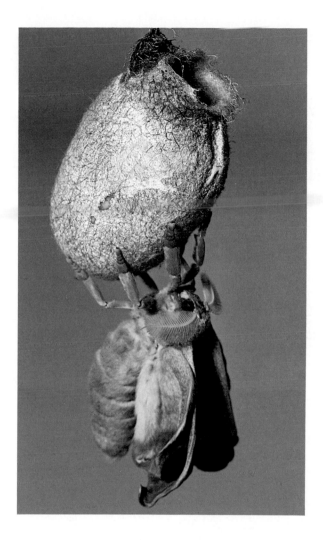

After the caterpillar in the cocoon completes its metamorphosis into a silk moth, its salivary glands secrete cocoonase. This enzyme is a member of the serine protease family, with a specificity like trypsin. It cleaves at lysine and arginine subunits in the thin strands of silk fibroin protein that are the principal structural material of the surrounding cocoon. Once again, the reaction mechanism of the enzyme depends on a critical serine side chain. Gradually, as the enzyme carries out its reaction, the thick overlying mesh of fibers is completely digested away in one place, and the moth emerges. The point to be made about cocoonase is that a serine protease has come under temporal control, and can therefore be used as an agent to carry out a specific stage in a developmental program.

of blood clotting introduced us to a new aspect of enzyme evolution. They retained the catalytic triad—serine, histidine, and aspartic acid—that characterizes all of the members of the family, but, in addition, these enzymes had extra domains not involved in catalysis. These domains allowed the chymotrypsin family to evolve from a set of simple digestive enzymes into complex regulatory proteases. The lock-and-key interactions formed by the extra domains targeted the enzymes to the physiological sites where they were to function. For example, thrombin and the blood clotting enzymes were directed to platelet membrane material and, later, plasmin and tPA to fibrin blood clots—their site of action.

In sum, we have seen how, amino acid by amino acid and domain by domain, a simple proteolytic enzyme can accumulate the evolutionary changes necessary to carry out increasingly diverse and complex physiological functions.

7

The Cooperative Action
of Enzymes

A neuron, the basic unit of the brain, is seen here growing on a microchip, the basic unit of the computer. Both objects perform somewhat the same information-processing function. However, the enzyme-produced and enzyme-operated neuron is more powerful, and it is the neuron that has created the microchip.

E nzymes have been guiding the activities of organisms for over three billion years, ever since the first living cells appeared in the primitive oceans. This great expanse of time has witnessed the continuous evolution of new life forms, whose activities are governed by new enzymes that arose from an original basic set of enzymes once responsible only for general growth functions. Over time, higher organisms developed, in which different cells acquired their own assigned responsibilities by relying on the activities of specialized set of enzymes. As this process was perfected, organisms evolved the capacity to carry out ever more complex physiological processes.

Perhaps the highest achievement thus far of this long evolutionary process has been the development of the human nervous system, which not only controls the rest of the body from a mechanical point of view, but also reaches out to understand itself and the universe. It is with a consideration of some of the enzymes that control the nervous system that we will now close our book.

The enzymes we have considered in our discussion so far are all members of a single family whose function is to carry out a physiologically diverse, although mechanistically related, class of chemical transformations. During our analysis of signal transmission in the nervous system, we will have the opportunity to examine several new types of enzymes, and take our first detailed look at catalysts that are not serine proteases. This will allow us not only to enlarge our field of vision about catalytic reaction mechanisms, but also to learn how a set of diverse enzymes, working together, can give rise to a complex physiological system. Specifically, we will explore the basic functioning of the nervous system in terms of the concerted activity of five prototype enzymes and related proteins. These enzymes are (1) a *pump* that redistributes ions between the inside and outside of the nerve cell; (2) a set of pores or *channels* that open and close to allow ions to enter or leave the cell; (3) an enzyme whose role is to *synthesize a signaling molecule* that enables the nerve cell to communicate with its target; (4) a *receptor protein* on the target cell that receives the signal; and (5) an enzyme responsible for *degrading the signaling molecule* and turning off nerve signal transmission when it is no longer needed. One of these five catalysts will turn out to be an enzyme of a type we have already encountered, for it is a member of the serine protease family. The others will represent new classes of enzymes. Because we have already developed a general understanding of protein structure and function, it will be easy to quickly grasp the reaction mechanisms of the new enzymes. At the same time, the nuances of their catalytic strategies will broaden our understanding of the ways enzymes work to guide the processes of life. And most important of all, we will see how an apparently complex physiological system in a higher organism can be broken down and understood in terms of the cooperative activities of a few key enzymes.

Santiago Ramón y Cajal, who first characterized the fundamental unit of structure and function in the brain, the neuron. An erratic and rebellious student early in life, he was apprenticed by his father first to a barber and then to a cobbler, with evidently good effect. His later progress as a medical student was distinguished and laid the foundation for an academic career, culminating in his professorship of histology and pathological anatomy at the University of Madrid.

The Neuron

While the other organs of the body were fairly well characterized in the last century, it was not until 1900 that the basic functional unit of the brain, the *neuron,* was identified. The great Spanish neuroanatomist Santiago Ramón y Cajal exposed brain tissue to a silver-containing mixture developed by his Italian counterpart, Camillo Golgi. This "stain" was known from studies with other tissues to enter random cells and build up crystalline deposits of silver chromate. As a result, individual cells were illuminated and their boundaries defined. When this procedure was applied to brain tissue, it became clear that, like the rest of the body, the nervous system was also composed of individual cells—and was not a vast interconnected network of tubes and chambers through which "brain secretions" flowed, as had been generally supposed. For his studies of brain structure, Ramón y Cajal earned a place in history as the founder of modern neurobiology, and he shared with Golgi the 1902 Nobel prize in medicine and physiology.

As the basic structural and functional unit of the brain, the neuron is a highly specialized cell. Its structure is uniquely suited to carrying out its primary function, the receiving and transmitting of signals. Like all cells in the body, neurons have a nucleus and surrounding cytoplasm. This area, the *cell body,* is the primary site of metabolic activities. But neurons differ from other cells in that they possess distinctive cellular extensions that either collect or transmit information. As shown in the photograph on this page, thousands of *dendrites* reach out like antennae from the cell body of a typical neuron. Each can receive the input of numerous other nerve cells. Also projecting from the cell body is a single large cylindrical fiber, the *axon,* which is the signal-transmitting part of the neuron. At its terminus, the axon divides into multiple branches, each of which can make contact with a different receiving neuron. The complexity of the nervous system results from the multiplicity of message-receiving dendrites on each neuron and the numerous message-distributing branches at the end of each axon. It is the networking of these billions upon billions of informational connections that allows such complex activities as memory and rational thought.

In overview, each neuron is like an on-off component in an electrical circuit—it either ''fires'' or it doesn't. The pattern of firing—in particular its frequency—reflects the intensity of the message.

The mechanism by which the nerve cell functions is ingenious—and enzymatic. Indeed, the neuron exemplifies the specialized, differentiated cell types that are the hallmark of higher organisms. Its remarkable capacities and high degree of specialization are a direct reflection of its unique set of enzymes.

Preparing the Nerve Cell for Action

When a neuron becomes active, a signal arises in the dendrites and cell body, and then travels down the axon, eventually passing to the next cell in the network. The key to understanding nerve cell function is to realize that long before the neuron is brought into play, it has been prepared—chemically set for firing. It is like a rifle that is ready and loaded.

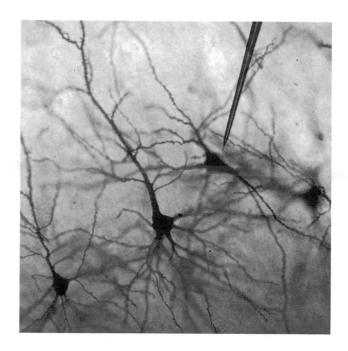

Three neurons, rendered visible by the injection of a dye, show clearly their multiple dendrite and axon projections.

The preparation of the nerve cell is carried out by a special type of enzyme—the first of the new enzymes we will analyze in this chapter. Unlike the protein catalysts we have encountered thus far, this enzyme is located in the cell membrane, where it functions to control the flow of ions between the inside and outside of the cell. Specifically, it transports sodium and potassium ions across the cell membrane—functioning quite literally as an enzymatic ''pump.'' The action of this pump will cause the inside of the neuron to become more negatively charged than the outside, and just this electrochemical difference is necessary for the neuron to fire.

The *sodium-potassium pump* is a relatively large protein that completely spans the lipid membrane. Its outer surface is in contact with the environment and its inner surface with the cell cytoplasm. The job of the pump is to continuously import potassium into the cell and export

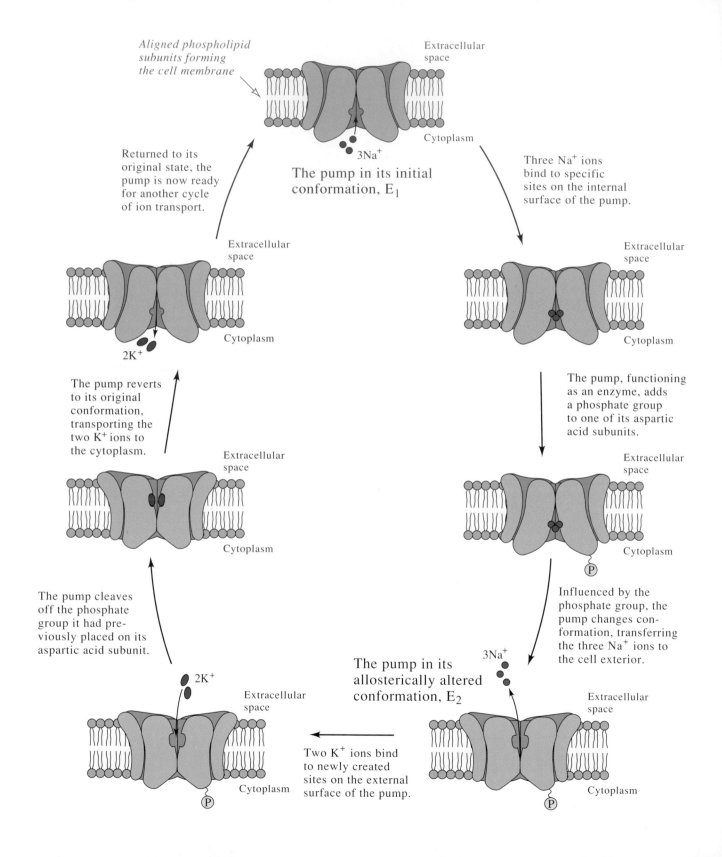

Aligned phospholipid subunits forming the cell membrane

Extracellular space

Cytoplasm

3Na⁺

The pump in its initial conformation, E_1

Returned to its original state, the pump is now ready for another cycle of ion transport.

Three Na⁺ ions bind to specific sites on the internal surface of the pump.

Extracellular space

Cytoplasm

Extracellular space

Cytoplasm

2K⁺

The pump reverts to its original conformation, transporting the two K⁺ ions to the cytoplasm.

The pump, functioning as an enzyme, adds a phosphate group to one of its aspartic acid subunits.

Extracellular space

Cytoplasm

Extracellular space

Cytoplasm

P

The pump cleaves off the phosphate group it had previously placed on its aspartic acid subunit.

Influenced by the phosphate group, the pump changes conformation, transferring the three Na⁺ ions to the cell exterior.

3Na⁺

The pump in its allosterically altered conformation, E_2

2K⁺

Extracellular space

Cytoplasm

P

Two K⁺ ions bind to newly created sites on the external surface of the pump.

Extracellular space

Cytoplasm

P

sodium. Its mechanism of action, which introduces us to an important new aspect of enzyme function, is shown on the opposite page. The transport cycle begins as the pump, in its initial conformation, E_1, binds three sodium ions (Na^+) to specific sites located on its internal (cytoplasmic) surface. Then, sensing that the Na^+ sites are "loaded," the pump swings into action. In fact, the pump is an enzyme, but it does not act chemically on the ions to be transported—but rather on itself in such a way as to cause ion transport. Specifically, the pump catalyzes the addition of a phosphate group to the side chain of one of its own aspartic acid subunits. This phosphate is derived from an important cellular molecule (adenosine triphosphate, ATP), which contains three phosphate groups in a high-energy configuration that allows their ready transfer to other molecules. After becoming the recipient of a bulky, negatively charged phosphate group, the pump is no longer stable in its original conformation, E_1, and is induced to undergo a change in shape to a new conformation, E_2. The consequence of this change in protein structure, or "allosteric" change, is that the three sites for binding sodium now face the outside of the cell. Moreover, they have a weakened affinity for the bound Na^+ ions, which are released into the extracellular fluid.

A reversal of these events allows the pump to import potassium into the cell. In its new form (E_2), the pump displays two sites on its outer surface that strongly bind potassium ions (K^+). As soon as the potassium sites are filled, the pump again functions as an enzyme; it cleaves off the phosphate group that it had previously placed on its aspartic acid subunit. Without the phosphate group, the pump reverts to its original conformation (E_1), relocating the K^+ ions to the inside of the cell and releasing them into the cytoplasm. The $E_2 \rightarrow E_1$ allosteric change also regenerates the sites that bind Na^+ ions, and returned to its original state, the pump is ready for another cycle of action.

Originally the suggestion of the biochemist Oleg Jardetzky, the proposed mechanism for the operation of the sodium-potassium pump is now supported by considerable data. It is uncertain only to the extent that X-ray crystallographic studies are not yet available to demonstrate the central postulate of the model—that the pump undergoes an allosteric change in protein structure upon the enzymatic addition and removal of the phosphate group, so as to cycle the Na^+ and K^+ binding sites between the inside and outside of the cell.

The $E_1 \rightarrow E_2$ allosteric change in the sodium-potassium pump is a clear example of a chemical reaction that, considered in isolation, will not proceed spontaneously because it yields a product (E_2) that has a higher energy level than the starting molecule (E_1). To make such "uphill" reactions possible, energy must be introduced into the system. *The energetically unfavorable reaction is carried out in association with a second reaction that is energetically favorable.* In many cases, the second reaction is the removal of a phosphate group from ATP, an energy-rich molecule made during the metabolism of glucose or during photosynthesis. The favorable energy change associated with the release of the components of ATP to a lower energy state compensates for the unfavorable character of the first reaction (for example, the allosteric change in the pump), so that the net generation of products satisfies the requirement of moving the total system to a lower energy level. Reactions that are inherently energetically unfavorable include not only the functioning of such enzymes as the pump, but also all biosynthetic reactions—in which larger molecules are built from smaller ones—such as the synthesis of proteins, nucleic acids, carbohydrates, and lipids. In all these cases, in one way or another, the energy of ATP is used to drive the reaction.

The sodium-potassium pump is found in all body cells, and its activity accounts for about 20 percent of the energy consumed by the cell. By redistributing ions, it maintains the internal environment of the cytoplasm and protects the cell against osmotic swelling. In nerve cells, the sodium-potassium pump is especially abundant, and its activity may account for as much as 70 percent of the energy used by the cell. The prevalence of the sodium-

The essence of the reaction mechanism of the sodium-potassium pump lies in the ability of the protein to interconvert between two conformations, shifting the ion-binding surface alternately between the inside and the outside of the cell. This allosteric change is brought about by the sequential addition and removal of a phosphate group, whose chemical character influences the shape of the protein. With each cycle of phosphorylation and dephosphorylation, the pump transports three Na^+ ions out of the cell and brings two K^+ ions in.

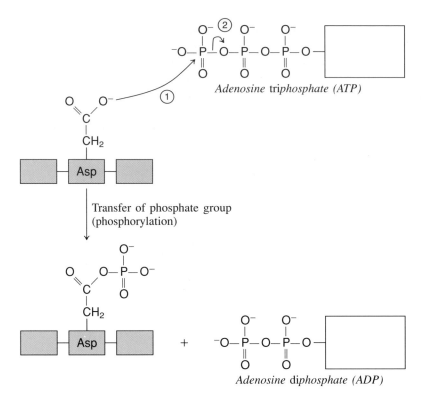

The addition of a phosphate group to a specific aspartic acid subunit in the sodium-potassium pump is the event inducing its change in conformation. Phosphorylation and dephosphorylation play an important role in the control of many enzymes.

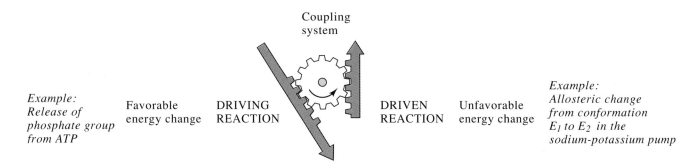

An energetically unfavorable reaction can be made to happen if it is carried out in association with a second, energetically favorable reaction. The favorable energy change of the second reaction compensates for the unfavorable character of the first reaction.

potassium pump reflects the fact that the nerve cell uses the pump not merely to maintain its internal ionic environment, but also to prepare the cell for signaling activity.

As the sodium-potassium pump transports three Na^+ ions out of the cell for every two K^+ transported in, and as some K^+ again leaves the cell due to membrane leakage, an imbalance of electrical charge develops across the cell membrane. The inside of the neuron becomes electrically negative, while the outside becomes electrically positive. This separation of electrical charges is reflected in a voltage differential of -0.07 volts, or -70 millivolts (mV), across the cell membrane. It is this *membrane potential,* or *resting potential,* that prepares the neuron for action.

Firing the Nerve

The actual firing of the neuron—its transmission of a signal—results from a temporary disruption in the cell's resting potential. This is brought about by the second of the five types of enzymes that operate the nervous system—a collection of special *ion channels* in the cell membrane. These channels are made of protein, and they open and close via an allosteric change caused by a fluctuation in the neuron's resting potential. This leads to a redistribution of Na^+ and K^+ ions between the inside and outside of the neuron—an electrochemical change that will constitute the nerve-cell signal.

The signaling process begins with stimulatory changes that first develop in the neuron's dendrites and cell body. These result in a gradual erosion of the -70-mV voltage differential across the cell membrane. When this reduction reaches a threshold level, a critical event occurs. In the area of the axon adjacent to the cell body, the membrane channels for Na^+ suddenly undergo an allosteric shift in conformation in response to the change in the local electrical environment. These "voltage-controlled" channels open and allow Na^+ ions that are present in relatively high concentration outside the nerve to rush to the inside. This influx of positive charge leads to a complete collapse of the normal -70-mV resting potential in the proximal part of the axon. In fact, the local electrical polarization is generally reversed; enough

charges usually pass inward to develop a 20-mV positive state in this segment of the axon.

This *depolarization* of the axon segment is a transient event, lasting only about a thousandth of a second, and is rapidly reversed. Almost as soon as depolarization has occurred, the sodium channels spontaneously close, and the membrane becomes resistant to the entry of further

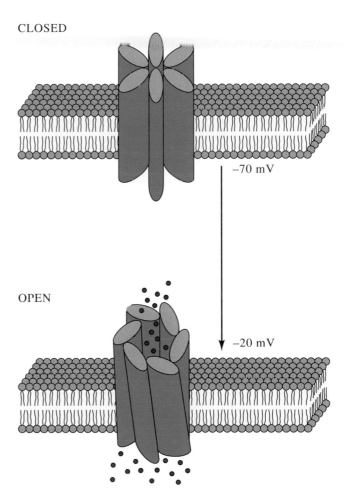

CLOSED

−70 mV

OPEN

−20 mV

A local reduction of the voltage differential may induce a sodium or potassium channel to change conformation from one allosteric form to another, thus opening the channel and leading to the free passage of ions. This change is presumably caused by key electrically charged amino acids in the channel protein, which shift their orientation in response to the changed electrical field.

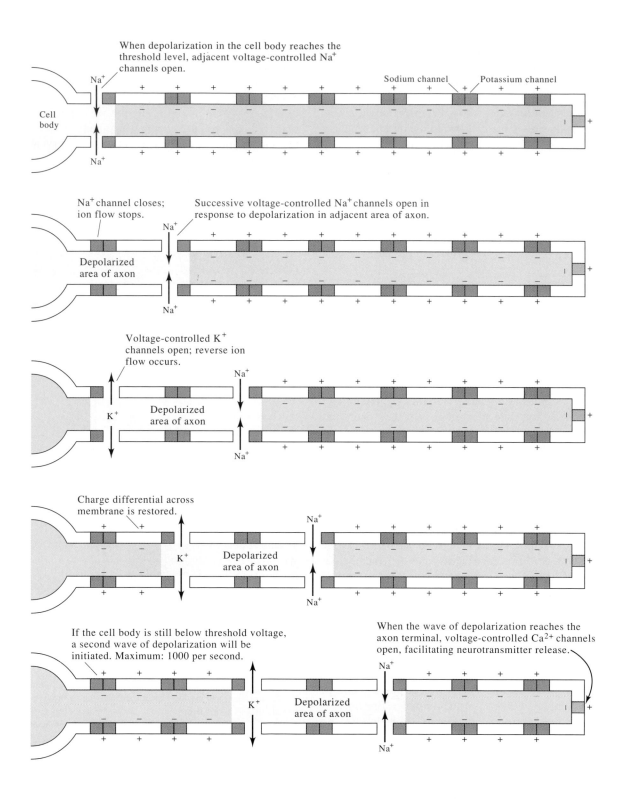

When depolarization in the cell body reaches the threshold level, adjacent voltage-controlled Na$^+$ channels open.

Na$^+$

Cell body

Sodium channel   Potassium channel

Na$^+$

Na$^+$ channel closes; ion flow stops.

Successive voltage-controlled Na$^+$ channels open in response to depolarization in adjacent area of axon.

Na$^+$

Depolarized area of axon

Na$^+$

Voltage-controlled K$^+$ channels open; reverse ion flow occurs.

Na$^+$

K$^+$

Depolarized area of axon

Na$^+$

Charge differential across membrane is restored.

Na$^+$

K$^+$

Depolarized area of axon

Na$^+$

If the cell body is still below threshold voltage, a second wave of depolarization will be initiated. Maximum: 1000 per second.

When the wave of depolarization reaches the axon terminal, voltage-controlled Ca^{2+} channels open, facilitating neurotransmitter release.

Na$^+$

K$^+$

Depolarized area of axon

Na$^+$

Na$^+$ ions. At the same time, a second set of channels opens fully and allows the free passage of K$^+$ ions. This results in a strong outward flow of K$^+$ due to the high concentration of potassium inside the cell and to like-charge repulsion between the concentrated positively charged Na$^+$ and K$^+$ ions. As the K$^+$ ions leave, the positive charge inside the proximal segment of the axon begins to diminish until, after a few thousandths of a second, the original -70-mV resting potential across the membrane is largely restored.

The two-stroke cycle of membrane depolarization (due to the opening of sodium channels and the influx of Na$^+$) and membrane repolarization (due to the secondary opening of potassium channels and the diffusion of K$^+$ back out of the cell) is called an *action potential*—the change in electrical potential associated with nerve action.

The action potential represents a perturbation of the normal -70-mV charge differential across the cell membrane. Initially, it occurs at a specific site in the neuron—the beginning of the axon, adjacent to the cell body. By itself, a single action potential would be of no consequence. What gives it significance is that the very way in which it is generated allows it to ignite other, similar action potentials that affect adjacent areas of the axon. This spread, or apparent movement, of the original action potential results from the special voltage-controlled property of the Na$^+$ and K$^+$ channels. Sensing the nearby voltage change caused by the original action potential, they, in turn, open and allow the influx of sodium—and then the efflux of potassium—in their part of the axon. This event extends the original action potential to the adjacent area of the axon and, through its repetition, progressively down to the end of the nerve cell. The process can be viewed as a chain reaction in which each successive depolarization–repolarization event leads to an exactly similar depolarization–repolarization in progressively more distal portions of the axon. The spread of the action potential to the end of the axon represents a full nerve signal, or *nerve impulse*.

A nerve impulse sweeps down the neuron from the cell body to the tip of the axon as successive groups of Na$^+$ and K$^+$ channels open and close.

It is interesting to note that although the ion channels that are involved in nerve firing do not make or break covalent chemical bonds—the traditional definition of an enzyme—they have many of the essential characteristics of enzymes. For instance, (1) ion channels are proteins with precise three-dimensional structures that are related to their function; (2) they serve as biological catalysts that carry a molecule (ion) across an energy barrier (posed by the impermeability of the cell membrane) to a lower energy state without themselves being permanently changed in the process; (3) they show substrate specificity in that they are selective for the type of ion they transport; (4) their catalyzed reaction has a maximum velocity, V_{max}, reflecting their ability to become saturated with substrate (page 70); and (5) like enzymes such as the sodium-potassium pump they undergo allosteric changes that allow them to open in response to stimuli. We thus see that ion channels can be usefully thought of as a special class of enzyme and that the definition of an enzyme as an agent of chemical transformation in living systems should not be viewed too narrowly.

Neurotransmitters

The power of the nervous system is not derived from the firing of a single neuron, but rather from the transmission of nerve impulses through a sequence of cells that form a functional pathway. But, since the nervous system is composed of individual cells, separated from each other by precise boundaries, no physical connection exists to allow the direct transmission of impulses from one nerve cell to the next. The same difficulty arises when a nerve cell activates its other potential target, a muscle cell. In both cases, the continued propagation of a nerve signal after it reaches the end of an axon poses a special problem—one, however, that is again solved by enzymes—or more precisely a pair of enzymes that synthesize and degrade a key signaling molecule.

The general solution to the problem of crossing the *synapse*, the space between two nerve cells, was first set forth in 1921 by Otto Loewi, a German physiologist. By studying the way in which the vagus nerve controls heart

THE DISCOVERY OF NEUROTRANSMITTERS

Neurotransmitters were discovered in a classic experiment by the German physiologist Otto Loewi. Loewi studied the vagus nerve, which originates in the brain and sends its axons to a number of major organs, including the heart. His experiment was designed to determine whether the vagus regulated the contractions of heart muscle directly, or indirectly by releasing a chemical substance that acted as a secondary signaling agent. As shown in the diagram, the experiment involved the use of two isolated hearts, connected to each other by a tube containing a solution of physiological salts. The tube served as a liquid bridge between the two hearts, allowing their chemical communication. Using electrical current, Loewi artificially stimulated the vagus nerve still attached to one of the hearts (the donor) and within seconds saw a change in the rate of its contraction. Then, about a minute later, the second, recipient heart (connected to the first by the liquid bridge) adjusted its contraction rate. Since there was no electrical connection between the two hearts, it appeared evident that a chemical substance must have been released by the original stimulated vagus nerve. This substance would have perfused through the tissue fluid of the first heart and then passed across the liquid bridge into the second heart.

This precedent-setting work strongly suggested the existence of chemical *neurotransmitters*—molecules that would carry signals between nerves and the cells they affected. The neurotransmitter hypothesis was firmly established when, five years later, Loewi isolated the substance released by the vagus nerve. Upon chemical analysis, it proved to be the small molecule *acetylcholine*.

In Loewi's experiments the affected cells were heart muscle cells, but it could be inferred that neurotransmitters would also exist within the central region of the brain, allowing one nerve cell to communicate with another. The important general point is that when one nerve cell communicates with another, the initial electrical type of signal is converted into a chemical signal, which, after crossing the synapse, is converted back into an electrical signal in the target cell.

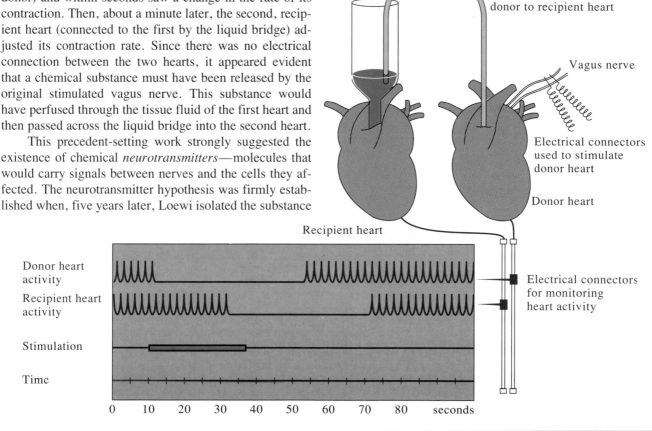

Liquid bridge from donor to recipient heart

Vagus nerve

Electrical connectors used to stimulate donor heart

Donor heart

Recipient heart

Donor heart activity

Recipient heart activity

Stimulation

Time

Electrical connectors for monitoring heart activity

0   10   20   30   40   50   60   70   80   seconds

beat, he found that, *when a nerve impulse reaches the end of an axon, it causes the release of a chemical, which then crosses the intercellular space and activates the target cell.* In essence, the original electrical signal is converted into a chemical signal, which is then converted back into an electrical signal. Loewi went on to identify the chemical released by the vagus nerve as the small molecule *acetylcholine*—and this molecule thus became the first *neurotransmitter,* work for which he was awarded a Nobel prize in 1936.

$$CH_3-\overset{\displaystyle O}{\overset{\displaystyle \|}{C}}-O-CH_2-CH_2-\overset{\displaystyle CH_3}{\overset{\displaystyle |}{\underset{\displaystyle |}{\overset{+}{N}}}}-CH_3$$
$$\underset{\displaystyle CH_3}{}$$

Acetylcholine

Acetylcholine is now known to be the transmitting agent between many neurons, in organisms ranging across a broad spectrum from insects to higher vertebrates. In addition, acetylcholine serves as the neurotransmitter at the junction between nerve cells and muscle cells (neuromuscular junctions).

The intricate structure of the nervous system is made even more complex by the existence of a multiplicity of neurotransmitters. These link the 10^{12} neurons to each other and to other target cells, such as muscle, through at least 10^{14} synaptic connections. More than 50 neurotransmitters are now known, ranging in size from small molecules to peptides with as many as 30 amino acids. Some exert a stimulatory influence on their target cells (causing depolarization), others an inhibitory influence (hyperpolarizing the target cell); some are released easily, others only upon repeated axon depolarization; some are short-lived, others more persistent. As a result of this profusion of neurotransmitters, the flexibility and sophistication of the nervous system is almost infinitely increased.

Crossing the Synapse

The site where neurotransmitters work is the *synapse*—the junction between a neuron and its target cell. An example

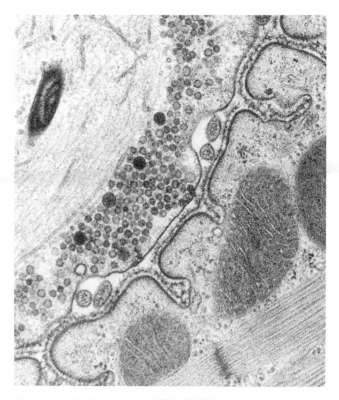

A neuromuscular junction magnified 100,000 times and observed in the electron microscope. In the upper left can be seen the axon terminal of the nerve cell, which contains hundreds of small synaptic vesicles filled with neurotransmitter molecules. Upon their release, the neurotransmitters diffuse across the pitted synaptic cleft (which runs diagonally across the middle of the micrograph) and activate the target muscle cell—whose energy-producing mitochondria and parallel arrays of contractile fibers are clearly visible at the lower right.

of a synapse magnified 100,000 times in the electron microscope is shown on page 239. At this magnification, we cannot view either participating cell in its entirety, but we do obtain a detailed image of the critical area—the axon terminal of one neuron and the corresponding dendrite of another neuron (or, in other cases, the surface of a target muscle cell). The membranes of the *presynaptic* and *postsynaptic* cells are separated by a specialized intercellular space filled with complex proteins and carbohydrates that form a connective matrix.

Neurotransmitters are synthesized in the presynaptic

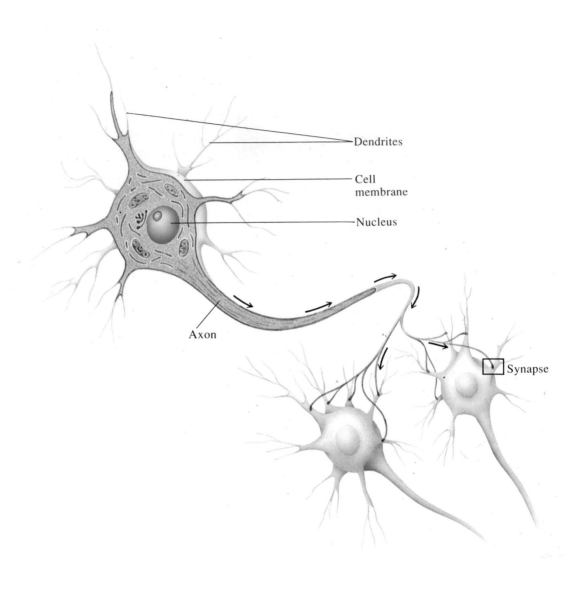

Above: A nerve cell sends signals down an axon, which then branches and passes the signals to the dendrites of other nerve cells at specialized junctions called synapses. *Facing page:* A close-up view of the synapse marked with a box. The electron micrograph shows two axons, seen in cross section, containing hundreds of small synaptic vesicles. Below and slightly to the left of the axons is a dendritic branch of an adjacent neuron. The thickened border between the axon and dendrite membranes marks the exact position of the synapse. The schematic diagram below the micrograph shows, how, when a nerve impulse reaches the end of an axon, the wave of depolarization triggers the release of small molecules that serve as neurotransmitters. Membrane-bound vesicles filled with thousands of neurotransmitter molecules move to the surface of the cell and, fusing with the cell membrane, release their contents into the synaptic cleft. The neurotransmitter then diffuses across the synaptic cleft to reach receptor sites on the target cell.

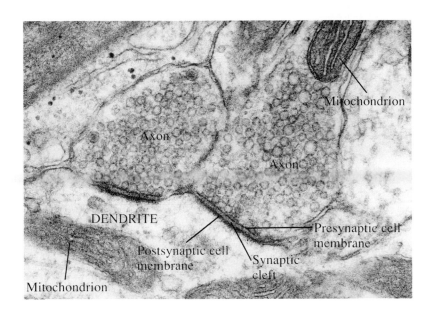

Mitochondrion

Axon

Axon

DENDRITE

Presynaptic cell
membrane

Postsynaptic cell
membrane

Synaptic
cleft

Mitochondrion

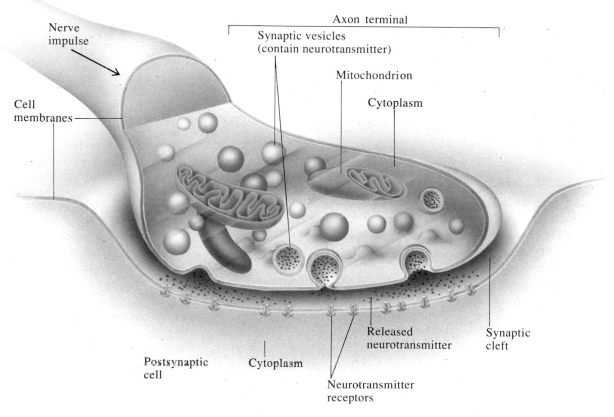

Axon terminal

Nerve
impulse

Synaptic vesicles
(contain neurotransmitter)

Mitochondrion

Cytoplasm

Cell
membranes

Postsynaptic
cell

Cytoplasm

Released
neurotransmitter

Synaptic
cleft

Neurotransmitter
receptors

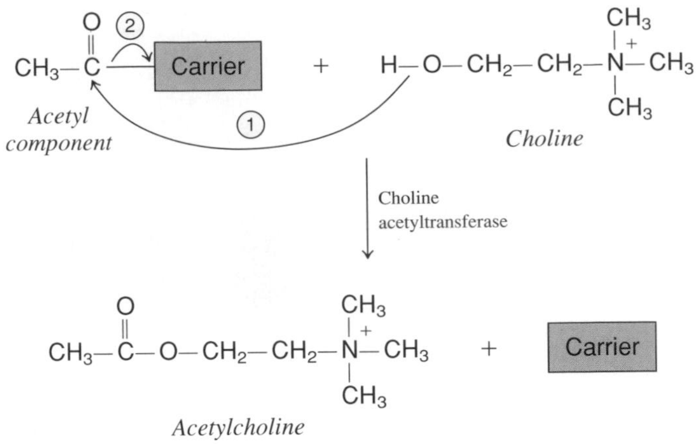

Acetylcholine is assembled from two common cellular metabolites, choline and an acetic acid component, by the enzyme choline acetyltransferase. The carrier molecule, to which the acetic acid is attached, serves the same function as ATP in providing an input of energy for this otherwise energetically unfavorable synthetic reaction.

cell by enzymes present in the cell cytoplasm. These constitute the third of the five prototype enzymes necessary for nerve cell function. For example, the enzymatic synthesis of acetylcholine is shown above. It is formed from two simpler compounds: an *acetic acid* component, derived from the metabolic breakdown of glucose, and *choline,* which is ordinarily attached to lipids and used in the construction of membranes. These two building blocks are sutured together by the enzyme *choline acetyltransferase* to yield the finished neurotransmitter. The synthesis of acetylcholine is a clear example of how enzyme evolution leads to cell specialization. Choline acetyltransferase is an enzyme found only in nerve cells, and its development has allowed these cells to draw upon two basic cellular metabolites to synthesize a new compound specifically useful in neuron function.

A precise sequence of events occurs as a nerve impulse reaches the end of an axon and the neurotransmitter is released. When the wave of depolarization arrives at the axon terminal, the loss of the -70-mV differential causes the opening of voltage-sensitive calcium channels, which then permit the entry of Ca^{2+} into the neuron. The elevated level of calcium ions in the cytoplasm promotes the fusion of neurotransmitter-containing synaptic vesicles with the presynaptic cell membrane. As the vesicles move to the cell surface, their membranes fuse with the overlying cell membrane, thereby discharging several thousand neurotransmitter molecules each into the synaptic cleft. The events leading to neurotransmitter release are illustrated on pages 238–239.

Receiving the Neurotransmitter Signal

The chemical signal carried by the neurotransmitter is received by specialized proteins embedded in the membrane of the postsynaptic cell. Thousands of such *neurotransmitter receptor proteins* line the dendrites and cell body of the target cell and bind neurotransmitter molecules through a lock-and-key interaction. The receptors are more than just simple binding proteins. They must be able not only to receive the signal carried by the neurotransmitter but also to function in such a way as to translate the signal into an appropriate response by the postsynaptic cell. In the well-studied case of the neuromuscular junction, the interaction of acetylcholine with its receptor leads to an allosteric change in the adjacent part of the protein so that it opens and becomes a channel for a variety of positively charged ions. The net effect is the entry of positive ions into the cell and a partial loss of the voltage differential across the target cell membrane. The neurotransmitter receptor, viewed as an ion channel, represents the fourth of the major nervous system–specific enzymes we will encounter in this chapter.

It is important to note that receptor channels are *ligand-controlled,* or controlled by small molecules, in contrast to the voltage-controlled character of the Na^+ and K^+ channels in the axon. This property allows each receptor channel to make its individual contribution to the depolarization of the dendrites and cell body of the postsynaptic cell without influencing, or being affected by, other neighboring channels.

The result of neurotransmitter binding is a gradual entry of positively charged ions into the target cell depending on the amount of neurotransmitter stimulation. For example, the membrane potential in the postsynaptic cell may change from -70 mV to a less polarized state, say, -60 mV. This change in voltage differential is in a stimulatory direction, but it is not sufficient to trigger the opening of the all-important voltage-controlled Na^+ channels that occur along the length of the axon. In fact, if the initial depolarization only reaches this sub-threshold level, it will eventually disappear—dissipated, for example, by the action of the sodium-potassium pump. However, if enough stimulatory signals are received by the neurotransmitter receptors, the outcome will be different. The function of the cell body and its associated dendrites is to perform a numerical summation of the many input signals that the neuron receives. If the individual small depolarizations add together sufficiently, the resultant membrane potential will fall below -20 mV and trigger the allosteric change that leads to the opening of the sentinel voltage-sensitive sodium channels at the beginning of the axon. This, of course, sets off the chain of sequential depolarizations that sweep down to the end of the axon and constitute the nerve impulse. Thus, only if a threshold depolarization of the dendrites and cell body occurs will the axon give a response. It is the "all-or-nothing" nature of the response of an individual nerve cell that makes it like an "on-off switch" in an electrical circuit.

The acetylcholine receptor is the target for several interesting compounds that inhibit the nervous system, in

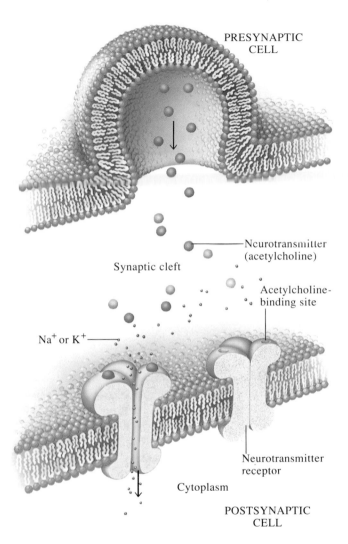

PRESYNAPTIC CELL

Neurotransmitter (acetylcholine)

Synaptic cleft

Acetylcholine-binding site

Na^+ or K^+

Neurotransmitter receptor

Cytoplasm

POSTSYNAPTIC CELL

Neurotransmitter signals are received by special receptor proteins located on the membrane of the target cell. The arrival of acetylcholine at its receptor induces an allosteric change, converting the receptor into an open channel for positively charged ions. The inward flow of ions begins to depolarize the target neuron or muscle cell, and when the depolarization reaches a threshold level, galvanizes the cell into action.

particular the part of the system that controls muscle contraction. One such compound, curare, was derived from an herb by South American Indians, who used it for centuries as a poison for their arrows. The active component of curare evolved to defend the plant against animal predators. It does this by competing with the neurotransmitter acetylcholine for binding to its receptor. Thus, after reaching a neuromuscular junction, curare inhibits the neurotransmitter-induced depolarization of the postsynaptic cell. The victim not only is paralyzed but also may die directly of respiratory failure. The toxin of the cobra snake, like curare, functions by blocking the acetylcholine receptor and inducing paralysis at neuromuscular junctions.

Turning the Neurotransmitter Off

We now come to the last major stage in the transmission of a nerve signal—terminating the influence of a neurotransmitter after it has carried the nerve impulse across the synapse to another cell in the system. Specifically, the depolarization signal must be switched off so that the dendrites and cell body of the postsynaptic cell (or the surface of a target muscle cell) can be returned to their normal −70-mV resting state. This brings us to the last of our prototype enzymes needed for nerve cell function—a *neurotransmitter processing enzyme*. For many nerves, the inactivation of the neurotransmitter is achieved by an enzyme of a type we are already quite familiar with, for it is a member of the serine protease family. This enzyme is *acetylcholinesterase,* which cleaves the neurotransmitter back to its original building blocks, acetic acid and choline.

The reaction mechanism of acetylcholinesterase closely resembles that of chymotrypsin. Like other serine proteases, the enzyme is inactivated by diisopropyl phosphofluoridate (DPF), which becomes covalently attached to a uniquely reactive serine subunit. The enzyme normally uses this serine to attack the peptidelike ester bond in the neurotransmitter, as diagramed on the opposite page. As in the case of chymotrypsin, the reaction proceeds in two steps. In the first, a covalent acyl–enzyme intermediate is formed, with the immediate release of one part of the substrate, choline. Then, in a second step, the acyl–enzyme intermediate is hydrolyzed, using the elements of water, to release acetic acid and regenerate the free enzyme. Formally, the enzyme is classified as a "ser-

ine esterase," rather than a serine protease, because of the particular structure of the chemical group it attacks:

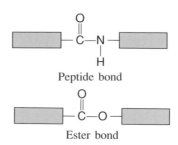

Peptide bond

Ester bond

The detailed structure of the active site of acetylcholinesterase is still under study, and the outcome is uncertain. The active site clearly contains a serine that carries out the nucleophilic attack on the substrate, and this serine is embedded in a stretch of homologous subunits (see p. 242). But it has proved impossible to align the rest of the amino acid sequence with those of other serine proteases. In particular, the expected histidine near position 57 and the aspartic acid near position 102 do not appear with their homologous regions. It is possible, of course, that these amino acids, which assist serine in catalysis, are contributed from other parts of the protein chain, whose three-dimensional folding pattern may be different from that of the typical serine protease. But if it were to turn out that the overall architecture of acetylcholinesterase bears no resemblance to that of chymotrypsin, then the two enzymes would represent an example of *convergent evolution,* in which proteins arrive at the same reaction mechanism starting from different ancestral proteins. In fact, recent evidence indicates that acetylcholinesterase, while not very homologous to the known serine proteases, *is* homologous in amino acid sequence to another protein known as *thyroglobulin.* Remarkably thyroglobulin is not an enzyme at all. Rather it is a protein made in the thyroid gland that serves as a precursor from which the hormone thyroxine (involved in the general control of body metabolism) is excised. We can only assume that acetylcholinesterase and thyroglobulin evolved from a common ancestral protein, let us say involved in thyroxine production. At

The inactivation of acetylcholine by acetylcholinesterase terminates signal transmission and stimulation of the postsynaptic cell.

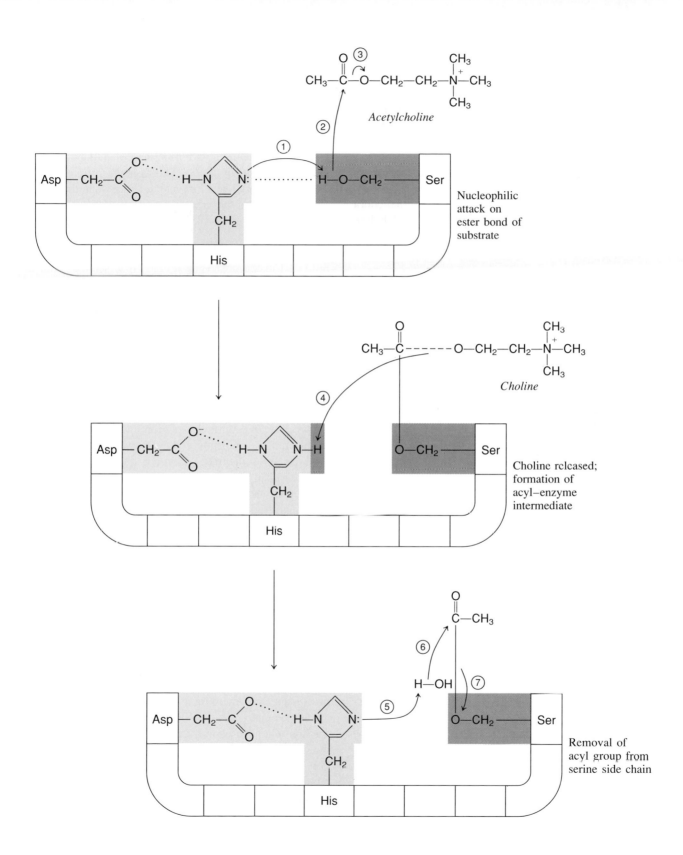

Acetylcholine

Nucleophilic
attack on
ester bond of
substrate

Choline

Choline released;
formation of
acyl–enzyme
intermediate

Removal of
acyl group from
serine side chain

Homology between Acetylcholinesterase and Related Serine Proteases in the Region of the Catalytic Triad

Enzyme	Active-site serine
Chymotrypsin	Gly—Asp—Ser—Gly—Gly—Pro—Leu
Trypsin	Gly—Asp—Ser—Gly—Gly—Pro—Val
Elastase	Gly—Asp—Ser—Gly—Gly—Pro—Leu
Thrombin	Gly—Asp—Ser—Gly—Gly—Pro—Phe
Plasmin	Gly—Asp—Ser—Gly—Gly—Pro—Leu
Acetylcholinesterase	Gly—Glu—Ser—Gly—Gly—Ala—Ser
	Ala*

	Active-site histidine
Chymotrypsin	Val —Thr—Ala—Ala—His—Cys—Gly
Trypsin	Val —Ser—Ala—Gly—His—Cys—Tyr
Elastase	Leu—Thr—Ala—Ala—His—Cys—Ile
Thrombin	Leu—Thr—Ala—Ala—His—Cys—Leu
Plasmin	Leu—Thr—Ala—Ala—His—Cys—Leu
Acetylcholinesterase	— — — — — — —

	Active-site aspartic acid
Chymotrypsin	Thr —Ile —Asn—Asn—Asp—Ile—Thr
Trypsin	Tyr —Leu—Asn—Asn—Asp—Ile—Met
Elastase	Ser —Lys—Gly—Asn—Asp—Ile—Ala
Thrombin	Asn—Leu—Asp—Arg—Asp—Ile—Ala
Plasmin	Phe —Thr—Arg—Lys—Asp—Ile—Ala
Acetylcholinesterase	— — — — — — —

*Acetylcholinesterase has an extra amino acid, alanine, between the catalytic serine and glycine.

some point, this protein evidently acquired the ability to bind to the transition state intermediate formed during the breakdown of acetylcholine. As we saw on pages 178 and 190, the act of binding can provide a basis for enzymatic catalysis by stabilizing the transition state. Over time, catalytic amino acids are introduced adjacent to the binding site, and their ability to promote electron rearrangements increases by many orders of magnitude the effectiveness of the new enzyme. Thus a serine-based active site not unlike the catalytic triad characteristic of the classic serine proteases ultimately arises as a result of convergent evolution. This is in marked contrast to the *divergent evolution* clearly evidenced by the other serine proteases, including trypsin, elastase, thrombin, plasmin, and plasminogen activator.

Like the other enzymes we have discussed with a complex regulatory function, such as thrombin and plasmin, acetylcholinesterase has, in addition to its serine protease–like domain, another protein domain that targets the enzyme to its site of action. This domain, acquired during the evolutionary development of acetylcholinesterase, serves to anchor the catalytic portion of the enzyme to heparan sulfate, one of the types of complex carbohydrate fibers that form an important part of the connective-tissue matrix in the synaptic cleft. After the enzyme has been produced in the postsynaptic cell, it is secreted into the synaptic space and becomes immobilized at just the site that is appropriate for cleaving acetylcholine after a nerve impulse.

We have saved for last a particularly interesting feature of acetylcholinesterase. The enzyme has an extraordinarily high *turnover number* of 25,000 reactions catalyzed per second, meaning that it cleaves an acetylcholine molecule in 40 millionths of a second. Indeed, it is possible to calculate, based on the speed with which molecules are known to move through solution, that only the inward diffusion of the substrate and the outward diffusion of the products limit the rate at which the enzyme works— implying that its active site reacts with acetylcholine as soon as the neurotransmitter binds to the enzyme. If so, this high efficiency of its activity leads us to the conclusion that acetylcholinesterase has reached the highest degree of evolution possible in terms of its inherent catalytic power.

An alternative, but equally impressive, possibility is that the enzyme uses a combination of a very efficient catalytic active site and an "electrostatic funnel," which literally pulls the positively charged acetylcholine substrate into the active site faster than diffusion itself would allow. In any event, the high turnover number allows the timely termination of the neurotransmitter signaling process and the rapid restoration of the resting potential of the postsynaptic cell membrane. Because the postsynaptic cell membrane can recover its −70-mV polarization within a fraction of a millisecond, the synapse is able to transmit a thousand impulses per second. Whereas only a small fraction of the acetylcholinesterase present in the synapse is required when the nerve is transmitting impulses at low frequencies, most of the enzyme molecules are brought into play to sustain transmission at high rates of firing.

═══ ALZHEIMER'S DISEASE ═══

The entire physiology of the body is operated by its collection of enzymes and other proteins, and when these proteins malfunction, serious diseases are the result. Alzheimer's disease is an example of such an illness occurring in the nervous system.

The pathological lesions that characterize the disease, the Alzheimer's "plaques," are clearly visible as the brown patches in the center of the photograph. Over a period of years, tens of thousands of such plaques develop in the brains of Alzheimer's patients, and, at a slower rate

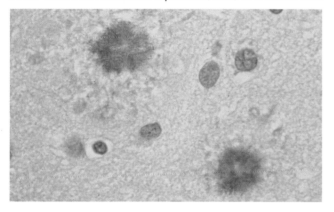

A thin section of tissue from the brain of an Alzheimer's patient as it appears upon microscopic examination.

and to a lesser degree, in the brains of normal aged individuals. The lesions arise in selected areas and are especially significant in the hippocampus—the part of the brain in which memory is established (through the multiplicity of interneuronal connections). As the lesions develop, there is extensive cell death and thus a disruption of signal transmission by neurons, including those neurons that use acetylcholine as a neurotransmitter. A progressive loss of memory, senile dementia, ensues. First there is a loss of the ability to establish short-term memory and then an erosion of the capacity to maintain already-established, long-term memory.

Current molecular biological studies show that the Alzheimer's plaque is composed of aberrant deposits of two normal body proteins that are important components of enzyme systems. One is an antiprotease (page 207), antichymotrypsin, and the other, present in greater amounts, is a fragment of nexin II, a second antiprotease believed to be directed against trypsinlike enzymes. How these abnormal protein deposits develop, how they cause cell death, and, particularly intriguing, why individuals who have three copies of chromosome 21 (Down's syndrome) develop Alzheimer's disease by age 40, are among the current topics of research aimed at understanding the molecular basis of Alzheimer's disease.

Morphine, Neurotransmitters, and Serine Proteases

Two well-known compounds, *morphine* and *heroin*, exert their physiological effect by disrupting the enzyme-mediated signaling system in the brain. These compounds provide an important insight into the way in which external agents can influence the body.

Morphine is a major constituent of the seeds of the opium poppy, accounting for as much as 10 percent of the weight of the plant. It is a remarkable small molecule whose effect on the brain is mood-altering and euphoric. Because of its ability to suppress pain, morphine has played an important role in medicine for centuries. It engenders a feeling of quiet warmth and well-being, followed by somnolence and sleep. It is used in the most serious of medical illnesses, and no better pain-suppressing agent has been found in one hundred years of research.

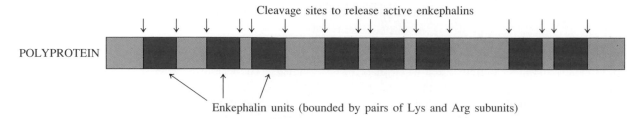

Cleavage sites to release active enkephalins

POLYPROTEIN

Enkephalin units (bounded by pairs of Lys and Arg subunits)

Enkephalins are released from their precursor polyprotein by a serine protease with trypsinlike specificity.

The tremendous demand for morphine has led to a method for increasing its strength. A simple modification of the natural compound—the chemical addition of two acetyl groups—generates the related molecule heroin, which is soluble in lipid membranes, enters the brain more easily, and is therefore more effective. It is the rapid uptake of heroin into the brain following injection that causes the sudden euphoric high or "rush," and accounts for the drug's misuse.

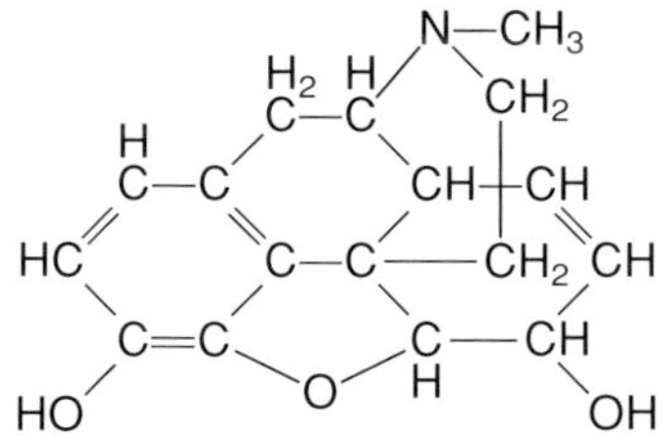

Morphine

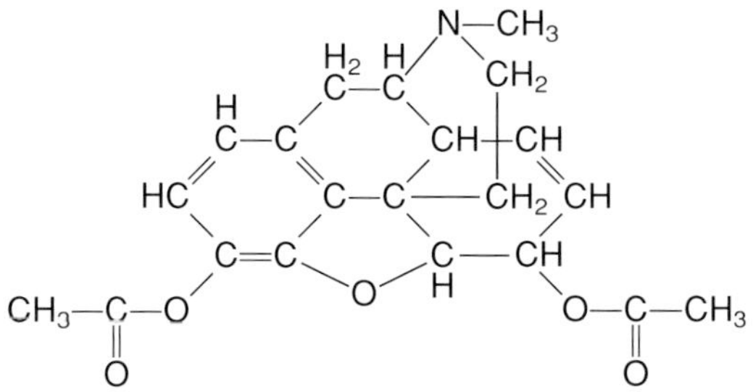

Heroin

Why should such unusual compounds as morphine and heroin have such a pronounced effect on the body? It is clear from even a cursory examination of their structures that morphine and heroin are not among the normal cellular building blocks—the amino acids, nucleotides, lipids, or sugars. Nonetheless, both molecules are structurally related, in part, to the amino-terminal portion of a pair of short peptides.

Tyr—Gly—Gly—Phe—Met
Tyr—Gly—Gly—Phe—Leu

These are the *enkephalins*—two very short proteins five amino acids long that are identical except for their last subunit.

These molecules are produced by certain neurons in the brain at times of great exertion or stress. They are used by these cells as neurotransmitters. Their role is to inhibit adjacent neurons, thereby intercepting nerve signals that would otherwise be perceived as pain. To accomplish the suppression of pain, the enkephalins interact with receptors (membrane ion channels) in the target neurons so as to make their depolarization more difficult. The enkephalins are derived from a large "polyprotein" containing about 100 amino acids. This precursor protein contains eight enkephalin units at various places, with each five-amino-acid unit being bounded by a pair of either lysine or arginine subunits. These positively charged amino acids serve as a signal when the enkephalins are to be released from the precursor polyprotein. In the enkephalin-producing

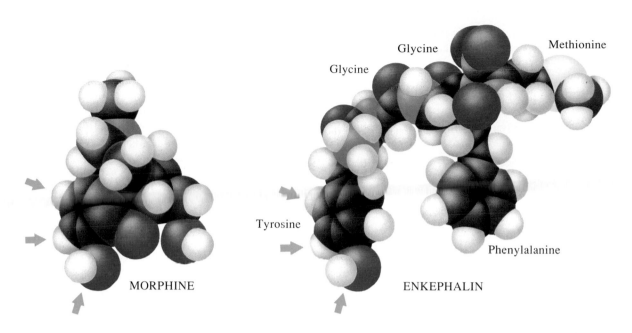

Glycine

Glycine

Methionine

Tyrosine

Phenylalanine

MORPHINE

ENKEPHALIN

The three-dimensional structures of the enkephalins and morphine have similarities, which are indicated by arrows.

neurons, a serine protease with trypsinlike specificity (whose discovery is so recent that it does not yet have a name) is brought into play to make the necessary cleavages and release the biologically active peptides. Although the *processing enzyme* is analogous to trypsin, it is prevented from acting as a general digestive enzyme within the cell because it is "compartmentalized." That is, the enzyme is confined in the endoplasmic reticulum (page 93) and its associated Golgi apparatus, the cellular sites of synthesis for proteins destined for secretion, such as the enkephalins.

The ability of morphine and heroin to mimic the physiological effect of the enkephalins is due to a specific aspect of their structure. As shown in the diagram above, one surface of the morphine molecule (marked with arrows) is very similar in structure to the N-terminal tyrosine subunit that occurs in both enkephalins. It is this similarity in structure that allows morphine—after it has diffused into the synaptic space between neurons in the brain—to trigger the same response as the enkephalins and mitigate

anxiety and pain. But whereas in the body the normal enkephalins are rapidly degraded by proteolytic digestion after their use as neurotransmitters (perhaps by a chymotrypsin-like enzyme that cleaves after their tyrosine or phenylalanine subunit, or by another type of protease that removes amino acid subunits one at a time from the end of a peptide), morphine and heroin, because of their different structure, remain intact. They are not substrates for the enkephalin-degrading enzymes and cannot be similarly removed. Their effect is consequently more prolonged and severe and can lead to death.

As to why a plant should produce morphine, the answer cannot be given with certainty. Nonetheless, a reasonable hypothesis can be advanced. The clue is the location of the morphine in the seed pod of the poppy plant. While many plants produce toxic substances to ward off predators (for example, the herb that produces curare), it may seem astonishing that a plant would produce a substance that is likely to encourage its consumption. The reason may lie in the plant's method of reproduction.

Many plants encase their seeds in a fruit filled with carbo-hydrates. At the appropriate time, the fruit will ripen and the carbohydrates will be converted to a soft, sweet collection of sugars. Animals eating the fruit (for example, birds eating berries) will ingest not only the sweet, fleshy fruit, but also the seeds upon which the reproduction of the plant depends. The seeds are designed to survive their journey through the digestive tract, and they will eventually be deposited on the ground at some distance from the parent plant. Overall, the plant has used the fruit as a mechanism for the dissemination of its seeds in such a way as to increase its geographical distribution. The fact that opium is located in the seed pod of the poppy plant suggests that the plant may be following the same strategy.

Perspective

Our discussion of enzymes and signal transmission in the nervous system has provided three major insights. First, we have encountered several new types of enzymes, particularly the membrane ion pumps and channels, whose ability to undergo allosteric change broadens our knowledge of the *mechanism* of enzyme action. Second, we have seen enzymes whose function is to manufacture, process, and degrade neurotransmitters, which have introduced us to the way in which signal molecules are used to *regulate* physiological processes. Finally, we have seen, through the example of morphine and heroin, how an organism's enzyme systems can be affected by exposure to *environmental agents*.

We will now consider two final topics related to the nervous system, one as fundamental as breathing and the other as complex as human emotion.

Nerve Gases

The major neurotransmitter-processing enzyme, acetyl-cholinesterase, is an Achilles' heel of the human species, and for fifty years the lethal arrow has been available to the major armies of the world. Nerve gases, by blocking acetylcholinesterase, are among the most fearful of weap-ons, potentially able to unleash a degree of savagery not yet seen in warfare. They are thus another example of external agents that disrupt the enzymes of the nervous system.

Nerve gases enter the body by inhalation or by absorption through the skin. Then, perfusing through body tissues, they reach their targets, the cells of the nervous system. The attack on these cells brings on a wide variety of symptoms, including dimming of vision, the filling of the bronchial passages with mucus, bronchial constriction, intense sweating, uncontrollable vomiting, convulsions, paralysis, and respiratory failure. Death from nerve gas poisoning can occur within minutes and is caused by an inability to breathe, leading to asphyxia. Even if the dose is received slowly through the skin, death can still occur after a period of several hours because the damage caused is cumulative and essentially irreversible.

The molecules that function as nerve gases are small, relatively simple compounds that can easily be made in a chemistry laboratory anywhere in the world from reagents generally available to industrial chemists. Sarin is an American nerve gas, and closely related to Sarin is Soman, the standard Russian nerve gas. Both compounds belong to the class of molecules known as organophos-phates and owe their discovery to routine experiments in organic chemistry laboratories designed to study the reactivity of various types of molecules.

Sarin   *Soman*

The biochemical explanation of the deadly effects of these compounds will strike a familiar chord—oddly enough, from our discussion of chymotrypsin. It will be recalled that biochemists were able to locate the active site of chymotrypsin in part because of the behavior of a small molecule known as diisopropyl phosphofluoridate (DPF). This compound proved to be a potent inhibitor of chymo-trypsin and led to the identification of Serine-195 as the crucial amino acid involved in the enzyme's reaction mechanism. When chymotrypsin attempted to initiate a

Not creatures from another planet, but men from our own civilization, forced into an other-worldly competition with their own body chemistry. French soldiers in gas masks.

cleavage reaction using the organophosphate as a pseudosubstrate, Serine-195 became permanently attached to the small molecule (page 167). It will also be recalled that the laboratory synthesis of DPF proved to be unexpectedly dangerous to the chemists carrying it out, causing shortness of breath and related effects.

Now that we know that there is an entire family of serine protease enzymes and, in particular, that acetylcholinesterase is a member of the this family, the reason for the respiratory distress caused by DPF becomes apparent. The molecule's real target in the body is not chymotrypsin or other serine proteases such as those involved in blood clotting, but in fact the very acetylcholinesterase that plays such an important role in the transmission of nerve impulses at synaptic junctions. Just as in the case of chymotrypsin, the active-site serine in acetylcholinesterase initi-

ates a nucleophilic attack on the DPF, displacing its fluoride and leaving the rest of the molecule covalently bonded to the serine side chain. Because this aberrant complex is hydrolyzed very slowly in comparison with the normal acyl–enzyme intermediate, the enzyme becomes effectively blocked. As more and more acetylcholinesterase becomes inactivated, the degradation of neurotransmitter molecules falls to a low level.

Nerve gases work in exactly the same way as diisopropyl phosphofluoridate, except that they are much more effective in blocking the active-site serine. Like DPF, they always contain a fluorine or similar group that is easily displaced and a residual part that becomes linked to the active-site serine of the enzyme. Their more powerful effect comes from their greater reactivity and from their enhanced stability once attached to the enzyme. At the interneuronal and neuromuscular junctions they affect, these compounds cause the buildup of acetylcholine and the overstimulation of target cells. Death results from asphyxia, because the diaphragm is one of the muscles innervated by nerves that use acetylcholine.

The differences in structure between various nerve gases (for example, the amount of bulky hydrocarbon material) give the gases different physical and chemical properties. These differences affect (1) the general reactivity of the compound, (2) the permanence of its attachment to the serine in the active site of the enzyme, and (3) the volatility of the compound—whether it is more likely to remain in the gas form or to settle as a surface film. For example, Sarin has about the same volatility as water and would be released as a vapor cloud. Related nerve gases with more hydrocarbon material, such as Agent VX, have a volatility approaching that of heavy lubricating oil and would be disseminated as a spray of liquid droplets, forming a direct-contact hazard and persisting on the ground or other surfaces for days.

Nerve gases have never actually been used in war. Even though they had been developed by the Germans prior to and during World War II, Hitler refrained from using them, perhaps because of his experience as a victim of the much simpler mustard gas of World War I, and perhaps because he assumed, incorrectly, that the Allies had similar weapons and would retaliate in kind. Nonetheless, the possible use of nerve gases has always been taken seriously, especially by NATO and the Soviets during the Cold War period. The rationale is that such weapons could stop an invading army when conventional firepower is unavailable, without resort to nuclear weaponry.

The lethal dose for Sarin is about 1 milligram (which is approximately equivalent to the amount of salt one shakes out of a salt shaker). It can be calculated that the size of current stockpiles is sufficient to create a lethal aerosol over thousands of square miles of territory. Although nerve gases are designed to be used against armies in the field, their inevitable drift downwind carries the potential for causing immense civilian casualties.

What stands between the civilized world and this catastrophe? One line of defense has been international treaties. The use of poison gas and other chemical weapons is prohibited by the 1925 Geneva protocol. But some nations, such as the United States, Great Britain, France, Russia, and China, regard the treaty as a no-first-use agreement and reserve the right to retaliate in kind if such weapons are used against them. And, of course, terrorists know no bounds. Perhaps more important is the fact that in recent years an increasingly effective defense against nerve gases has been developed. Gas masks and protective clothing have been designed that intercept the organophosphate compounds and neutralize them before they work in the body. The most recent gas masks, together with a four-pound disposable overgarment worn along with gloves and boots, absorb nerve gases in an inner layer of activated charcoal and can reduce the concentration of such chemical warfare agents in breathed-in air about 100,000-fold. In addition, portable scrubbers can apply decontaminating agents to surfaces that soldiers would touch, such as door handles, gunsights, and machine controls. These devices have increased the capacity of soldiers to survive and even operate effectively on contaminated battlefields.

Protective equipment would have to be used immediately, and therefore nerve gas detectors have been developed. One fascinating device is an enzyme-based machine invented by the British. It contains molecules of acetylcholinesterase enzyme immobilized on a solid surface but still active; they are continuously perfused with a solution of an acetylcholine-like substrate. The machine constantly monitors the breakdown of the substrate. If a nerve gas is released into the environment, it will inhibit the acetylcholinesterase in the machine (just as it would the enzyme in the human body). As the breakdown of the tester substrate falters, the machine senses this and triggers an alarm. It is the extensive knowledge of the detailed enzymology of the nervous system, particularly the serine-based enzyme acetylcholinesterase, that has made the development of this machine possible.

The new methods of defense, which have cost billions of dollars to develop and deploy, are effective. While unprotected troops would suffer casualties from chemical weapons that would be similar to those from conventional weapons, this is not the case for an attack on prepared troops. Soldiers wearing antichemical protective equipment are in fact far more vulnerable to conventional attack than they are to attack with nerve gases. Thus, even though nerve gases might still be used by terrorists who would view these weapons as inexpensive and effective against civilian populations, it is increasingly unlikely that they would become part of an escalating global conflict.

Enzymes and Emotion

We have seen the lethal effects of high concentrations of chemicals that inhibit acetylcholinesterase, but what happens when such compounds are encountered in a sublethal dose? It has become increasingly evident that important aspects of physiology and even behavior can be significantly affected—with results ranging from abdominal and respiratory distress to "a violent mind." This point is illustrated by the rather remarkable case of David Garabedian.

For many centuries, a system of medicine was in place based on the four "humors," or body fluids. Imbalances in these "humors" were supposed to govern health and disposition. Clockwise from the bottom left: Too much yellow bile, or choler, makes an angry master; a melancholy patient suffers from an excess of black bile; blood arouses the sanguine nature of a lutenist to play to a maiden; and a woman dominated by phlegm is unemotional and resists the advances of her suitor. Today we are beginning to understand these emotional responses in terms of enzymes and neurotransmitters.

David Garabedian once worked for the Old Fox Lawn Care Company in Massachusetts. On a March afternoon in 1983, he arrived at the home of Eileen Muldoon to spray her lawn with insecticides. After knocking on the front door and receiving no response, he set out to inspect the front lawn and work through some calculations in order to come up with a cost estimate for the job. At about that time, according to Garabedian, he experienced an urge to urinate and walked around to the back of the house to do so. Mrs. Muldoon unexpectedly came upon this scene and was understandably annoyed. Garabedian tried to apologize but to no avail. She turned away and this set in motion a tragic chain of events.

> She was very angry at me and began yelling. I became confused. I tapped her on the shoulder to say "I'm sorry." But she just turned around and gouged me in the face. That's when I grabbed her at the back of the neck, and I just remember squeezing and then both of us falling off the ledge. At that point a drawstring fell out of my jacket. I picked it up and put it around her neck. I just started squeezing.

In a rage, Garabedian dislodged several stones from a nearby wall and used these to repeatedly bludgeon Mrs. Muldoon.

Garabedian then recalls coming to his senses: "I didn't know how I could tell my father about this awful thing." He drove back to the chemical plant where he worked, but told no one of the murder or even of visiting the Muldoon house. He went home and sat down with his father and said nothing. He went to bed. In the middle of the night, the police arrived and he was arrested. At the trial that followed, David Garabedian—a man who had never evinced a violent temper—was charged with first-degree murder.

The case for the defense was based on a plea of temporary insanity. More specifically, it was asserted that the temporary insanity had resulted from chemical poisoning of the nervous system, the first time that such a defense had been used in a criminal trial.

In its modern form, the insanity defense arose around 1843 in connection with the case of Daniel McNaghten in England. A clearly deranged man, McNaghten was convinced that he had been singled out for persecution by the government. He attempted to shoot the British prime minister, instead killing the prime minister's secretary. McNaghten was found not guilty by reason of insanity, and the "McNaghten Rules" became established. For a person to be excused from punishment for a crime based on an insanity defense,

> it must be clearly proved that, *at the time of committing the act,* the party accused was labouring under such defective reasoning *from disease of the mind* as not to know the nature and quality of the act he was doing.

Such a defense, when successful, leads not to release, but to indefinite confinement in a mental hospital.

The temporary insanity defense has never been popular with the public. For example, after the trial of McNaghten, a satirical poem was published in the *Times* of London, containing these lines:

> Ye people of England! Exalt and be glad!
> For ye're now at the will of the merciless mad.
> They're a privileged class whom no statute controls,
> Whose murderous charter exists in their souls.

The opposition to the temporary insanity defense arises because the defense runs counter to our deeply held attitudes about personal responsibility, a bulwark of society that cannot be given up lightly. But does each of us, in fact, have the same emotional stability and the same capacity for rational behavior, and, more particularly, do we have this capacity at all times? Such fundamental questions about personal responsibility are not new, but the approach we will take to them—based on our discussion of enzymes, neurotransmitters, and the brain—is quite recent.

The basic question is whether a person who commits a crime is always responsible for his actions. The defense for David Garabedian claimed that, in fact, he was not responsible for the murder of Mrs. Muldoon. He had been driven to murder—not by Mrs. Muldoon, nor by some weakness in his own character, but by chemicals that can affect the mind.

The case for the defense was formulated by Dr. Peter Spencer, a toxicologist and professor of neuroscience at the Albert Einstein College of Medicine in New York, and Dr. David Bear, a neuroscientist at the University of Massachusetts. It was evident that Garabedian was not permanently "mad." But the defense asserted that his behavior had become progressively erratic during the period leading up to the fatal encounter.

> David had become unusually fatigued, irritable, tense, and impatient. On one occasion he uncharacteristically argued with and struck his younger sister. He had repeated episodes of abdominal pain and diarrhea, and a five- to ten-pound weight loss occurred during the first three weeks of March. He complained that his skin itched and that he was urinating more frequently and with a great sense of urgency. His father noted that his son was salivating excessively and complaining of headaches.

The explanation that the defense then put forward attempted to attribute Garabedian's temporary insanity to his exposure to a specific chemical—a chemical that had literally "poisoned" his mind as well as his body.

The chemical at issue was a lawn care product called Dursban, an important constituent of the insecticides used by Garabedian. He encountered it on a daily basis, both while mixing and spraying the agents. The important property of Dursban is that—like diisopropyl phosphofluoridate and other organophosphates—it interferes with nerve-signal transmission within the brain by inactivating acetylcholinesterase. As we have seen, this enzyme is normally responsible for terminating the action of acetylcholine, but in the presence of an inhibitor, such as Dursban, the neurotransmitter remains active within the synapse. The crucial question thus becomes, Could such a chemical, by prolonging or exaggerating nerve firing, cause changes in *behavior?*

A great deal is known about the consequences of exposure to anticholinesterase chemicals. They produce a spectrum of ill effects that reflect the degree to which en-

zyme and neurotransmitter activity in the nervous system is disturbed. With the most powerful agents, lethal respiratory paralysis occurs (as in the case of nerve gases). But for less potent agents, such as those deliberately designed for use as pesticides, the effects are less acute. In general, depending on the amount of exposure and the potency of a particular anticholinesterase compound, many impairments result, such as fatigue, loss of appetite, abdominal cramps, diarrhea, increased sweating and salivation, and urinary frequency (because acetylcholine is one of the neurotransmitters involved in controlling the bladder). These are in fact the very symptoms that David Garabedian experienced. The question was, Could these disturbances of physiological systems be legitimately extended to include *psychological* behavior?

Evolutionary pressures have led to the development in all animals of the ability to mount an aggressive response. But the parts of the nervous system that control this response also carefully modulate it through several intermediate levels of increasing severity, from disapproval to disagreement, then to confrontation, and finally violence. In a normal individual, the programed neuronal response for aggression is expressed only at the level that is appropriate. The result is an inherent balance between the nature of the provocation and strength of the response. Numerous experiments with animals, however, have indicated that damage or artificial electrical stimulation to specific parts of the brain—particularly to an area known as the hypothalamus—can cause individuals to become highly excitable and easily triggered into aggressive behavior.

Taking into account Garabedian's conduct in the period leading up to the crime and considering the body of knowledge that has accumulated about how the nervous system is organized, the defense formulated its scenario for the murder as follows:

> Let us imagine what might have been happening inside David's brain that day. The nerve cells in the hypothalamus are firing signals at their usual, low rate. When Mrs. Muldoon walks up, the "encounter" takes place. Nerve cells in the hypothalamus that have been resting— idling, firing very slowly—suddenly begin firing extremely rapidly.

Such an increase in firing would cause a greater amount of acetylcholine to be released into the synapses. But now the problem arises. Since the enzyme that normally inactivates acetylcholine is inhibited by the insecticide chemical, the neurotransmitter is not cleaved and begins to build up.

Suddenly the circuit does something it has never done before, and a tremendous emotional reaction follows. There is a feeling that can only be described as overwhelming rage. The chemical poisoning combined with the environment [the stress of the encounter] may have produced a terrible event that neither alone could have produced. The brain was fooled by a substance from outside the body that mimicked the action of its normal signals so that, in effect, a chemical led to his anger. This was not deliberate rage, but a mistake in the brain based on chemical poisoning.

If the real villain is the insecticide, then there may be no human villain.

In essence, the defense argued that Garabedian did not have the capacity to control his behavior because his brain had been, quite literally, poisoned. An imbalance in neurotransmitter release and breakdown had resulted in either an overstimulation of neurons involved in aggression, or an inhibition of neurons normally involved in suppressing aggressive behavior.

In fact, this defense proved unsuccessful. The jury decided that Garabedian was guilty of first-degree murder. There were two factors that probably tipped the scales. First, the defense, for all the logic employed, could only present indirect and circumstantial evidence that Garabedian's behavior at the time of the crime had been influenced by the chemicals he worked with. Second, other workers had been exposed to these chemicals and had not committed murder.

Clearly the case is fraught with ambiguity. There is no objective way to know for certain whether Garabedian is responsible for the murder of Mrs. Muldoon or whether his actions were overwhelmingly governed by chemical agents. The more general question is whether this and

other cases contain important clues to human behavior—indicating that there is a delicate balance of neuronal activity within the brain that can be inadvertently disrupted by environmental agents.

Garabedian is currently serving a life sentence for murder, but he has hopes of receiving a new trial. This hope is based, in part, on increasing evidence that anticholinesterase chemicals may indeed cause abnormally aggressive behavior. Among the new evidence is a bizarre experience with a similar type of chemical, but in this case murder did not result, and there is no reason for the participants to have exaggerated or invented any aspect of the behavior that they ascribe to the particular chemical.

Dr. Claude Lechene is a professor of medicine at Harvard.

I had a large dog who was heavily infested with fleas. When nothing else worked I started using a very potent insecticide. Since I also had a cat, I sprayed the chemical on him as well. Within days, the cat, who was usually very gentle, began chasing mice and birds. He would drag them into the house, free them, and then run after them and kill them. Soon almost every room of my house was filled with rows of animals. The cat was continuing to go outside and bring in more animals, which he then killed. The cat was becoming a murderer.

Dr. Lechene told a companion about these unusual events and was then astonished to hear that his own behavior had recently become much more aggressive—in a way that is relevant to the behavior of David Garabedian. He was told that his mood had become extremely changeable and that he was constantly picking fights over trivial matters.

When she said that, I realized that something was terribly wrong with me—the cat— even the dog. All three of us sleepy all the time. Then there was the increased aggression by me and the cat. When I checked into the composition of the insecticide I found that it contained an acetylcholinesterase inhibitor.

The crucial point—in terms of the Garabedian case—was that, when Lechene stopped using the insecticide, his symptoms and the cat's promptly disappeared. The "involuntary intoxication" was over. The "temporary insanity" was gone.

The specific outcome of the Garabedian case is not the major issue. What is more generally important is the developing awareness that various aspects of emotional behavior—not only acts of extreme violence, but also perhaps the much more common nuances of our personality—may reflect imbalances in neurotransmitters and the enzymes that control them. To a greater or lesser extent, variations in mood in the average person, which rarely approach the extreme of criminal behavior, may nonetheless be significantly influenced by the many chemical compounds we encounter every day.

Final Perspective

We have traveled a long distance from the place we started. A century ago the sciences of physics and chemistry were already highly developed and well articulated. But the biological sciences were, to a significant degree, still caught in the grip of vital forces. That world is gone forever. In our century, an almost unparalleled and productive period of research has entirely transformed the landscape of biology. Most important has been the emergence of enzymes as the molecular machines that bring about the chemical transformations that occur in living systems.

The science of biochemistry was born with the goal of recovering enzymes from cells and determining how they work at the molecular level. We have seen this goal fulfilled. As our window onto the world of enzymes, we chose a particularly remarkable enzyme, chymotrypsin. Focusing on this enzyme, we saw the three great themes of biology—anatomy, physiology, and evolution—translated into molecular terms. In the case of anatomy, we found that our model enzyme was a protein that exerts its remarkable influence over the molecules of the cell by selectively forming a lock-and-key interaction with the molecule it wishes to change. When we then explored what happened after the enzyme bound its target molecule, we were led to the heart of the subject of enzyme physiology—the molecular mechanism of enzyme action. We found that the catalytic power of the enzyme lay in its extraordinary architectural organization. Adjacent to the area where it bound to the substrate in a lock-and-key fashion, the enzyme brought into play a number of its amino acid subunits. The interaction of these amino acids with the target molecule not only tugged at the substrate, helping it pass across the high-energy transition state barrier, but also provided powerful but highly localized acid and base catalysts to promote the electron rearrangements needed for the chemical transformation of the substrate. Indeed, the synergy obtained from the cooperative use of several amino acid side chains at close range accomplished, in a controlled and sophisticated way, the same effect as is achieved when the random agencies of heat or acid/base drive reactions in the nonliving world. The subtle but powerful mechanisms enzymes use are so successful that physiologically important chemical reactions can be made to occur 1,000,000,000-fold more rapidly than they would otherwise proceed. This tremendous rate enhancement lifts a selected group of chemical reactions into the range where they can sustain the processes of life.

Enzymes are remarkable not only for their catalytic powers, but also for their ability to evolve. They are like a set of machine tools that are undergoing continuous improvements. Discrete changes in amino acid subunits, occurring randomly and chosen for their usefulness by natural selection, alter the substrate binding site so that the enzyme can bring its catalytic power to bear on different target molecules. Thus we saw the humble digestive enzyme chymotrypsin evolve into an entire family of enzymes that occur throughout the body. All of these enzymes are variations on the same theme, but they work on different substrates and are subject to different mechanisms of control. Their evolution has allowed the cell to venture into chemical transformations involved in processes ranging far afield from digestion—into areas as diverse as blood clotting and neurobiology.

What emerges above all else from our discussion of enzymes is an enhanced appreciation of the unity of life at the underlying molecular level. In the microcosm of enzymes and biochemistry, recurrent basic themes bind together all living things and all living processes. The family of chymotrypsin-like enzymes, reaching out as it does to embrace such diverse processes as digestion, blood clotting, and signal transmission in the nervous system, illustrates the power of what may ultimately prove to be only a small number of major enzyme families. The very simplicity of this idea suggests that life is not so much the miracle of a large number of improbable events, but the ingenious interplay of a few very good ideas.

Further Readings

LUBERT STRYER: *Biochemistry*, 3d ed., W. H. Freeman, New York, 1988. An outstanding general biochemistry text, comprehensive and well written.

CHRISTIAN DE DUVE: *A Guided Tour of the Living Cell*, Scientific American Library, New York, 1984. Well written, with an emphasis on the structures of the cell.

HELENA CURTIS AND SUE N. BARNES: *Biology*, Worth, New York, 1989. The leading general text on biology, interestingly written and beautifully produced.

ALAN FERSCHT: *Enzyme Structure and Mechanism*, W. H. Freeman, New York, 1985. An advanced treatment of enzymes and reaction mechanisms.

JOSEPH S. FRUTON: *Molecules and Life*, Wiley-Interscience, New York, 1972. The best comprehensive overview of the development of the science of biochemistry.

HERBERT C. FRIEDMANN: *Benchmark Papers in Biochemistry: Enzymes*, Hutchinson Ross, Stroudsburg, Penn., 1981. A collection of classic papers translated and commented upon by the author.

ROBERT KOHLER: *Journal of the History of Biology* 4:35–61, 1971; *Journal of the History of Biology* 5:327–353, 1972; *Isis* 64:181–196, 1973. Three papers analyzing the seminal work of Eduard Buchner.

JAMES SUMNER AND JOHN NORTHROP: The chemical nature of enzymes, Nobel Lectures, 1946, *Nobel Lectures in Chemistry*, Elsevier, Amsterdam, 1964. John Northrop: Biochemists, biologists, and William of Occam, *Annual Review of Biochemistry* 30:1–10, 1961.

SIDNEY ALTMAN AND TOM CECH: RNA enzymes, Nobel Lectures, 1989.

DAVID BLOW: The structure of chymotrypsin, in P. D. Boyer (ed.), *The Enzymes*, vol 3, 1971, pp. 185–212. Paul Sigler, David Blow, Brian Matthews, and Richard Henderson: *Journal of Molecular Biology* 35:143–164, 1968. David Blow, Jens Birktoft, and Brian Hartley: *Nature* 221:337–340, 1969. Thomas Steitz and Robert Schulman: *Annual Review of* *Biophysics and Bioengineering* 11:419–444, 1982. Joseph Kraut: *Annual Review of Biochemistry* 46:331–358, 1977. The data and the debate about how chymotrypsin works.

PAUL CARTER AND JAMES WELLS: *Nature* 332:564–568, 1988. Dissecting the catalytic triad of a serine protease.

RICHARD LERNER AND ALFONSO TRAMONTANO: Catalytic antibodies, *Scientific American*, 258(3):58–70, March 1988. Transition state stabilization and the evolution of enzymes.

HANS NEURATH: Evolution of the proteolytic enzymes, *Science* 224:350–357, 1984. Robert Stroud: A family of protein-cutting proteins, *Scientific American*, 231(1):74–88, July 1974.

RUSSELL DOOLITTLE: Fibrinogen and fibrin, *Scientific American*, 245(6):126–135, December 1981. A good overview. Also, George Stamatoyannopoulos, Arthur Nienhuis, Philip Leder, and Philip Majerus: *The Molecular Basis of Blood Diseases*, W. B. Saunders, Philadelphia, 1987, 534–721. An advanced discussion of the molecular biology of blood-clotting enzymes.

DÉSIRÉ COLLEN, HENRI LIJNEN, PETER TODD, AND KAREN GOA: *Drugs* 38:346–388, 1989. Joseph Loscaizo and Eugene Braunwald: Tissue plasminogen activator, *New England Journal of Medicine*, 319:925–931, 1988. Review articles on plasminogen activator (tPA) and its use as a thrombolytic agent in the treatment of heart attacks.

MATTHEW MESELSON AND JULIAN PERRY ROBINSON: Chemical warfare and chemical disarmament, *Scientific American*, 242(4):38–47, April 1980. An overview of the biochemistry and politics surrounding the nerve gas problem.

RICHARD RESTAK: *The Mind*, Bantam Books, New York, 1988. The chapter on the violent mind discusses the Garabedian case and many others where behavior is affected by damage to the nervous system.

CARMELA ABRAHAM AND HUNTINGTON POTTER: Alzheimer's disease: Recent advances in understanding the brain amyloid deposits, *Biotechnology* 7:147–153, 1989.

Sources of Illustrations

Paintings by Tomo Narashima; line drawings by Vantage Art, Inc.

Prologue   *p. 2:* Kevin Schafer.   *p. 3:* Bob and Clara Calhoun/Bruce Coleman.   *p. 4:* Lennart Nilsson, *A Child Is Born,* Dell, 1986.   *p. 6:* Lennart Nilsson, *The Body Victorious,* Delacorte, 1987.   *p. 9:* Kunsthistorisches Museum, Vienna.   *p. 10:* Biophoto Associates/Science Source, Photo Researchers.   *p. 12:* Eric Grave/Photo Researchers.   *p. 14:* Manfred Kage/Peter Arnold.   *p. 17:* E. Bernstein and E. Kairinen, Gillette Research Institute.   *p. 18:* Muriel Voter Williams.

Chapter 1   *p. 20:* The Pasteur Institute.   *p. 22:* Rijksmuseum van Oudheden.   *p. 24:* Mauritshuis, Le Hague, Scala/Art Resource.   *p. 25:* The Metropolitan Museum of Art; purchase, Mr. and Mrs. Charles Wrightsman Gift, 1977 (1977.10).   *p. 27:* Giraudon/Art Resource.   *p. 30:* (left and right) The Deutsches Museum.   *p. 32:* Reunion des Musées Nationaux.   *p. 33:* The Pasteur Institute.   *p. 34:* (left) David Scharf/Peter Arnold. (right) Manfred Kage/Peter Arnold.   *p. 36:* The Deutsches Museum.   *p. 39:* The Deutsches Museum.   *p. 40:* Bettmann Archive.   *p. 44:* The Royal Swedish Academy of Sciences.   *p. 48:* The Deutsches Museum.   *p. 52:* The Deutsches Museum.

Chapter 2   *p. 54:* Municipal Museum, The Hague, Holland/Cordon Art B.V.   *p. 56:* Camera Hawaii.   *p. 62:* From Alfred L. Lehninger, *Principles of Biochemistry,* Worth, 1982.   *p. 65:* The Deutsches Museum.   *p. 69:* The University of Birmingham.

Chapter 3   *p. 72:* Alexander McPherson.   *p. 81:* The Deutsches Museum.   *p. 82:* Cornell University Archives.   *p. 83:* After Lubert Stryer, *Biochemistry,* 3d ed., W. H. Freeman, 1988.   *p. 84:* Alexander McPherson.   *p. 85:* Alexander McPherson.   *p. 86:* (left and right) Rockefeller Archive Center.

Chapter 4   *p. 90:* Polygen Corporation.   *p. 92:* (top) After Stryer, *Biochemistry,* 3d. ed.   *p. 93:* (left) Biophoto/Photo Researchers. (right) Keith R. Porter/Photo Researchers.   *pp. 98, 99, 101, 102, 103, 104, 105, 106:* After Alfred L. Lehninger, *Biochemistry,* Worth, 2d ed., 1975.   *p. 108:* Laboratory of Molecular Biology, Cambridge University.   *p. 114:* After David Blow, in *The Enzymes,* vol. 5, 3d ed., Academic Press, 1973.   *p. 115:* After G. N. Ramachandran et al., Biochemistry, *Biophys. Acta* **359:**298–302, 1974.   *p. 117:* Linus Pauling.   *p. 118:* After Strycr, *Biochemistry,* 3d ed.   *p. 119:* After Robert W. McGilvery, *Biochemistry: A Functional Approach,* 2d ed., W. B. Saunders, 1979.

p. 120: Dr. Mary Osborn, Max Planck Institute.   *p. 122:* From Stryer, *Biochemistry,* 3d ed.   *p. 123:* Muriel Voter Williams, Photo/Nats.   *p. 138:* After Geoffrey Zubay, *Biochemistry,* Addison-Wesley, 1983.   *p. 139:* (top) Thomas Steitz (bottom) Polygen Corporation.   *p. 140:* After Richard E. Dickerson and Irving Geis, *The Stucture and Action of Proteins,* Harper & Row, Benjamin/Cummings, 1969.   *p. 144:* After Dickerson and Geis, *The Structure and Action of Proteins.*   *p. 149:* Christian Anfinsen.   *p. 150:* After Stryer, *Biochemistry,* 3d ed.   *p. 151:* Mauritshuis, Le Hague, Scala/Art Resource.

Chapter 5   *p. 156:* Royal College of Physicians.   *p. 171:* Jens Berktoft.   *p. 172:* After Dickerson and Geis, *The Structure and Action of Proteins.*   *p. 189:* Table based on data in Paul Carter and James Wells, *Nature* **332:**564–568.

Chapter 6   *p. 192:* Mia Tegner, Scripps Institute of Oceanography.   *p. 194:* Darwin Museum, Down House, courtesy of the Royal College of Surgeons of England.   *p. 196:* After Dickerson and Geis, *The Structure and Action of Proteins.*   *p. 205:* Originally prepared in modified form for J. David Rawn, *Biochemistry,* Niel Patterson Press/Carolina Biological Supply Co., 1988.   *p. 206:* From Stryer, *Biochemistry,* 3d ed.   *p. 208:* Alfred Owczarcak/Taurus Photos.   *p. 209:* Boehringer Ingelheim International GmbH, Lennart Nilsson.   *p. 210:* (electron micrograph) Dr. Henry Slayter.   *p. 218:* Abbott Laboratories.   *p. 221:* Lennart Nilsson, *Behold Man,* Little Brown, 1978.   *p. 224:* Genentech, Inc.   *p. 225:* Muriel Voter Williams.

Chapter 7   *p. 226:* Synaptek Scientific Products, Inc./SPL/Photo Researchers.   *p. 228:* Ateneo de Madrid.   *p. 229:* Fritz Goro.   *p. 236:* After W. A. Bain, Method of demonstrating humor transmission of effects of cardiac vagus stimulation in frogs, *Quarterly Journal of Experimental Physiology and Cognate Medical Sciences* **22:**269–274, 1932.; copyright with Sir Edward Sharpey Shafer's Trustees.   *p. 237:* John E. Hauser, Washington University School of Medicine.   *p. 238:* After David H. Hubel, *Eye, Brain, and Vision,* W. H. Freeman, 1988.   *p. 239:* Alan Peters, Sanford L. Palay, and Henry deF. Webster, *The Fine Structure of the Nervous System,* Oxford Press, 1991.   *p. 245:* Dr. Huntington Potter and Dr. Carmella Abraham.   *p. 247:* After Solomon H. Snyder, *Drugs and the Brain,* W. H. Freeman, 1986.   *p. 249:* Roger-Violet.   *p. 251:* Zentralbibliothek Zürich.

Index

Other books in the Scientific American Library Series

POWERS OF TEN
by Philip and Phylis Morrison and the Office of Charles and
Ray Eames

HUMAN DIVERSITY
by Richard Lewontin

THE DISCOVERY OF SUBATOMIC PARTICLES
by Steven Weinberg

FOSSILS AND THE HISTORY OF LIFE
by George Gaylord Simpson

ON SIZE AND LIFE
by Thomas A. McMahon and John Tyler Bonner

THE SECOND LAW
by P. W. Atkins

THE LIVING CELL, VOLUMES I AND II
by Christian de Duve

MATHEMATICS AND OPTIMAL FORM
by Stefan Hildebrandt and Anthony Tromba

FIRE
by John W. Lyons

SUN AND EARTH
by Herbert Friedman

ISLANDS
by H. William Menard

DRUGS AND THE BRAIN
by Solomon H. Snyder

THE TIMING OF BIOLOGICAL CLOCKS
by Arthur T. Winfree

EXTINCTION
by Steven M. Stanley

MOLECULES
by P. W. Atkins

EYE, BRAIN, AND VISION
by David H. Hubel

THE SCIENCE OF STRUCTURES AND MATERIALS
by J. E. Gordon

SAND
by Raymond Siever

THE HONEY BEE
by James L. Gould and Carol Grant Gould

ANIMAL NAVIGATION
by Talbot H. Waterman

SLEEP
by J. Allan Hobson

FROM QUARKS TO THE COSMOS
by Leon M. Lederman and David N. Schramm

SEXUAL SELECTION
by James L. Gould and Carol Grant Gould

THE NEW ARCHAEOLOGY AND THE ANCIENT MAYA
by Jeremy A. Sabloff

A JOURNEY INTO GRAVITY AND SPACETIME
by John Archibald Wheeler

SIGNALS
by John R. Pierce and A. Michael Noll

BEYOND THE THIRD DIMENSION
by Thomas F. Banchoff